Automotive Emission Control

Automotive Emission Control

William L. Husselbee

RESTON PUBLISHING COMPANY, INC.
A Prentice-Hall Company
Reston, Virginia

Library of Congress Cataloging in Publication Data

Husselbee, William L.
   Automotive emission control.

   Includes index.
   1. Automobiles–Pollution control devices. 2. Automobiles–Motors–Exhaust gas. I. Title.
TL214.P6H87   1984   629.2'528   83-17845
ISBN 0-8359-0173-4

©1984 by Reston Publishing Company, Inc.
A Prentice-Hall Company
Reston, Virginia 22090

All rights reserved. No part of this book may be reproduced in any way or by any means without permission in writing from the publisher.

10 9 8 7 6 5 4 3 2 1

PRINTED IN THE UNITED STATES OF AMERICA

Contents

Preface, xiii

Chapter 1
Air Pollution and Smog—Their Causes and Effects, 1

What is Air Pollution?, 1
What Is Smog?, 3
Smog Versus Air Pollution, 3
Other Contributing Factors to Air Pollution and/or Smog, 3
Emissions from the Automobile—Hydrocarbons, 4
Cause of NOx Emissions, 6
Cause of Carbon Monoxide Emissions into the Atmosphere, 7
Cause of Particulate Emissions into the Atmosphere, 8
Cause of Oxides of Sulfur Emissions into the Atmosphere, 8
Air Pollution Research and Legislation, 9
Air Pollution Regulatory Agencies, 10
Emission Control Standards, 11
Summary, 12
Review Questions, 14

Chapter 2
Chemical Reaction of Gasoline Combustion Within an Automobile Engine, 17

Composition of Air, 17
Composition of Gasoline, 18
Operating Cycle of the Engine, 20
Factors Influencing Combustion Chemistry and Negative Emissions, 24
Summary, 30
Review Questions, 32

Chapter 3
Instruments Used to Measure Exhaust Emissions and Test Control Devices, 35

Vacuum Gauge, 35
Timing Light, 36
Tachometer, 37
Infrared Analyzer, 37
Propane Enrichment Equipment, 41
Vacuum Pumps—Hand Operated, 41
Summary, 43
Review Questions, 44

Chapter 4
Using Emission Control Test Equipment, 47

Testing the Engine with a Vacuum Gauge, 47
Checking Basic Ignition Timing and the Action of the Distributor Advance Units with a Timing Light, 51
Adjusting Idle Speed and Air/Fuel Ratio with a Tachometer, 53
Using the Infrared Analyzer, 55
Infrared Analyzer Maintenance, 58
Testing Exhaust Emissions with an Infrared Analyzer, 62
Causes of and Diagnosing Excessive Hydrocarbon Emissions, 71
Causes and Diagnosis of Excessive CO Emissions, 76
Other Test Procedures with the Infrared Analyzer—Accelerator Pump, 78
Carburetor Power Circuit Test with the Infrared Analyzer, 78
Carburetor Adjustments with the Infrared Analyzer and Propane Enrichment Equipment, 79
Using a Portable Vacuum Pump, 79
Summary, 80
Review Questions, 82

Chapter 5
Crankcase Emission Control Systems, 85

Need for an Engine Ventilation System, 85
Early Ventilation Systems, 86
Positive Crankcase Ventilation Systems, 87
Types of PCV Systems, 87
Retrofit Crankcase Emission Systems, 96
Summary, 97
Review Questions, 98

Chapter 6
Testing and Servicing PCV Systems, 101

Symptoms of Improper PCV System Operation, 101
Maintenance of a PCV System, 102
PCV System Inspection, 102
PCV System Testing, 103
Replacing a PCV Hose, 107
Cleaning the PCV Passages in the Carburetor, 107
Replacing a PCV Valve, 107
Replacing a PCV Inlet Filter in the Air Cleaner Housing, 108
Cleaning and Servicing the Filter Inside the Oil Filler Cap, 109
Summary, 110
Review Questions, 110

Chapter 7
Evaporation Emission Control Systems, 113

Function, 113
Types of EEC Systems, 113
EEC System Design, 114
EEC System Operation, 127
Types of Purging Methods, 128
Summary, 132
Review Questions, 133

Chapter 8
Evaporation Emission Control System Service Inspection of the EEC System, 137

Checking the Condition of Other System Components, 137
Testing the Operation of an EEC System with an Infrared Analyzer, 139
EEC System Component Replacement, 140
Summary, 142
Review Questions, 143

Chapter 9
Engine Modification Systems, 145

Function of an Engine Modification System, 145
Engine Modifications, 146
Combustion Chamber Shape Effects on Emissions, 150
Stratified-Charge Combustion Chambers, 151
Reducing Emissions by Altering Combustion Chamber Area, 153
Jet Air System, 153

Carburetor Modifications, 156
Electronically Controlled Carburetors, 165
Carburetor-Assist Devices, 166
Thermostatically Controlled Air Cleaners, 174
Spark Advance Control Devices, 177
Increased Operating Temperatures, 177
Vacuum-Operated Heat-Riser Valves, 178
Summary, 180
Review Questions, 181

Chapter 10
Testing and Servicing Engine Modification Systems, 185

Adjusting a Jet Air Valve, 186
Adjusting the Air/Fuel Ratio with an Infrared Analyzer, 186
Carburetor Adjustments Using Propane Enrichment Equipment, 191
Propane Enrichment Adjustment Procedures, 192
Testing Choke Delay Valves, 198
Dashpot Adjustment, 199
Adjusting an Antidiesel Solenoid, 199
Adjusting an Air-Conditioning Solenoid, 201
Testing and Servicing a Typical Thermostatically Controlled Air Cleaner, 201
Testing the Vacuum Motor Diaphragm, 202
Removing and Replacing the Air Cleaner Temperature Sensor, 204
Summary, 205
Review Questions, 206

Chapter 11
Spark Timing Control Systems, 209

Spark Timing Effects on Emissions, 209
Types of Spark-Timing Control Devices, 211
Summary, 228
Review Questions, 230

Chapter 12
Testing Spark-Timing Control Systems, 233

Testing Dual-Diaphragm Distributors, 233
Testing Temperature-Sensitive Vacuum Control Switches, 235
Testing and Servicing Spark-Delay Valves, 238
Testing an OSAC Valve, 239
Testing and Adjusting a Vacuum Control (Deceleration) Valve, 239
Testing a Typical TCS System, 241
Testing a Typical Electronic Spark Control System, 242

Summary, 246
Review Questions, 246

Chapter 13
Air Injection Systems, 249

The Design of the Air Injection System, 250
Injection System Operation, 255
Air Injection Systems Used with Catalytic Converters, 255
Air Injection System Used with Dual Catalytic Converters, 263
Aspirator-Type Air Injection Systems, 266
Pulse Air Injection Used with Multiple Catalytic Converters, 268
Summary, 270
Review Questions, 271

Chapter 14
Testing Air Injection Systems, 273

Testing a Basic Air Injection System with an Infrared Analyzer, 274
Checking Air Injection System Components, 274
Testing a Chrysler Catalytic Air Injection System, 277
Testing a Ford Catalytic Air Injection System, 279
Testing a Typical General Motors Catalytic Air Injection System, 287
Checking a Typical Aspirator Air Injection System, 283
Summary, 284
Review Questions, 285

Chapter 15
Exhaust Gas Recirculation Systems, 287

Basic EGR System Design, 288
Vacuum Sources Used to Operate an EGR Valve, 290
EGR System Modulating Devices, 292
Specific EGR Systems—Chrysler Corporation, 295
Ford EGR Systems, 299
General Motors EGR Systems, 308
Summary, 308
Review Questions, 311

Chapter 16
Servicing EGR Systems, 313

General EGR System Service, 313
Inspecting, Testing, and Servicing Specific EGR Systems—Chrysler
  Corporation, 315
Testing Ford EGR Systems, 321

Testing a General Motors EGR System—Ported Vacuum Control System, 332
Summary, 334
Review Questions, 335

Chapter 17
Catalytic Converter Systems, 337

Converter Location, 337
Function and Types of Catalyst Materials, 339
Types of Catalytic Converters, 339
Specific Catalytic Converter Systems—Chrysler, 345
Catalytic Converter Protection Subsystem, 347
Ford Catalytic Converter Systems, 350
General Motors Catalytic Converter Systems, 354
Summary, 356
Review Questions, 357

Chapter 18
Catalytic Converter Service, 359

Factors Reducing Service Life of Catalytic Converters, 359
Service Precautions, 360
Converter System Inspections, 360
Testing Catalytic Converters, 363
Testing the Converter Efficiency with an Infrared Analyzer, 364
Catalytic Converter Service, 365
Catalytic Converter Replacement, 366
Summary, 368
Review Questions, 369

Appendix—Answers to Review Questions, 371

Index, 375

Preface

For over 30 years now, many portions of this country have been subjected to an abnormal phenomenon, namely air pollution. The problem first became apparent during World War II, and most people thought it would disappear once the war effect was over. However, this was not the case.

Although there are many types of natural air pollutants, these are not of much concern, basically because there is very little which can be done to curb their causes. This leaves us with man-made air pollutants, which we create and permit to enter and contaminate the atmosphere. These contaminators are not only unsightly but are hazardous to humans, animals, and plant life.

In studies made at the California Institute of Technology in the 1950s, it was discovered that one of the primary causes of the air pollution problem (known as smog) in the Los Angeles basin and similar situated areas was the direct result of certain harmful emissions from motor vehicles. These particular studies and ones later funded by the government led to the passage of state and Federal legislation, which aimed at the reduction of harmful emissions from motor vehicles.

In the beginning, the Federal emission standards were not too difficult for the automotive industry to meet; however, with the passage of time, the standards did become more stringent. As a result, emission control systems have become more complex year by year. Now, every automobile and many trucks come factory-equipped with a very sophisticated emission-control system, which functions along with the other engine systems to reduce exhaust emissions to a relatively insignificant amount, considering what they were even 20 years ago.

Although it is up to automotive manufacturers to initially design, test and produce the emission control systems to reduce a vehicle's level of pollution, it is up to automotive and truck technicians in the field to test and service these devices. In many cases, a mechanic will check the serviceability of the emission-control systems as a part of a complete tune-up. But many states now require that a vehicle's emission-control system be checked yearly for serviceability before the owner can obtain its annual registration. In either case, it does require skilled technicians not only to test these devices, but also to diagnose and repair them when a malfunction occurs.

Initially training and updating mechanics to perform emission-control system testing and service has always been a problem since the installation of the first devices. Most automotive manufacturers do provide training centers for their technicians in order to keep them informed on changes or modifications to emission control devices. But a mechanic, who works in an independent garage or service center, is not as fortunate. This person, of his own accord, has to attend a community college or trade school, or read a current text to obtain working knowledge of the various types of emission systems found on the different types of domestic and foreign-built vehicles.

This text, *Automotive Emission Control,* assists in the training of mechanics by acting as a valuable reference source. The book presents many facts regarding the sources of atmospheric pollution from motor vehicles, and explains how the industry is minimizing or eliminating this pollution by the installation of emission control devices.

In regard to the devices themselves, this book explains in detail the function, design, and operation of past as well as current emission systems. Moreover, it provides detailed, step-by-step instruction on the testing and servicing of these types of emission systems.

Persons other than automotive technicians will also find this text helpful. For example, the interested consumer can find answers to questions on vehicle pollution and what can be done to a personal vehicle to eliminate the problem. Furthermore, the decision makers in industry as well as in government will find in this text a review of what has been done and is being done to reduce pollutants from motor vehicles.

To perform its overall purpose, this book is divided into chapters on theory, followed by chapters on how to perform specific procedures and/or repairs. A chapter on theory, one which explains the design and principles of operation, such as Chapter 15 on EGR systems, covers the basic concepts that underlie this particular system's operation. In other words, a theory chapter within the text explains each component of a given system and describes its operation.

Following close behind a theory chapter is a "how to" chapter, such as Chapter 16 on EGR system service. In a "how to" chapter, this book presents up-to-date troubleshooting, service, and repair of components discussed within the theory chapter. Thus, the book provides the reader with a firm conceptual understanding of the theory behind the many emission-control systems, as well as with the testing and repair techniques necessary to maintain the devices.

The book also explains the many types of test equipment used to perform the various checks on emission-control systems. The kinds of equipment included in this text are the vacuum gauge, vacuum pump, tachometer, timing light, infrared analyzer, and propane enrichment tool.

Because of the complex nature of the material contained in the various chapters of this book, there are also several methods provided to assist the reader in learning the material. For instance, at the end of each chapter is a summary, which condenses the main concepts within the unit into a number of sentences. Also, after the chapter summary are a number of chapter review questions, which assist the reader in determining just how well he remembers the material contained in each chapter. The answers to these questions are found in the Appendix located at the back of the text.

William L. Husselbee

Chapter 1

Air Pollution and Smog—Causes and Effects

During World War II, there appeared over many metropolitan areas in the United States a very disturbing phenomenon, namely air pollution. The average citizen firmly believed that this condition would cease when the war effort was over. However, this was not the case. In some cities, of course, air pollution from factories that produced war materials did decrease; but the problem still existed and steadily grew in intensity due to other sources.

Since this period air pollution has gained more and more public attention. Originally, ecology groups were the driving force in bringing the air pollution problem to the public's attention. However, the main reason air pollution has become a topic of concern is that it has become a national problem.

Although air pollution is not a problem in all areas of the United States, certain cities such as Los Angeles, California, do experience severe air contamination. The problem proliferated in the years following World War II as a direct result of the increase in population, number of motor vehicles, factories, urban areas, and power sources.

What Is Air Pollution?

Up to this point, the words "air pollution" have been mentioned several times. *Air pollution* is the introduction of any contamination into the atmosphere in an amount large enough to injure human, animal, or plant life. There are, in fact, many different types and causes of air pollution. However, these contaminants do fall into two general groups, natural and man made.

Natural Air Pollution

Natural air pollution has always existed in one form or another. For instance, when the volcano Mt. St. Helens erupted, it threw millions of tons of ash and smoke into the air. Prevailing winds then carried the contaminants for hundreds of miles. Another type of common air pollution is dust, stirred up into the atmosphere by winds.

Trees and plants give off a gaseous form of air pollution. Conifers (evergreen trees) give off, for example, terpene hydrocarbons, and rotting vegetable matter in forests emits methane hydrocarbons. *Hydrocarbons* basically are compounds made up of hydrogen and carbon atoms.

The point is that natural air pollution has always been with us in one form or another. This type of air pollution is a product of nature, and there is no known way to control it.

These natural contaminants do alter the air we breathe. Perfectly pure air, for example, contains approximately 78 percent nitrogen, 21 percent oxygen, and 1 percent other inert, harmless gases. However, this pure air is not only very rare today but probably has been so for many centuries.

Man-Made Air Pollution

Man-made air pollution is any form of contaminant that we create and permit to enter the atmosphere (Fig. 1-1). Air pollutants of this type are not only unsightly but are hazardous to human, animal, and plant life.

There are, in fact, many sources of man-made air pollution that are controllable. One common cause, which most people are familiar with, is in the form of chemical fumes and smoke from the exhaust stacks of large factories and power plants. Another source of air pollution comes from incinerator fires and the chimneys of wood-burning stoves or fireplaces in residential homes.

In many areas of the country legislation now controls the amount of the above-mentioned air contaminants. In other words, large manufacturers and power companies must now have control devices on their smokestacks to limit the amount of harmful emissions. Also, in many areas, open trash or incinerator fires are no longer permitted.

Another man-made source of air pollution is the motor vehicle. Studies made at the California Institute of Technology in the 1950s, which attempted to find the causes for the severe air pollution problem in the Los Angeles basin, found that some of the by-products of the combustion process in the automobile engine formed new substances in the air. Specifically, these by-products are hydrocarbons (HC) and oxides of nitrogen (NOx). These combustion by-products along with carbon monoxide (CO) are now known to be significant con-

Figure 1-1. Air pollution over heavily populated areas is the result of such factors as fumes and smoke from factories along with the emissions from motor vehicles (Courtesy of Ford Motor Co. of Canada Ltd.).

tributors not only to the Los Angeles pollution problem but also to that of many other large metropolitan areas.

What Is Smog?

Smog is a special kind of air pollution. The word "smog" is a combination of the words "smoke" and "fog," and it was the first label attached to this special form of air pollution.

Photochemical smog occurs when hydrocarbons and oxides of nitrogen react with sunlight (Fig. 1–2). As previously stated, both of these invisible gases are present in very small amounts in the normal automobile engine exhaust. When combined in the atmosphere with sunlight, they react to form the eye-irritating haze known as photochemical smog. In other words, smog is not smoke; it is not fog; and it is not a combination of the two. Smog is the particular end product of a complex photochemical reaction.

Smog Versus Air Pollution

Smog is only the end result of a specific and rather uncommon type of air pollution that occurs in only a few locations. That is,

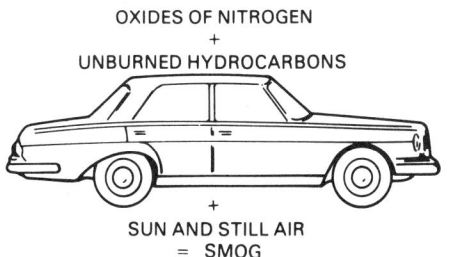

Figure 1–2. Smog occurs when hydrocarbons and oxides of nitrogen unite in the presence of sunlight.

smog only occurs when significant quantities of hydrocarbons and oxides of nitrogen unite in the presence of sunlight. Air pollution is a much more general and common problem affecting many areas both day and night. In other words, in most locations the problem isn't smog, but rather it is air pollution. For this reason, enacted federal laws affecting motor vehicle emissions are primarily concerned with controlling air pollution.

Other Contributing Factors to Air Pollution and/or Smog

Several contributing factors increase the air pollution and/or smog problem over many metropolitan areas. One of these is, of course, the very design of the city itself. That is, tall buildings and relatively narrow streets tend to limit air movement, thus trapping and concentrating contaminants in the air. This causes the levels of certain pollutants to climb above safe maximum limits.

However, air pollution and/or smog does not always remain over major cities and surrounding urban areas. Pollution can move great distances by atmospheric conditions such as the wind. Wind currents can move air pollutants 40 to 100 miles from major metropolitan areas. Thus, the air over certain towns in rural areas, although far removed from the city, may actually contain unsafe quantities of certain contaminants.

The air pollution situation occurring in Los Angeles, California, is a good example of another contributing factor. That is, Los Angeles itself is in a basin, or valley, with the Pacific Ocean to the west and the mountains to the east. When the winds come up from the mountains, they blow the air pollution out over the ocean, making the air over the city reasonably clean. But when there is no wind, the air is still, and the smoke from industry, along with automobile pollutants, does not blow away. Thus, the mountains trap the air pollution over the city; and the dirty, still air covers Los Angeles like a blanket that gradually builds up to a thick, smelly layer of air pollution or smog.

Under normal conditions in this area, the warm air near the ground would rise to the cooler layers of the atmosphere. This would carry the air pollutants up and away from the city. However, Los Angeles and similarly located cities are subject to a weather condition called "inversion." Under this condition, a relatively warmer layer of air hovers above the basin in which the city lies. This inversion layer shuts off the movement of ground-level air and keeps the air pollution near the ground. This, of course, worsens the air pollution condition.

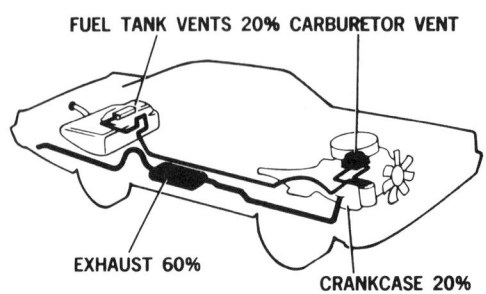

Figure 1-3. Sources of HC emissions from an automobile (Courtesy of Chrysler Corp.).

Emissions from the Automobile—Hydrocarbons

Hydrocarbons, a compound composed of hydrogen and carbon, can be emitted from three different areas of the automobile (Fig. 1-3). For example, about 60 percent of the average automobile HC emissions come out within the exhaust gases. Another 20 percent enters the atmosphere from the engine's crankcase, while an additional 20 percent discharges from the fuel tank and carburetor vents.

Hydrocarbons themselves are not our main concern because they are the prime source of the energy we use (in the form of gasoline) to operate the automobile. What is of concern are the hydrocarbons that pass through the combustion process unchanged and exit through the exhaust into the atmosphere. Once these hydrocarbons escape, they are no longer of any use to us and we just regard them as air pollutants or emissions.

For many reasons, 100 percent, or complete combustion, of the hydrocarbon fuel does not occur within the internal combustion engine. Consequently, some unburned hydrocarbons are exhausted into the atmosphere. One of the causes of these unburned HC emissions is known as *quenching*, a phenomenon which occurs when a flame in the combustion chamber approaches any relatively cool metal surfaces. The flame does not burn close to these cool surfaces; consequently, it goes out or is "quenched" by the cool areas. As a result, this leaves a small amount of unburned fuel or hydrocarbons in this area, which passes out of the combustion chamber during the exhaust stroke.

Figure 1-4 shows typical quench areas within a combustion chamber. Note the quench area formed by the closed or nearly closed space near the end of the chamber. In this particular engine, another quench area exists between the top of the piston and the first compression ring. In

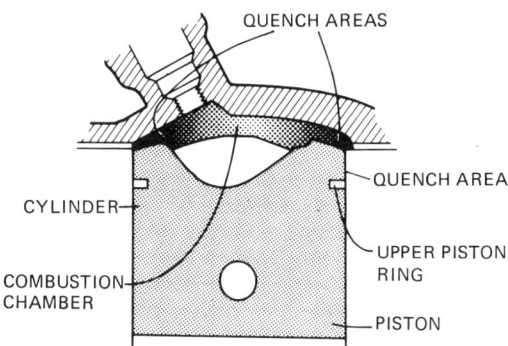

Figure 1-4. Typical quench areas within a combustion chamber.

either area, the flame will not burn. Therefore, a small amount of unburned fuel from both areas remains after the combustion process is over.

Another cause of unburned hydrocarbons in the exhaust is combustion chamber deposits. These deposits are porous. Therefore, as the piston comes up on the compression stroke, it forces some fuel into these deposits. The absorbed fuel never burns, and it comes out of these deposits late in the power stroke or during the exhaust stroke where it passes into the atmosphere.

An engine in poor mechanical condition will also emit excess hydrocarbons. If, for example, the compression in any given cylinder is low due to faulty valves or rings, the air/fuel charge within that cylinder will not sustain the compressed state necessary for good combustion. Consequently, during its exhaust stroke, large quantities of unburned hydrocarbons will pass out of the chamber and into the exhaust system.

A malfunction within the ignition system can also cause a significant increase in HC emissions. For instance, if any part of the ignition system (a spark plug, a plug wire, or the contact points themselves) becomes defective, the end result would be a weak spark or no spark at all. This would cause either incomplete combustion or a total misfire. In either case, there would be an increase in the emission of unburned hydrocarbons from the affected cylinder.

The overall temperature of the air/fuel mixture will also have an effect on unburned HC emissions. If, for example, the fuel and air temperatures are too low, this results in poor mixing and distribution of the air/fuel charge. Both of these factors can result in excessive concentrations of either rich or lean mixtures within an intake manifold. When these rich or lean concentrations move into a given cylinder, they will not burn evenly, resulting in the emission of unburned hydrocarbons. In other words, at higher temperatures, the air/fuel mixture more easily changes to a vapor. A totally vaporized air/fuel mixture travels easily through the manifold runners without losing any of the heavier hydrocarbon particles. Therefore, the vaporized air/fuel charge reaching the cylinders is more uniform in quantity and will burn evenly.

Another cause of unburned hydrocarbons in the exhaust is the quantity of air/fuel mixture reaching the individual combustion chambers. For example, if too rich a mixture, one containing more fuel than necessary, reaches the combustion chambers, there will not be sufficient oxygen present to allow complete combustion. Consequently, unburned hydrocarbons contained in the partially burned mixture escape through the exhaust and into the atmosphere.

If the carburetor feeds too lean a mixture into the chamber, the fuel mixture becomes so diluted with air that it will not burn properly. In other words, the fuel vapor particles spread out too much to properly transfer the combustion flame through the chamber. Therefore, the flame goes out before it burns all the fuel. This action also results in unburned hydrocarbons escaping into the atmosphere.

Excessive exhaust gas dilution of the air/fuel mixture is also a cause for unburned hydrocarbons in the exhaust. This particular negative condition occurs during periods of high intake manifold vacuum such as at engine idle or deceleration. This dilution of the air/fuel mixture itself results in a charge that may not burn completely or can cause a complete misfire.

Negative Effects of Hydrocarbon Emissions

There are, in fact, several negative effects due to excessive amounts of hydrocarbons in the atmosphere. As stated earlier, the chemical combination of hydrocarbons and oxides of nitrogen in the presence of sunlight result in photochemical smog.

However, all the hydrocarbons do not combine in this manner to form smog. Some heavier hydrocarbons, the aromatics, remain in the air and act as an irritant to the eyes. Finally, one of these heavier hydrocarbons, benzopyrene, is also thought to be a carcinogen, a cancer-causing agent.

Cause of NOx Emissions

The second emission to be dealt with here, but not the least in importance, is of oxides of nitrogen (Fig. 1–5), a result of high temperature and pressure as found in the internal combustion engine. The high

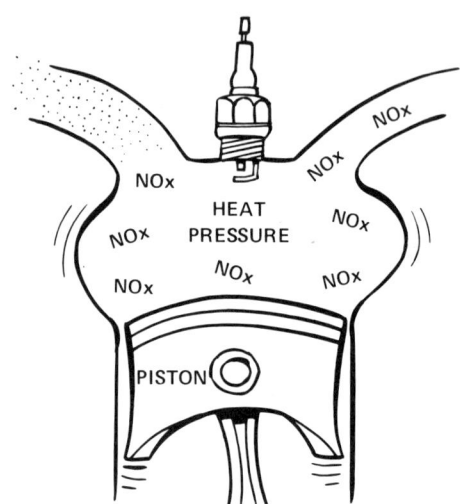

Figure 1–5. Oxides of nitrogen form in the combustion chamber under high temperature and pressure.

temperatures necessary to form NOx, above 2,500 °F, only occur during some phases of engine operation.

NOx is the symbol representing any combination of nitrogen and oxygen. The "x" symbolizes any number of oxygen atoms present in the compound.

When an engine is operating, atmospheric air moves through the carburetor where it mixes with fuel before entering the engine. Under normal conditions the combustion process consumes the oxygen from the air while the nitrogen passes out of the exhaust virtually unchanged.

During the combustion process, temperatures may exceed 4,500 °F. At temperatures above approximately 2,500 °F, oxides of nitrogen form very rapidly from the nitrogen and oxygen in the air. Consequently, the formation of NOx is dependent upon temperature, and any variable that causes an increase in temperature above about 2,500 °F will cause an increase in NOx emissions.

Effects of NOx Emissions in the Atmosphere

As mentioned earlier, NOx is one of the compounds that form photochemical smog. The oxides of nitrogen have a composition of about 97 to 98 percent nitric oxide (NO) and about 2 to 3 percent nitrogen dioxide (NO_2). Nitric oxide itself is a colorless gas. But when exhausted into the atmosphere, it combines with oxygen (O_2) to form nitrogen dioxide (NO_2), a gas with a brownish color. Nitrogen dioxide then combines with certain active hydrocarbons in the presence of sunlight to form smog.

One of the negative aspects of this combination is the formation of ozone (O_3). The sunlight breaks apart some of the nitrogen dioxide to form nitric oxide and oxygen. This combination of the single oxygen atom with diatomic oxygen forms ozone.

Not only is ozone an odorous gas, one of the smells associated with smog, it also has other detrimental effects. For example, ozone acts as an irritant to lung tissue and to the eyes. In addition, ozone rapidly deteriorates rubber products and affects the growth rate of some crops and plants.

Cause of Carbon Monoxide Emissions into the Atmosphere

Carbon monoxide (CO) is another harmful emission produced by the automobile engine during its combustion process. Carbon monoxide is an invisible gas that results from incomplete combustion of the engine's air/fuel mixture (Fig. 1–6). If complete combustion of the fuel charge occurs, the process would produce carbon dioxide (CO_2) instead of carbon monoxide.

However, whenever the air/fuel ratio becomes richer than about 15:1, there is insufficient oxygen present in the charge to

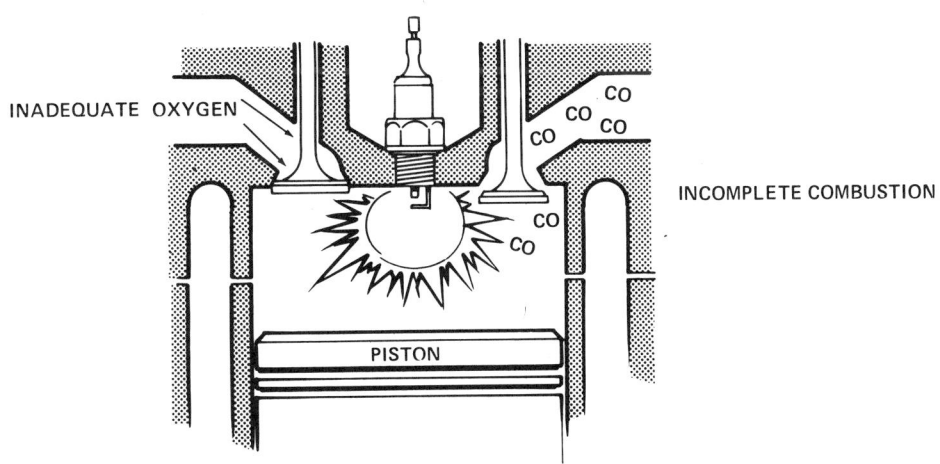

Figure 1–6. Carbon monoxide is the result of incomplete combustion of the air/fuel mixture.

complete the combustion process. This lack of oxygen results in an incomplete conversion of carbon (C) to carbon dioxide (CO_2). Moreover, an increase in CO emissions usually results in an increase in HC emissions. The reason for this is simply that there is insufficient oxygen available to completely burn all the air/fuel mixture.

Effects of Carbon Monoxide Emissions

Excessive amounts of CO emissions can create several problems. For instance, carbon monoxide itself is an odorless, colorless, poisonous gas. When people inhale carbon monoxide into their lungs, it transfers into the bloodstream. There it takes the place of oxygen within the red blood cells, resulting in a reduction of the oxygen supply to the body. This lack of oxygen can cause headaches, a reduction in mental alertness, and even death if the carbon monoxide concentration is high enough.

Another negative effect of carbon monoxide is that it also increases the rate at which photochemical smog forms, by accelerating the conversion of nitric oxide to nitrogen dioxide. In this case carbon monoxide speeds up this chemical reaction by combining with oxygen and the nitric oxide to form carbon dioxide and nitrogen dioxide. In other words $CO + O_2 + NO$ creates $CO_2 + NO_2$.

Cause of Particulate Emissions into the Atmosphere

The automobile engine also emits particulates from its exhaust into the atmosphere. Particulates are among some of the combustion by-products of the engine, and they may consist of calcium, ammonia, iron, hydrocarbons, carbon, lead, and sulfur. These kinds of emissions are known as *particulates* because they are actually small visible amounts of solids or liquids suspended in the atmosphere. This is in contrast to the other invisible emissions produced by an operating engine.

Particulate emissions primarily result from the use of hydrocarbon fuels and certain fuel additives. Carbon particulates, for example, originate from the hydrocarbon fuels and oils. The source of lead (Pb) particulates is tetraethyllead (Pb $(C_2H_5)4$) that manufacturers add to the fuel to raise its octane rating. However, with the advent of lead-free and low-lead gasoline this form of particulate problem is on the decline.

Effects of Particulate Emissions

There are primarily two negative aspects to particulate emissions. First, small, suspended particulates are one of the causes of the reduced visibility that accompanies photochemical smog. Second, authorities suspect lead particulates as being a health hazard in two ways: (1) the respiratory intake of airborne lead during breathing and (2) the contamination of food sources by the settling of lead in the soil.

Cause of Oxides of Sulfur Emissions into the Atmosphere

Another form of exhaust emission is that of oxides of sulfur (SOx), a recently discovered pollution problem that became quite evident with the introduction of catalytic converters on automobiles. (Catalytic converters will be discussed in Chapters 17 and 18.)

Oxides of sulfur result from the small amount of sulfur present in the gasoline itself. Under normal conditions the sulfur content of the fuel is less than 0.1 percent. However, this small amount of sulfur is emitted from an engine's exhaust into the atmosphere in the form of oxides of sulfur.

Effect of Oxides of Sulfur Emissions

The production of a sulfuric acid mist is the major problem arising from SOx emissions from motor vehicles. However, this particular problem primarily occurs with motor vehicles equipped with catalytic converters. The reason is that these converters provide an oxidizing atmosphere that enhances the formation of sulfuric acid from the sulfur compounds within the fuel.

In the catalytic converter sulfur dioxide (SO_2) converts to sulfur trioxide (SO_3) within the oxidizing atmosphere of the unit. In other words $SO_2 + O_2$ creates SO_3. Next, sulfur trioxide (SO_3) unites with water (H_2O) vapor (the water vapor results from the oxidation of hydrocarbons) and forms a sulfuric acid (H_2SO_4) mist.

This sulfuric acid mist is very corrosive. Consequently, it will, upon contact, deteriorate textiles, building materials, and vegetation. Moreover, it is a well-known fact that sulfuric acid in any form is very harmful to living tissue.

Air Pollution Research and Legislation

Once the problem of air pollution was apparent, it became the subject of intense research and investigation. As mentioned earlier, studies at the California Institute of Technology in the 1950s showed that some of the by-products of the combustion process reacted with sunlight to produce photochemical smog. As a result, California became the first state to pass laws to limit automotive emissions.

At first, the standards established by California in a given year became federal standards soon afterwards. For instance, beginning with the 1961 new-model automobiles, California required crankcase emission control devices. This same type of device became mandatory for the rest of the United States starting with the 1963 new-model automobiles.

The first federal air pollution research program began in 1955. Since that time, many bills such as the Clean Air Act of 1963 have become law. This act provided money to states for the development of air pollution control programs.

In 1965 Congress amended the 1963 Clean Air Act. This amendment authorized the setting of emission standards by the Department of Health, Education and Welfare (HEW). These standards were first applied nationwide for 1968 new-model automobiles and medium-sized trucks. However, in June 1968, HEW adopted the California standards for nationwide application, beginning with the 1970 and later-year vehicles.

Although the Clean Air Act was a new approach to the air pollution problem, many changes followed in 1970. These changes attempted to take a fragmented program and form it into a unified attack on pollution of all kinds. This new program resulted from the passage of the National Emissions Standard Act in 1970 as Title II of the Clean Air Act.

This particular act set HC and CO emission standards for 1972 through 1974 model year vehicles and projected them for the 1975 and later automobiles and light trucks. In addition, this bill also set NOx emission standards for the 1973–74 model year vehicles and projected the standards for the 1975 and later models. Finally, the standards set within this act represented a 90 percent reduction in hydrocarbons, carbon monoxide, and oxides of nitrogen below those set in the early 1970s.

Perhaps the most important legislation affecting the automotive service industry was the 1977 Clean Air Act amendments. These particular amendments provided direction to those states that could not meet air quality standards. The amendments themselves required the establishment of inspection/maintenance programs for areas not able to meet air quality standards for carbon monoxide and other oxidents by December 31, 1982.

The establishment of an inspection/maintenance program brings with it added responsibilities to the automobile repair shop. That is, when a particular vehicle is identified by the inspection program as having excessive emissions, it will be up to an automotive service facility to correct those problems. How effectively these repairs are made and how efficiently mechanics repair and tune any vehicle will have a large impact on the overall improvements in air quality.

Air Pollution Regulatory Agencies

Federal Agencies

As mentioned earlier, at first HEW was responsible for all air pollution regulation. However, in 1970, along with the passage of the clean air amendments, the Environmental Protection Agency (EPA) was established. This agency, part of HEW, is now the federal agency responsible for enforcing the Clean Air Act.

Relating to motor vehicle emissions, the EPA has two basic responsibilities. First, this agency can set interim standards. As previously stated, the Clean Air Act amendments of 1970 established standards representing a 90 percent reduction in hydrocarbons, carbon monoxide, and oxides of nitrogen below those standards first set in the early 1970s. However, these same amendments provided for the postponement of these standards by one year if the EPA determines that certain conditions listed in the act are met. In such an event the EPA must prescribe interim standards that reflect the greatest degree of control achievable within available technology. The original standards were postponed and interim ones set. Later, Congress acted to extend these interim standards and delay the implementation of the original lower requirements until a future date.

The second main responsibility of the EPA is monitoring the automobile manufacturers. A large portion of this monitoring results from the federal motor vehicle emission control program, made up of the following three elements:

1. A certification program—the EPA certifies all new vehicle prototypes to show that they meet federal emission standards.

2. Selective enforcement auditing—the random testing of selected vehicles right off the assembly line, which assures that production vehicles are meeting the federal emission standards.

3. Recall—if, through a surveillance program on vehicles in use, defects are

found that prevent a group of vehicles from meeting federal standards, the EPA can require the manufacturer(s) to recall these particular vehicles and correct the problems.

How EPA Certifies New Vehicles

As each new model year approaches, automotive manufacturers build prototype or emission data cars for the use of EPA. These particular vehicles are driven for 4,000 miles to stabilize the engine and the emission control systems before the certification test begins. However, vehicle manufacturers are responsible for conducting a 50,000-mile durability test of their own emission control systems.

The first step in the EPA test is to precondition the vehicle. This amounts to positioning it on a chassis dynamometer and permitting the vehicle to stand for 12 hours at an air temperature of 73 °F. The purpose of this is to prepare the vehicle for a simulated cold-start. The dynamometer will be necessary later on, for the actual test to simulate a driving cycle that represents urban driving conditions. Finally, instruments are also made ready to analyze the vehicle's exhaust for harmful pollutants.

The actual test sequence takes about 41 minutes. The first 23 minutes make up the cold-start driving test. The next 10 minutes represent a waiting or a hot-soak period. The final 8 minutes of the test is a hot-start test, representing a short trip in which the vehicle stops and starts several times while hot. If the emission test results for all prototype vehicles are equal to or lower than all the HC, CO, and NOx standards, EPA certifies the vehicle, and the manufacturer can offer it for sale to the public.

State Agencies

Many states now have their own air pollution control agency for enforcing emission standards. For example, California established the Air Resources Board (ARB) and Oregon the Department of Environmental Quality (DEQ). The authority of these agencies roughly parallels that of the federal EPA. However, although California's ARB and Oregon's DEQ operate under permission of the EPA, their authority extends only to those vehicles sold or brought into the respective state.

A state agency can have as many as three functions. First, similar to the EPA, a state operation may perform emission certification tests on prototype new-model motor vehicles. Second, a state agency can have the responsibility of establishing and monitoring inspection/maintenance programs for motor vehicles in areas that cannot meet air quality standards. Finally, a state office can establish emission control standards within its jurisdiction. However, although these state standards may be more strict, they can never be below those set by federal law.

Emission Control Standards

As early as 1968, the federal government and some states had already established emission control standards. Since that period, these standards have become more stringent in regard to the allowable emissions from the automobile and light- to intermediate-size trucks. Before 1968, for instance, an average automobile could produce HC emissions from the exhaust at a rate of 1,250 parts per million (ppm), with a CO level of about 6 percent of the total exhaust volume.

However, starting in 1968, a vehicle with a similar engine, could not emit more than 650 ppm of hydrocarbons and a 5.0 percent volume of carbon monoxide. Of course, now, all new automobiles and light trucks must not discharge from their exhaust more than 175 ppm of hydrocarbons and 0.5 percent carbon monoxide by volume to remain within current standards.

Obviously then, the federal standards and state emission regulations have changed a great deal since 1968. In some cases the standards were downgraded because at the time certain vehicles could not meet the requirements. But now, all new vehicles must meet the current standards in order to be sold in this country.

The main point to keep in mind is that these standards or regulations do exist. And a federal or state agency is responsible for supplying the garages and/or service centers with the current emission control standards or regulations.

Effect of Emission Controls on Fuel Consumption

Most manufacturers have met the strict emission control standards through the combined use of such approaches as engine modifications, retarded ignition timing, lower compression ratios, carburetor modifications, and exhaust gas recirculation. The combined effect of these basic changes has been to reduce the efficiency of the engine, resulting in a decrease in the miles per gallon (mpg) obtained by a vehicle on the highway. Although the exact loss of fuel economy due to emission controls does vary somewhat for different makes and models of vehicles, estimates of the average loss in fuel economy for the 1973-model automobiles was in the vicinity of 15 percent. In addition, these emission controls had the effect of lowering the driveability of the vehicle.

The 1975, and later, standards were much more stringent than earlier regulations. To meet the newer standards, manufacturers installed catalytic converters on their vehicles. This control device permitted adjustment of the engine systems to provide somewhat better fuel economy than the earlier models and improved the driveability of the vehicle.

Summary

1. The problem of air pollution became quite apparent during and after World War II.
2. Although air pollution is not a problem in all areas, certain cities do experience severe air contamination.
3. Air pollution is the introduction of any contamination into the atmosphere in an amount large enough to injure human, animal, or plant life.
4. Natural air pollution has always existed in one form or another.
5. Trees and plants give off a gaseous form of natural air pollution.
6. Man-made air pollution is any form of contaminant we create and permit to enter the atmosphere.
7. Legislation now controls the amount of air pollution from factories and power plants.
8. The automobile produces air pollution by emitting such substances as hydrocarbons, oxides of nitrogen, and carbon monoxide.
9. Photochemical smog occurs when

hydrocarbons and oxides of nitrogen react with sunlight.
10. Smog is only the end result of a specific and rather uncommon type of air pollution that occurs in only a few locations.
11. There are several factors that increase the air pollution problem over metropolitan areas.
12. Hydrocarbons, compounds composed of hydrogen and carbon, can be emitted from three different areas of the automobile.
13. Complete combustion of hydrocarbon fuel does not occur within the internal combustion engine for many reasons.
14. There may be several quench areas in an engine's combustion chamber.
15. The temperature of the air/fuel mixture will have an effect on unburned HC emissions.
16. Another cause of unburned hydrocarbons in the exhaust is the quantity of air/fuel mixture reaching the individual combustion chambers.
17. Exhaust gas dilution of the air/fuel mixture is also a cause of unburned hydrocarbons in the exhaust.
18. There are several negative effects resulting from excessive amounts of hydrocarbons in the atmosphere.
19. Oxides of nitrogen result from high temperatures and pressure, as found in the internal combustion engine, acting on the air passing through the chambers.
20. At temperatures above about 2,500 °F, oxides of nitrogen form very rapidly.
21. Oxides of nitrogen are one of the compounds that form photochemical smog.
22. A negative aspect of the combination of oxides of nitrogen and hydrocarbons is ozone.
23. Not only is ozone an odorous gas, it also has other detrimental effects.
24. Carbon monoxide is an invisible gas resulting from the incomplete combustion of the air/fuel mixture.
25. Carbon monoxide is an odorless, colorless, poisonous gas.
26. The automobile engine also emits particulates from its exhaust into the atmosphere.
27. Particulate emissions primarily result from the use of hydrocarbon fuel and certain fuel additives.
28. There are two negative aspects to particulate emissions.
29. Oxides of sulfur result from the small amount of sulfur present in the gasoline itself.
30. The production of a sulfuric acid mist is the major problem arising from SOx emissions from motor vehicles.
31. A sulfuric acid mist is very corrosive.
32. Once the problem of air pollution was apparent, it became the subject of intense research and investigation.
33. The first federal air pollution research program started in 1955.
34. The Clean Air Act became law in 1963.
35. Title II of the Clean Air Act became law in 1970.
36. The Clean Air Act amendments of 1977 established state inspection/maintenance programs.
37. The EPA is responsible for enforcing the Clean Air Act.
38. The EPA has two basic responsibilities regarding motor vehicle pollution.

39. Each year, the EPA certifies prototype new-model vehicles.
40. Many states now have their own air pollution control agency for enforcing emission standards.
41. A state emission agency can have as many as three functions.
42. By 1968, the federal government and some states had established emission control standards.
43. Federal and state emission laws have changed a great deal since 1968.
44. The combined effect of many emission control devices is the reduction in vehicle fuel economy and driveability.

Review Questions

The questions listed below will assist you in determining how well you remember the material contained in this chapter. Read each question carefully before choosing the answer. If you cannot answer the question, review the section in the chapter that covers the question.

1. Trees emit a gaseous form of pollution called _____.
   a. carbon monoxide
   b. hydrocarbons
   c. oxides of nitrogen
   d. oxides of sulfur

2. Smog occurs when sunlight reacts with hydrocarbons and _____.
   a. oxides of nitrogen
   b. carbon monoxide
   c. carbon dioxide
   d. water

3. _____ percent of the total hydrocarbon emissions from the automobile come out of the tailpipe.
   a. 40
   b. 50
   c. 60
   d. 70

4. One of the factors that affects HC emissions from a vehicle's exhaust is air/fuel _____.
   a. temperature
   b. pressure
   c. vacuum
   d. volume

5. Oxides of nitrogen form in an engine's combustion chamber when temperatures exceed _____ °F.
   a. 1,000
   b. 1,500
   c. 2,000
   d. 2,500

6. _____ is a very poisonous gas.
   a. nitric oxide
   b. nitrogen dioxide
   c. carbon monoxide
   d. carbon dioxide

7. The production of a sulfuric acid mist primarily occurs from motor vehicles equipped with _____.
   a. air injection pumps
   b. catalytic converters
   c. PCV systems
   d. dual exhausts

8. Congress passed the Clean Air Act in _____.
   a. 1961
   b. 1962
   c. 1963
   d. 1964

9. The EPA, relating to automotive emissions, has _____ main functions.
   a. two

b. three
  c. four
  d. five
10. Federal emission control standards first became effective in _____.
  a. 1965
  b. 1966
  c. 1967
  d. 1968

For the answers to these review questions, turn to the Appendix.

Chapter 2

Chemical Reaction of Gasoline Combustion within an Automobile Engine

As mentioned in Chapter 1, an automotive engine produces carbon dioxide and water as by-products of its combustion process. However, ideal conditions that promote this type of combustion do not occur very often. Therefore, the engine emits a number of harmful pollutants into the atmosphere.

In order to understand what happens inside the engine to cause incomplete combustion and the harmful emissions, the reader must know the makeup of the various constituents that together produce or create the burning process. These include the composition of the air and gasoline, along with the actual operating cycle of the engine.

Composition of Air

A great quantity of atmospheric air moves through an operating engine. This air mixes with fuel within the carburetor, and this combined mixture passes on into the engine combustion chambers, where it ignites and burns. Under ideal conditions, the carburetor supplies a mixture ratio of about 15:1; that is, 15 pounds of air for each 1 pound of gasoline (Fig. 2–1). As a result, each gallon of gasoline requires as much as 1,200 cubic feet of air for normal combustion in the engine.

At this point, the reader may wonder why the need for so much air in the mixture. The answer is simple when you consider the composition of the air from our atmosphere. The air we breathe and the air used in the automobile engine is a mixture of 21 percent oxygen, 78 percent nitrogen, and 1 percent other gases. However, the *oxygen* in the air alone supports the combustion of gasoline. Consequently, the engine must take in a great deal of air along with the fuel so that sufficient oxygen is available for adequate combustion.

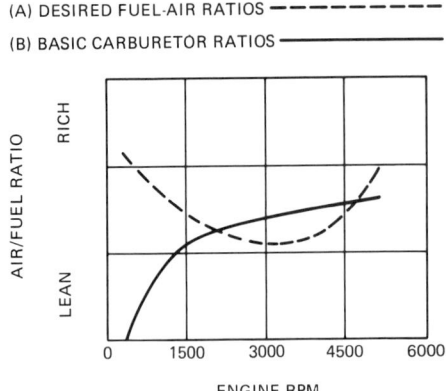

Figure 2-1. Ideal air/fuel ratio for complete combustion is about 15:1.

Normally, the *nitrogen* portion of the air passing through the engine is harmless. At low temperatures, for example, nitrogen is inert. That is, it will not unite chemically with anything, except under very special circumstances.

However, during the combustion process, temperatures can reach from 5,000 to 6,000 °F. At about 2,500 °F, some of the nitrogen within the air/fuel mixture unites with oxygen to form oxides of nitrogen (see Fig. 2-2). Although not very much of the total nitrogen inducted into the engine reacts with the oxygen, enough does to cause some severe pollution problems.

Composition of Gasoline

The second component necessary for combustion is fuel; in the automobile engine, this is most commonly gasoline.

OXYGEN (O) + NITROGEN (N) + HIGH TEMPERATURE = NITROGEN OXIDES (NO_x)

Figure 2-2. Formation of NOx emissions.

Gasoline contains literally hundreds of chemicals, the major proportion of which are known as hydrocarbons, compounds consisting of hydrogen and carbon atoms.

The typical hydrocarbons in gasoline may be N-pentane, 2-methylbutane, and benzene. Although these are long chemical names for the compounds, they are all hydrocarbons in that they contain only atoms of hydrogen and carbon.

For the purpose of clarifying the discussion of the combustion of hydrocarbons, these compounds are divisible into three groups. We can, for instance, divide them into straight-chain, branched-chain, and ring hydocarbons. Sometimes, the ring hydrocarbons are also known as aromatic hydrocarbons.

There is one important thing to understand about these hydrocarbons in reference to combustion. That is, it is the energy stored in the chemical bond within the hydrocarbons themselves that is released when these compounds burn in the presence of oxygen inside the combustion chamber.

Since the efficiency of any engine is partly dependent upon its compression ratio, the octane rating of a fuel must be correct for knock-free performance. The octane rating necessary for knock-free operation must be higher with increases in engine compression ratio. For instance, at a ratio of 4:1, the octane number necessary is about 60. For a compression ratio that is very high, such as 12:1, the required octane number is 102.

The *octane number or rating* is a numeral given a fuel. This number is nothing more than a measure of the antiknock properties of a particular gasoline. The higher the octane rating, the more resistant this fuel is to spark knock or detonation.

The octane rating of a fuel is based on a

comparison of its performance with other fuels in a variable compression ratio test engine. The test engine operates on a fuel, consisting of a mixture of two hydrocarbons, to its peak performance without detonating.

The first hydrocarbon in the test fuel is known as N-heptane; that is, a straight-chain compound. This has an assigned octane number of 0. The second hydrocarbon is known as a 2-2-4-trimethylpentane. This compound is sometimes referred to as an isooctane and is a branched-chain hydrocarbon with an octane number of 100.

It should be obvious then that one way to raise or lower the octane number of a fuel is to alter the mix of the hydrocarbons used. For example, fuel that has a low octane number will have more straight-chain hydrocarbons with low octane rating. On the other hand, a fuel with a high octane rating will have more branched or ring hydrocarbons that have high octane numbers. In fact, fuel with a very high rating can have as much as 40 percent of its total volume made up of ring hydrocarbons that have a higher octane than either straight- or branched-chain compounds. However, this method of raising a fuel's octane rating is rather expensive.

Another and less expensive way of increasing the octane number of a fuel is by adding small amounts of certain additives. One of the most common additives is tetraethyllead (TEL). This is usually known as the antiknock additive within the gasoline. Tetraethyllead is so named because there are four ethyl groups (1, 2, 3, and 4) together with an atom of lead to which attach the four ethyl groups. In the past, 2 to 5 grams of this material have been added to a gallon of gasoline in order to raise its octane number, with each gram raising the octane rating about five numbers.

As mentioned in Chapter 1, the use of TEL in fuel to raise its octane rating is on the decline because lead creates a form of air pollution. Therefore, federal regulations currently limit TEL content to 3.5 grams per gallon. In its place, some unleaded fuels have methylcyclopentadienyl manganese tricarbonyl (MMT), a catalyst-compatible improver, to further enhance its antiknock quality.

There are also small amounts of other additives in gasoline that serve useful purposes. For instance, some fuels have additives that are detergents, which act to keep an engine clean (Fig. 2-3). There can also be deposit modifiers, anti-icing agents, antirust agents, metal deactivators, and dyes.

Whatever goes into the engine within the fuel must come out in some form or other, even if the particular chemical does not perform any useful function in the engine itself. For example, gasoline may contain the impurities, sulfur and phosphorus. These compounds are in the original crude oil and rather difficult and ex-

PROPERTIES OF EXXON® GASOLINES

	EXXON 2000 (UNLEADED)	EXXON (LEADED)	EXXON EXTRA UNLEADED
Research octane number[a]	93	94	97
Motor octane number[a]	84	84	87
TEL content	Nil[b]	Approx. 2 grams/gal	Nil[b]
Approximate compression ratio served	8.2:1	8.2:1	8.2:1 and higher
Carburetor icing control		Adequate	
Volatility adjustments		Continually with season for each market area	
Detergent		For required carburetor deposit control	

[a]Lower for higher-altitude market areas.
[b]Trace lead contents are due to pickup in the distribution system.

Figure 2-3. Properties of Exxon® gasoline.

pensive to remove. These materials are not useful, and they do form certain substances inside the engine that create air pollution.

Operating Cycle of the Engine

The Intake Stroke

The final constituent necessary to complete the chemical process of converting the heat energy stored in the fuel into motion is the *operating cycle of the engine.* This cycle consists of four strokes of the piston within the various cylinders: the intake, compression, power, and exhaust stroke.

The intake stroke of a four-cycle engine begins with a piston at top dead center (TDC). Activating the starter causes the crankshaft to rotate in a clockwise direction (Fig. 2-4). The crankshaft, through the connecting rod, forces the piston to move downward in the cyclinder. This movement of the piston creates a vacuum or difference in pressure in the space above the piston.

The engine manufacturer times the intake valve action so that it opens automatically at or slightly before the piston starts down. Therefore, a mixture of air and gasoline rushes through the intake manifold and into the cylinder, pushed by atmospheric pressure from outside the engine. In other words, atmospheric pressure forces the air/fuel mixture into the cylinder in an attempt to raise and stabilize the lower pressure within the cylinder.

Atmospheric pressure results from the blanket of air under which we live. The atmosphere itself exerts pressure on our bodies and everything else that exists

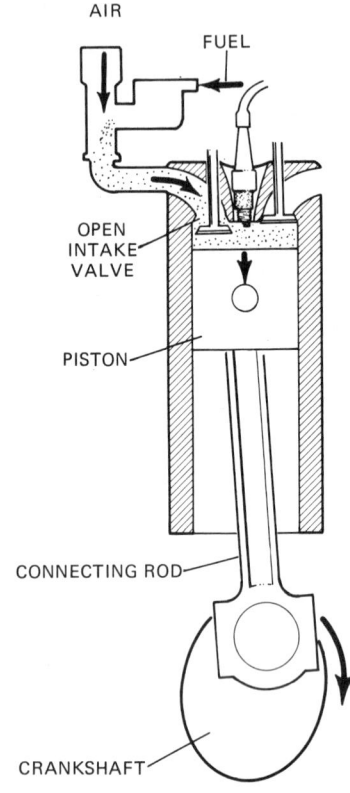

Figure 2-4. Intake stroke of a four-cycle engine.

around us (Fig. 2-5). This pressure results from the weight of the air itself. At sea level, for example, a cubic foot of air weighs about 0.08 pound or 1.25 ounces. This may not seem to be very much; however, the total blanket of air is over 50 miles thick.

The total weight or downward push of the atmosphere results in a pressure of 14.7 pounds per square inch (psi) at sea level. From sea level upward to plateaus and mountains this pressure decreases. The pressure reduction is due to a lowering of the volume of air at higher altitudes. Consequently, the total weight of the air decreases and so does atmospheric pressure.

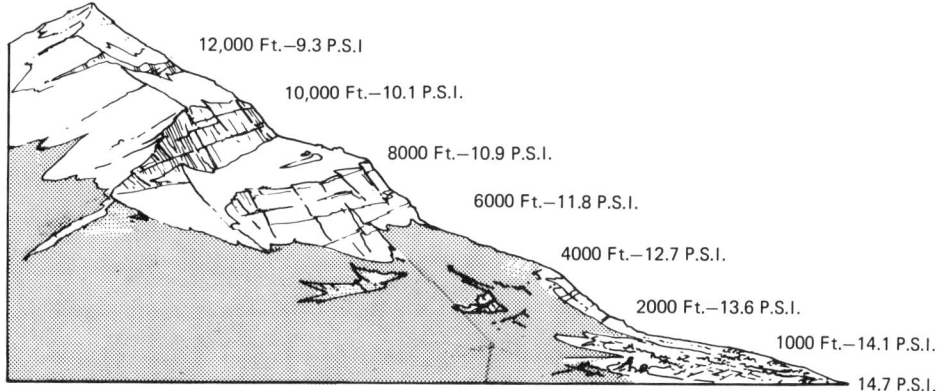

Figure 2-5. Weight of the air produces atmospheric pressure.

The arrows in Fig.2-4 illustrate the inrushing air/fuel mixture due to the effect of atmospheric pressure. Also notice that the exhaust valve remains closed during this downward stroke of the piston. This valve closure prevents the entering air/fuel charge from escaping through the exhaust port. After the piston reaches the bottom of the intake stroke, the cylinder is practically full of an air/fuel charge. The drawing of this charge into the cylinder during the downward movement of the piston constitutes the intake stroke of the engine.

During the intake stroke, the carburetor determines the air/fuel mixture. This complex mixture consists of all the chemicals contained in the gasoline, along with the air pulled from the atmosphere. Therefore, the complete mixture contains oxygen and nitrogen molecules, ring, straight-, and branched-chain hydrocarbons, and the additives that are in the fuel.

Compression Stroke

After the piston reaches bottom dead center (BDC) on the intake stroke, it moves upward again as the starter continues to turn the crankshaft in a clockwise direction (Fig 2-6). As the piston begins to move up, the intake valve closes; the exhaust valve remains closed. Now, since both valves are in the closed position, the piston compresses the air/fuel mixture to a pressure of about 100 to 150 psi into the small space between the top of the piston and cylinder head. The actual compression pressure varies somewhat from one engine design to another.

As the pressure of the mixture increases tremendously so does its temperature. As mixture temperature goes up, the straight-chain hydrocarbons have a tendency to begin to burn before the other types of hydrocarbons. If the fuel did not have an antiknock additive, there could be spontaneous combustion of the mixture.

Without the spark plug actually firing the mixture, it will explode rather than burn smoothly. This explosion sets up shock waves that strike the cylinders and pistons, causing the characteristic knocking or pinging sound. In addition, spontaneous combustion completely wastes the stored energy within the hydrocarbons.

The antiknock additive TEL prevents this from occurring by decomposing. TEL

Power Stroke

Just as, or slightly before, the piston reaches TDC on the compression stroke, a timed electrical spark appears at the spark plug gap (Fig. 2–7), igniting the air/fuel charge. The burning mixture begins to expand; almost immediately, the pressure within the combustion chamber above the piston increases to about four times the pressure before ignition occurred. This results in a pressure of between 400 to 600 psi in an average engine cylinder, with a total force applied to the top of the piston being around 2 tons.

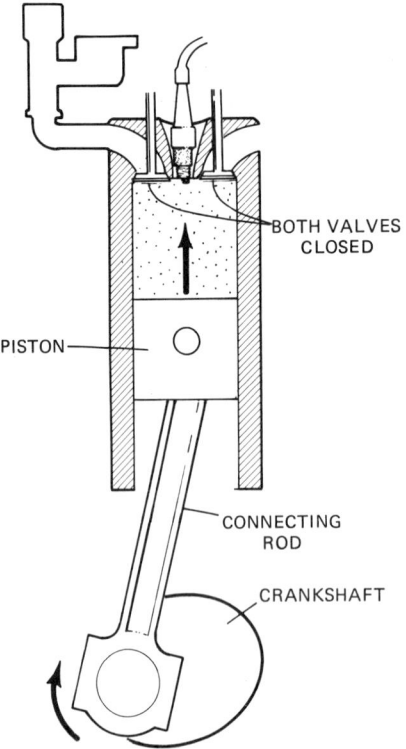

Figure 2–6. Compression stroke of a four-cycle engine.

breaks up to form four ethyl and lead groups. The lead groups then combine with the decomposing straight-chain hydrocarbons to form substances that do not burn as rapidly as would normal straight-chain compounds. As a result, the additive retards or slows down the combustion until the piston reaches the top, or nearly the top, of its compression stroke.

When the piston reaches TDC again during its upward travel, the compression stroke is over, and the crankshaft has now rotated 360° from its starting point. The air/fuel charge is now under full compression so that it will produce a great deal of power when the spark plug ignites it.

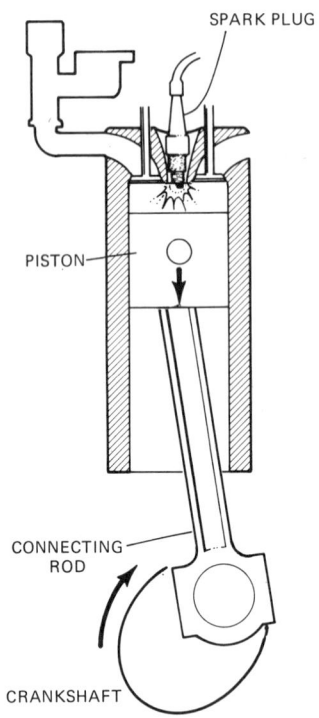

Figure 2–7. Power stroke of a four-cycle engine.

Notice also in Fig. 2–7 that both valves remain closed during the power stroke. This action assures that the total force of the expanding gas applies itself to the head of the piston. This tremendous force pushes the piston downward on the power stroke, causing the connecting rod to rotate the crankshaft. In other words, the force resulting from the expansion of the burning air/fuel mixture is now turning the crankshaft and not the starting motor.

Under ideal conditions, this combustion process produces some harmless by-products. As combustion begins, the straight-chain hydrocarbons, along with their attached lead groups, begin to burn. The branched-chain and ring hydrocarbons also start to burn and react with the oxygen. All three types of hydrocarbons form carbon dioxide, water, and heat during the combustion process.

Unfortunately, complete combustion does not always occur, partially due to the high temperatures inside the cylinders and to the nature of the fuel itself. Consequently, there are several negative side effects. For example, the lead compounds from the antiknock additive have now finished their job. At this point, the scavenging agents, included in the fuel specifically to remove the lead compounds do so by reacting with the lead to form gaseous compounds. The chemical names for these gases include lead chloride, lead bromide, and lead bromochloride.

The high combustion chamber temperatures can also cause several other problems. For instance, at $2,500°F$ the nitrogen contained in the air/fuel mixture begins to combine with oxygen. Although not very much of the nitrogen reacts with oxygen, enough does to create some severe emission problems in the form of the compound nitric oxide. The high temperatures can lead to some unusual chemical changes in the ring hydrocarbons contained in the gasoline. In this case these compounds can stick together, or they can fuse to form strange chemicals such as benzopyrene.

The burning of the remaining hydrocarbons in the fuel is also never quite complete for a number of complex reasons. For example, the air/fuel ratio normally used in automotive engines is too rich; this ratio is necessary so the engine can achieve maximum power. Therefore, the rich mixture does not have sufficient oxygen available to burn all of its hydrocarbons.

Thus, an overly rich air/fuel mixture results in incomplete combustion, and some hydrocarbons are left over. In addition, incomplete combustion gives rise to the formation of carbon monoxide. Remember, if an engine achieves perfect combustion of the hydrocarbons, the process produces carbon dioxide instead of carbon monoxide.

As mentioned in Chapter 1, quenching leaves some hydrocarbons unburned. This occurs because the area at the edge of the combustion chamber, cylinder, and piston is metal and thus conducts heat away. Since these components are somewhat cooler, the combustion process does not burn all the mixture adjacent to these areas; consequently, some unburned hydrocarbons remain that will later pass out of the cylinder during the exhaust stroke.

Exhaust Stroke

Near the end of the downward movement of the piston on the power stroke, the camshaft opens the exhaust valve, but the intake valve remains closed (Fig.2–8). By

CHEMICAL REACTION OF GASOLINE COMBUSTION WITHIN AN AUTOMOBILE ENGINE

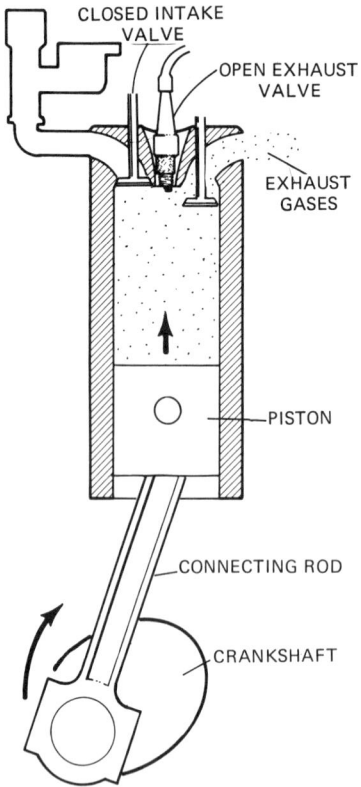

Figure 2–8. Exhaust stroke of a four-cycle engine.

this time, the crankshaft is approaching 540° of total rotation since the cycle began, but it is now turning due to the force applied to the piston by the burning air/fuel mixture.

Although much of the gas has expended itself driving the piston down, some pressure still remains as the exhaust valve opens. This remaining pressurized gas flows comparatively freely from the cylinder by way of the passage (port) opened by the exhaust valve. Then, as the piston again moves up the cylinder, it drives any remaining gases out of the cylinder, past the exhaust valve, and into the exhaust manifold. In other words, while the exhaust valve is open, the upward movement of the piston provides an effective method for expelling all waste gases from the cylinders and combustion chambers.

The exhaust gases contain a very complex mixture of materials and waste gases that pass out of the tail pipe and into the atmosphere. Common materials found in a vehicle's exhaust include unburned hydrocarbons, used hydrocarbons such as benzopyrene, carbon monoxide, carbon dioxide, nitric oxide, lead compounds of various types, unused nitrogen, water vapor, and heat waste. Chapter 1 has already discussed what occurs when most of these combustion by-products react within the atmosphere to produce various forms of air pollution.

Factors Influencing Combustion Chemistry and Negative Emissions

Compression

Many factors influence combustion chemistry and, consequently, the harmful emissions from the automobile engine. These factors include engine compression, air/fuel mixture, idle speed, basic ignition timing, and operating temperature.

As previously mentioned, hydrocarbons from gasoline become a pollutant whenever they escape from an engine in an unburned state. The best method of controlling HC emissions is to ensure complete combustion of the fuel charge. However, since we are dealing with an engine that cannot produce perfect combustion, other ways are necessary to assist in reducing emission levels.

One way is the control of compression pressure. *Compression* plays a role in achieving total combustion. If, for example, the fuel particles have the proper spacing during the compression stroke, there is a smoother transfer of the flame front as it moves throughout the combustion chamber. Ignition timing also plays an important part in controlling the proper compression pressure. Later we will discuss ignition timing and how it relates to controlling various forms of emissions.

Compression pressure directly relates to temperature, a determining factor in the combustion process. Higher temperatures, for instance, cause an increase in compression pressure and fuller burning of the air/fuel charge. This causes a reduction in unburned hydrocarbons in the exhaust gases.

However, higher temperatures also make NOx formations possible. Therefore, by reducing compression pressures, the combustion temperature is lower, thus reducing the formation of NOx emissions. In many cases, manufacturers can reduce both NOx and HC emissions by reducing the compression ratio. A reduced ratio is achieved by opening up or eliminating sharp corners and pockets in the combustion chamber that trap gas particles and prevent them from burning. These are the quench areas referred to earlier in the text.

But due to the overall increased engine temperatures of modern engines, reduced compression by itself does not present a complete answer to the NOx problem. In many cases, it is necessary to lower temperatures in ways other than just the reduction in compression pressure. These include the use of a modified camshaft or of exhaust gas recirculation (EGR). Both these items will be discussed in a later chapter.

Effects of the Air/Fuel Ratio on Emission Levels

The air/fuel mixture is another critical factor that can affect combustion chemistry and the amount of harmful pollutants. For example, if a mechanic turns an idle mixture adjustment screw out, the amount of fuel within the air/fuel ratio increases. Thus, the total mixture becomes richer (Fig. 2–9). Note in this illustration, that a rich setting on the mixture screw causes an increase in HC emissions. This enriched mixture contains less air and

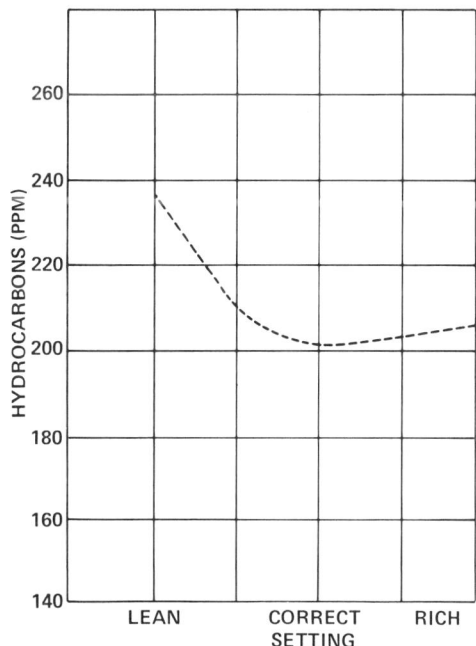

Figure 2–9. Effect on HC emissions due to changes in the air/fuel ratio at idle.

therefore a reduction in the oxygen available for complete burning of the air/fuel charge. Consequently, HC emissions increase.

Also note in Fig. 2–9 that if a technician would adjust the same mixture screw in enough (towards the lean side), the HC emissions increase drastically. This situation occurs because the fuel mixture has become so diluted, or thinned out, by too much air that the charge cannot burn completely or ignite at all. The common term given to this condition is *lean misfire*.

As pointed out, a lean misfire results in a large increase in HC emissions. This increase is due to the failure of the fuel charge to ignite or burn completely; consequently, there is an amount of raw fuel (hydrocarbons) that is emitted into the atmosphere from the vehicle's tail pipe. Finally, lean misfire can occur in one or more cylinders, and the condition can also vary from cylinder to cylinder while the engine is operating. This situation is a direct result of problems in the design of the intake manifold that create an uneven distribution of the air/fuel charge.

The air/fuel ratio also has a direct effect on CO emissions. When complete combustion occurs, one of its by-products is carbon dioxide (CO_2). However, anytime there is a lack of oxygen due to an overly rich air/fuel ratio, the amount of CO_2 produced by an engine decreases, and CO formations increase (Fig. 2–10). Note in this illustration how the carbon monoxide increases with the mixture adjustment screw turned out one-fourth of a turn. This increase occurs because there is insufficient oxygen to combine in a 2:1 ratio with the carbon by-products in the charge. Remember, carbon monoxide is easily converted to carbon dioxide with the addition of

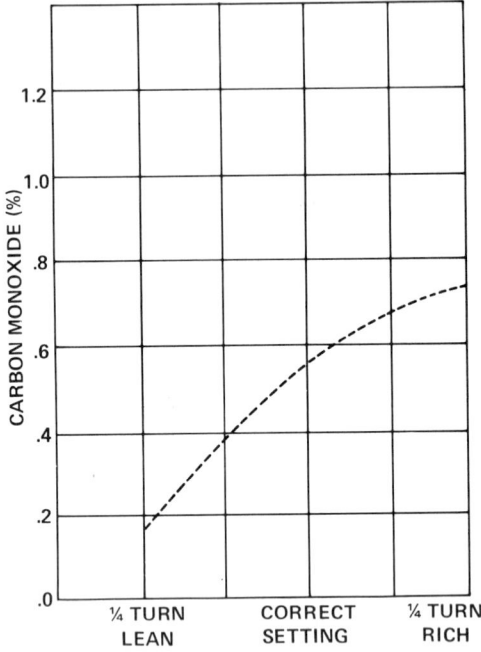

Figure 2–10. Effect on CO emissions due to changes in the air/fuel ratio at idle.

more oxygen. Therefore, fuel mixture adjustments at idle play an important role in emission control during this phase of engine operation.

On the surface it would seem that a lean mixture would be the answer to the CO problem. If carbon monoxide were the only emission presenting a problem, then a lean mixture would be the best corrective action to take. However, since carbon monoxide is only one of the three harmful pollutants from an automobile engine, a lean mixture is not the best overall solution. In fact, as stated before, a mixture that is too lean increases HC emissions while CO levels decrease.

Effect of Idle Speed on Emission Levels

Engine idle speed (revolutions per minute (rpm)) has a direct relationship to the amount of HC emissions (Fig. 2–11). Note in the drawing that when idle rpm is low, HC levels are higher. This results from the closure of the throttle valve at a low idle speed. A closed valve position starves the engine of the oxygen necessary to maintain the correct mixture for complete combustion. Consequently, a specified engine rpm is necessary to reduce HC emissions.

Since it is not possible to obtain complete control of carbon monoxide through idle mixture adjustments, other effective, alternate methods are necessary; one of these is idle speed (Fig. 2–12). Since CO levels are lower through an increase in an oxygen supply, a higher engine rpm is necessary to maintain a wider throttle valve opening. By keeping the throttle valves in a more open setting during idle rpm, an adequate supply of oxygen moves through the carburetor to the combustion chambers. As a result, there is a better control of CO levels. Note in Fig. 2–12 the relationship between the various idle speeds and the resulting CO emissions.

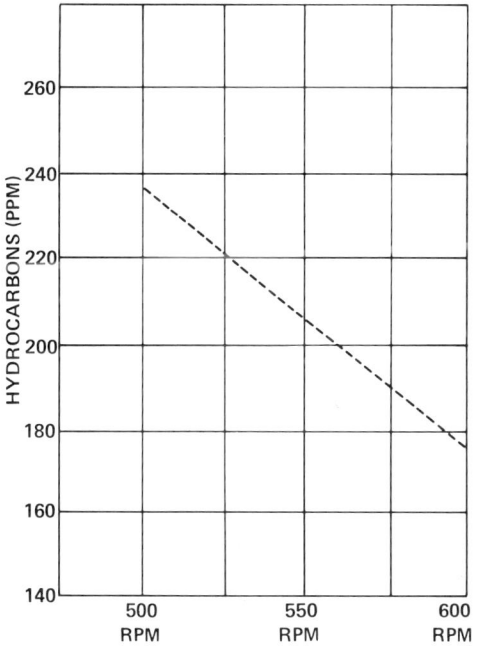

Figure 2–11. Effect on HC emissions due to changes in engine idle rpm.

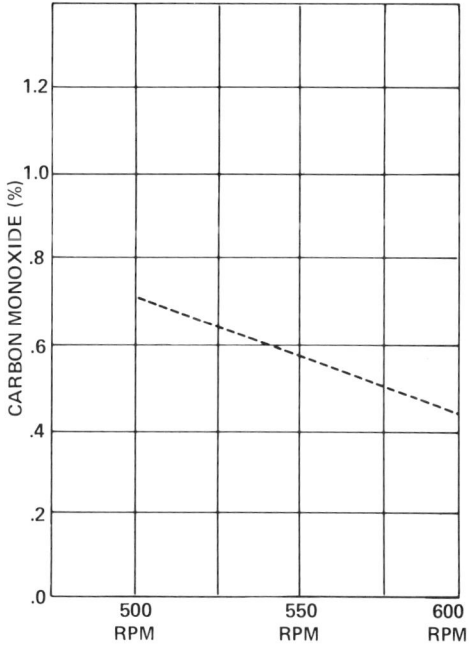

Figure 2–12. Effect on CO levels due to changes in engine idle speed.

Effect of Basic Ignition Timing on Emission Levels

Before the advent of emission control devices, most automotive engines operated on advanced basic ignition timing. *Advanced basic ignition timing* provided an arc at the spark plugs before the pistons reached TDC on their compression strokes. This action permitted combustion to begin before TDC and finish shortly after TDC.

This arrangement permits the various pistons to liberate the maximum amount of heat energy in the fuel during the combustion process. The advantages of this advanced timing are an increase in fuel economy and engine performance.

However, an engine with advanced ignition timing produces excessive HC emissions at idle. In this situation, the pistons convert more of the heat energy within the fuel into useful work too early in the engine's cycle, thus lowering the temperature of the exhaust gases. As these temperatures decrease, the oxidation of hydrocarbons in the exhaust manifold also lowers. As a result, an engine with advanced basic ignition timing emits more hydrocarbons out of its exhaust and into the atmosphere.

By retarding the basic timing, there is a slightly higher compression of the unburned air/fuel mixture prior to ignition, but a delayed burn time (Fig. 2–13). This delayed burn time keeps the exhaust gas temperatures higher, permitting the burning or oxidation of excess, unburned hydrocarbons to continue right out into the exhaust manifold and system. This lowers the amount of hydrocarbons emitted into the atmosphere but does reduce engine performance and increases fuel consumption.

Note in Fig. 2–14 that by advancing the ignition timing, CO emissions decrease. However, by doing so, the HC levels increase (Fig. 2–13). In other words, if a mechanic advances the timing as much as 2.5° over the recommended setting, CO

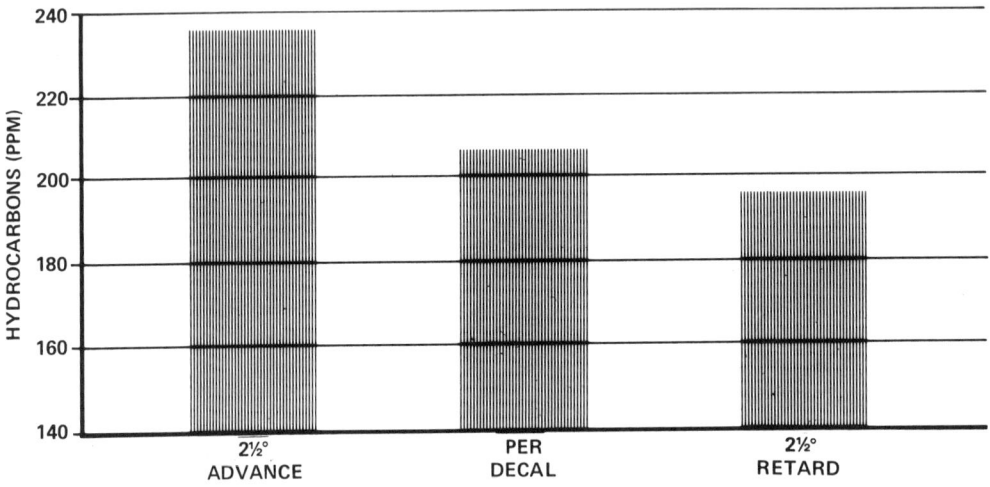

Figure 2–13. Effect on HC emissions with changes in basic ignition timing.

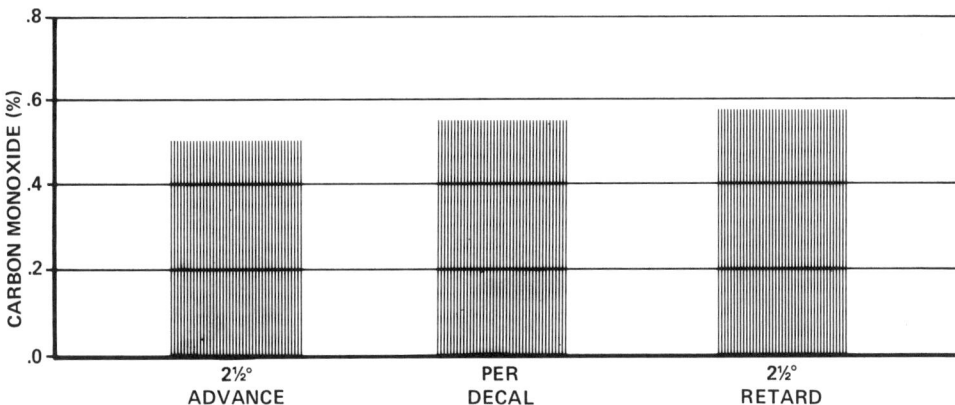

Figure 2-14. Effect on CO levels with changes in basic ignition timing.

levels decrease, but the HC emissions increase. It should be very clear then that advancing the basic ignition timing creates a larger problem; namely, the increase in HC emissions.

Another advantage to retarded timing is an increase in air flow through the carburetor. When any engine operates on retarded basic ignition timing, its idle speed is naturally lower than with advanced timing. This decrease in rpm is due to less fuel heat energy being converted into work by the piston, and more heat supplied to the oxidation process within the exhaust manifolds.

In order to raise the rpm up to the specified amount, the throttle valves must be in a somewhat wider, open postion. This new position permits additional air to pass through the carburetor and into the engine's combustion chambers. The additional air dilutes the amount of residual exhaust gases within the cylinder while supplying more oxygen for combustion. As a result, the engine idles faster with a leaner, more combustible mixture. Finally, the increased air flow also provides the oxygen necessary for the oxidation of both hydrocarbons and carbon monoxide and their emission levels therefore decrease.

Advanced timing also has a negative effect on NOx emissions. As stated before, advanced timing creates high combustion chamber temperatures and pressure. These factors increase NOx emissions. However, the timing has to be advanced somewhat during engine acceleration in order for the engine to develop sufficient power with reasonable fuel economy.

Since NOx emissions are not such a real problem during engine idle, it is only when the engine accelerates between idle and cruise that it emits increased NOx levels. The reason, of course, is that during these periods the engine is operating with an enriched mixture, a higher combustion chamber temperature, and cylinder pressure.

During cruise conditions the engine is running with a light load. Consequently, its mixture is leaner and its combustion chamber temperature and cylinder pressure are lower. So the ignition timing can be advanced for increased fuel economy during cruise without affecting NOx levels to a great extent.

Effects of Combustion Chamber and Air/Fuel Temperatures on Emissions

The temperature of the combustion chambers, along with the air/fuel mixture, has an effect on HC emissions. At higher engine and air/fuel temperatures, the mixture vaporizes more effectively. As the fuel becomes more vaporized, it travels more efficiently through the various runners within the intake manifold. Therefore, there is a greater combustion of the fuel in the engine's cylinders.

One reason for the higher operating temperature of an emission-controlled engine is the use of a thermostat with a higher opening temperature. The raising of the air/fuel temperature is due largely to the use of thermostatic air cleaners. These units preheat the air before it enters the carburetor, thus keeping the air at its most efficient temperature.

Effect on Hydrocarbon Emissions by Spark Plug Misfire

If a spark plug fails to fire for any reason, HC emissions will increase drastically (Fig. 2–15). Note in the diagram that when all the spark plugs are firing satisfactorily, the total HC emissions are about 200 ppm. However, when one spark plug misfires, the HC levels increase to approximately 2,200 ppm. For this reason, the components of the ignition, especially the spark plugs, must receive proper maintenance at the interval specified by the manufacturer.

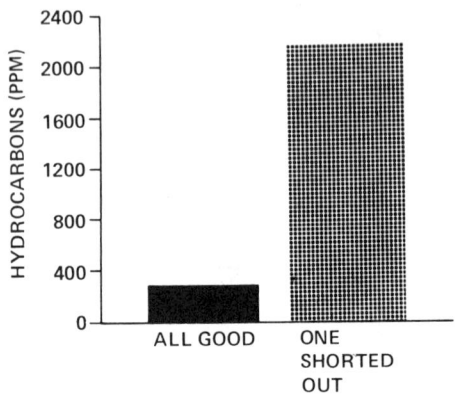

Figure 2–15. Effect on HC emissions by a spark plug misfire.

Summary

1. Ideal combustion conditions do not occur very often in an automobile engine.
2. Under ideal conditions the carburetor supplies a mixture of about 15:1 to the engine.
3. The air used by an engine consists of 21 percent oxygen, 78 percent nitrogen, and 1 percent other gases.
4. The oxygen in the air supports the combustion of gasoline.
5. At about 2,500°F, some nitrogen unites with oxygen to form oxides of nitrogen.
6. Gasoline contains literally hundreds of chemicals, most of which are hydrocarbons.
7. Hydrocarbons are divisible into three groups.
8. The octane rating of a fuel must be correct for knock-free performance in a given engine.

9. The octane rating, given a fuel, represents its antiknock properties.
10. The octane rating of a fuel is based on a comparison of its performance with other fuels in a test engine.
11. One way to raise or lower the octane number of a fuel is to alter the mix of its hydrocarbons.
12. Another way of increasing a fuel's octane number is by the addition of small amounts of certain additives such as TEL.
13. There are also small amounts of other additives in gasoline.
14. Gasoline may contain the impurities sulphur and phosphorus.
15. The operating cycle of the engine is necessary to complete the chemical process of converting the heat energy stored in the fuel into motion.
16. The purpose of the intake stroke is to create a vacuum in the cylinders that causes atmospheric pressure to force an air/fuel charge into the combustion chambers.
17. Atmospheric pressure results from the weight of the blanket of air under which we live.
18. Atmospheric air pressure is 14.7 psi at sea level.
19. During the intake stroke, the carburetor determines the air/fuel mixture.
20. During the compression stroke, with both valves closed, the piston compresses the air/fuel mixture to a pressure of about 100 to 150 psi.
21. As the pressure of the mixture increases during the compression stroke, its temperature increases as well.
22. The antiknock additive in the fuel prevents spontaneous ignition of the fuel during the compression stroke.
23. With both valves closed, a timed electrical spark occurs at the spark plug gap, slightly before the piston reaches TDC on the compression stroke.
24. During the power stroke, the force resulting from the burning air/fuel charge turns the crankshaft.
25. Under ideal conditions, the combustion process produces carbon dioxide and water.
26. Complete combustion in the engine does not occur due to high temperatures inside the cylinders and the nature of the fuel itself.
27. High combustion chamber temperatures can cause NOx emissions as well as unusual changes in ring hydrocarbons.
28. An overly rich air/fuel mixture causes both HC and CO emissions.
29. Quench areas in the combustion chambers cause HC emissions.
30. The upward movement of the piston, while the exhaust valve is open, provides an effective method of expelling all waste gases from a cylinder.
31. The exhaust gases contain a very complex mixture of materials and waste gases.
32. The compression of an engine plays a role in achieving total combustion.
33. High engine temperatures promote NOx formations.
34. The air/fuel mixture is a critical factor that can affect combustion chemistry and the amount of harmful pollutants.
35. A lean misfire results in a large increase in HC emissions.

36. The air/fuel ratio has a direct effect on CO emissions.
37. Engine idle speed has a direct relationship to both CO and HC emissions.
38. An engine with advanced ignition timing produces excessive HC emissions at idle.
39. CO levels also vary with changes in basic ignition timing.
40. Retarded ignition timing affects NOx formation.
41. The temperature of the combustion chamber and the air/fuel mixture affects HC emissions.
42. If a spark plug fails to fire, HC emissions increase drastically.

Review Questions

The questions listed below will assist you in determining how well you remember the material contained in this chapter. Read each question carefully before choosing the answer. If you cannot answer the question, review the section in the chapter that covers the question.

1. The portion of the air necessary for combustion is _____.
   a. nitrogen
   b. oxygen
   c. hydrogen
   d. sulfur
2. NOx formation begins at about _____ °F.
   a. 2,500
   b. 2,000
   c. 1,500
   d. 1,000
3. An engine with a compression ratio of 12:1 requires a fuel with an octane rating of _____.
   a. 98
   b. 100
   c. 102
   d. 106
4. Atmospheric pressure at sea level is _____ psi.
   a. 13.0
   b. 13.7
   c. 14.0
   d. 14.7
5. The average pressure in the combustion chamber during the power stroke is _____ psi.
   a. 100 to 150
   b. 200 to 400
   c. 400 to 600
   d. 600 to 800
6. An overly rich air/fuel mixture creates _____ emissions.
   a. CO
   b. CO_2
   c. NOx
   d. HC_2
7. One way to reduce HC emissions is to reduce the _____ areas in the combustion chambers.
   a. detonation
   b. hot
   c. cool
   d. quench
8. Engine idle speed has a direct relationship to the amount of _____ emissions.
   a. NOx
   b. HC

c. CO_2
d. water vapor

9. An engine with retarded ignition timing produces more _____ emissions.
   a. NOx
   b. CO
   c. HC
   d. CO_2

10. If a spark plug misfires, _____ levels increase drastically.
    a. NOx
    b. HC
    c. CO
    d. CO_2

For the answers to these questions, turn to the Appendix.

Chapter 3

Instruments Used to Measure Exhaust Emissions and Test Control Devices

Before discussing the function, design, operation, and servicing of the many types of emission control systems, it is first important that the reader be familiar with the various pieces of equipment used by the industry to test these devices. These include such items as the vacuum gauge, timing light, tachometer, infrared analyzer, propane enrichment equipment, and vacuum pump.

Vacuum Gauge

A *vacuum gauge* is one of the most useful instruments that a mechanic utilizes on an engine to detect and locate the cause of mechanical problems. When an engine is operating, the gauge itself provides a quick analysis of the engine's mechanical condition through the vacuum effect inside the intake manifold, without the need of removing the spark plugs for the purpose of checking cylinder compression. In other words, the movement of the vacuum gauge needle provides a very good indication of the compression within the cylinders, which may be good or bad depending on the condition of the rings, valves, or head gaskets.

A vacuum gauge may be a separate instrument or incorporated into the console of an engine analyzer (Fig. 3–1). In either case, the gauge has a scale marked off in divisions ranging from 0 to 30 inches of mercury (Hg) or 0 to 76 millimeters Hg.

The movement of the needle on the gauge scale during engine operation actually indicates the difference in pressure or vacuum that exists between the inside of the intake manifold and the atmosphere. The amount of vacuum shown on the scale (the difference in pressure) results from the action of all the engine's pistons functioning on their respective intake strokes.

Since atmospheric pressure varies with

Figure 3-1. Typical vacuum gauge.

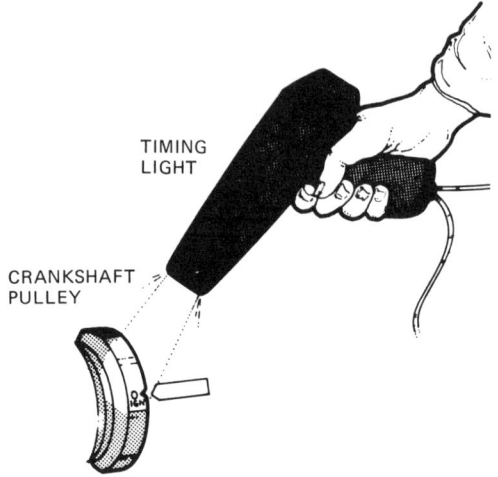

Figure 3-2. A timing light is necessary, in most cases, to check ignition timing.

changes in altitude, all vacuum gauge readings are dependent on altitude. Consequently, there is a gauge adjustment necessary for altitudes above sea level. For instance, for every 1,000 feet (305 meters) above sea level, the vacuum gauge reads low by 1 inch (25.4 millimeters) Hg. As a result, it is necessary to add 1 inch Hg to the gauge reading at that altitude.

Timing Light

A timing light can also be a separate unit or a component of an engine analyzer. In either situation, the *timing light* is nothing more than a stroboscopic light used to check ignition timing on most engines (Fig. 3-2). In operation, the light flashes each time the number-one or designated spark plug fires.

A mechanic uses this flashing light to view the position of the engine's timing marks for one of two reasons. First, the flashing light indicates the relative position of the timing marks so that a technician can check and adjust basic ignition timing at engine idle. Second, a mechanic can use the timing light flashes to check the operation of the vacuum and centrifugal advance mechanisms of the distributor at various engine speeds.

Although a standard timing light indicates whether the vacuum or centrifugal advance mechanisms are functioning to advance ignition timing, it does not show the actual number of degrees of total advancement. Therefore, many manufacturers produce timing lights that incorporate a timing advance unit (Fig. 3-3). These units are also referred to as powered timing lights.

A *powered timing light* cannot only test basic ignition but also quickly measure the number of degrees of total advancement provided by the vacuum and centrifugal advance mechanisms. In other words, the powered advance unit of the light measures the amount of total advancement of the spark due to the action of both advance units of the distributor or electronic control devices.

To accomplish this task, the timing light itself has two controls, a timing light switch and rotating indicator control (Fig. 3-3). The timing light switch is nothing

Figure 3-3. Controls of a powered timing light of a Sun 1115 engine analyzer.

more than an on-off switch for the unit. The rotating indicator control has a number of degree markings used to measure the amount of spark advance.

When this powered timing light is functioning, it electronically moves the engine's timing marks in relationship to the movement of the indicator control on the light. For example, when a technician switches on the timing light, aims it at the timing marks, and turns the indicator control, the timing mark on the rotating crankshaft pulley moves until it lines up with the zero mark on the stationary timing scale. Then, a mechanic measures the amount of timing advance by noting which degree marking on the rotating control lines up with the stationary mark on the control-unit housing. This process, of course, is accomplished at given engine speeds, and the results compared to specifications supplied by the manufacturer.

Tachometer

A *tachometer* measures engine speed in rpm. A mechanic uses this device for setting idle speed to specifications as well as testing many of the emission control devices at given engine rpm.

The tachometer shown in Fig. 3-4 mounts on the console of a Sun 1115 engine analyzer. This instrument measures crankshaft revolutions (rpm) from 0 to 12,000.

The instrument itself has two scales. The upper single red scale has divisions that measure from 0 to 1,500 rpm. The blue lower scale has division markings from 0 to 6,000 above the heavy curved line, and markings indicating 0 to 12,000 rpm below the line. When this instrument is in operation, engine speed and the position of several controls on the analyzer determine which scale and indicator light will activate.

Infrared Analyzer

An *infrared (IR) analyzer* measures the amount of hydrocarbons and carbon monoxide in a vehicle's exhaust (Fig. 3-5).

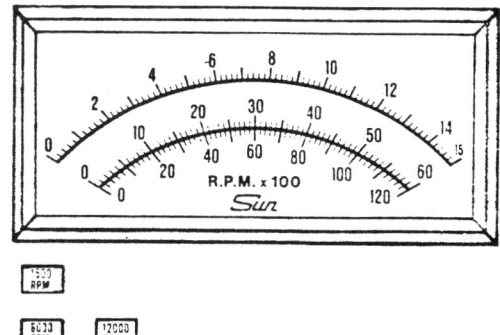

Figure 3-4. RPM X 100 tachometer of the Sun 1115 engine analyzer (Courtesy of Sun Electric Corp.).

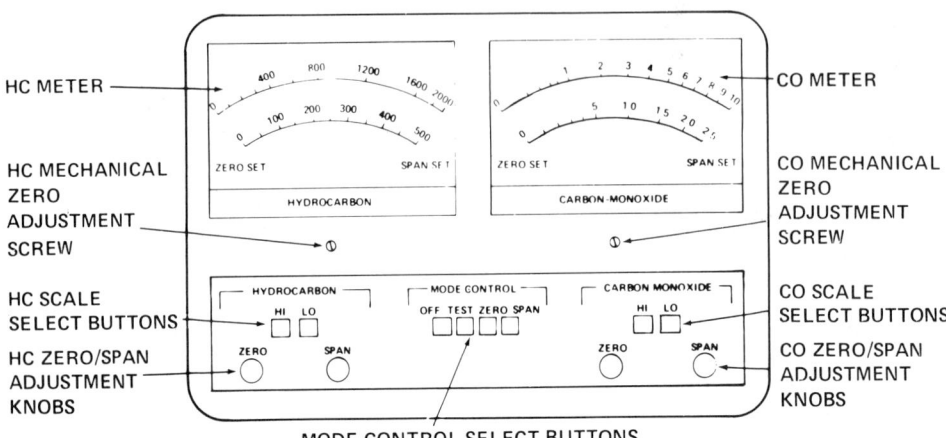

Figure 3-5. An infrared analyzer accurately measures the amount of HC and CO emissions in a vehicle's exhaust.

As mentioned earlier, both of these compounds are harmful air pollutants. Hydrocarbons, present in a vehicle's exhaust, are unburned gasoline. If an engine does not burn up all the fuel during its combustion process, raw gasoline passes out of the tail pipe and registers as hydrocarbons on one of the analyzer's meters.

Several factors cause an abnormally high reading on the HC meter. For instance, a fouled spark plug, defective spark plug wires, or burned valves increase the HC reading simply because the combustion process in the affected cylinder did not consume all the fuel. In addition, an excessively rich air/fuel mixture from the carburetor raises the HC levels. However, this condition shows up more apparently on the CO meter than on the HC meter.

The *hydrocarbon meter* on the analyzer measures HC emissions in parts-per-million (ppm) or grams per mile (g/mi). The automotive industry uses ppm or g/mi as the measurement for HC levels because there are much smaller amounts of hydrocarbons in a sample of a vehicle's exhaust. One ppm, for example, is equal to 0.0001 percent of the total exhaust sample. Finally, typical specifications for the allowable HC emissions range from as high as 900 ppm for early model automobiles to a low of 175 ppm for a current-model vehicle.

The infrared analyzer's CO meter shows the richness or leaness of the air/fuel ratio by the position of the needle on the scale. The richer the mixture, for instance, the higher the needle deflects, indicating a greater percentage of carbon monoxide in an exhaust sample. If a vehicle has a closed choke valve, restricted air cleaner, or extremely high carburetor float levels, the analyzer will show this by its CO meter needle swinging over into the high numbers on the scale.

A lean mixture, on the other hand, produces the lowest CO emission readings on the analyzer. However, if an air/fuel charge is too lean, the engine misfires intermittently or totally. This misfire may not be heard, but it is definitely there whenever the analyzer shows a low CO reading and a high HC indication or a HC needle that fluxuates between the low and high ends of the scale.

The analyzer measures the amount of carbon monoxide in the exhaust, using a percentage figure. A given specification table may show a percentage range for CO levels from 0.05 percent for current automobiles to as high as 6 percent for early model vehicles.

The infrared analyzer measures the amount of HC and CO emissions from a vehicle's exhaust by comparing a sample of its content with the surrounding air. The unit does this by pulling in two samples, one from shop air and the other from the vehicle's exhaust by means of a probe inserted into the tail pipe. These samples then pass into two separate tubes.

From the tail pipe probe, the exhaust sample moves through the water separator and primary filter, secondary filter, and finally into the sample cell (Fig. 3–6). As the gases pass through the sample cell, the HC and CO molecules in the sample will absorb a certain amount of infrared waves. The amount of this absorption depends upon the concentration of carbon monoxide and hydrocarbons within the sample tube of the cell.

The metal-shielded heat source creates the infrared waves. The unit (the infrared energy source) emits a total spectrum of infrared radiant energy waves through both the sample and reference cells.

A rotating, segmented chopper disc interrupts the constant spectrum of infrared waves from the infrared energy source. This interruption of the flow of waves creates a pulsating ac signal that is later amplified and rectified to a dc signal, used to activate both the HC and CO meters.

After moving through the chopper, the infrared waves enter both the sample and reference cells. Once through both cells, the waves move into filters which exclude all of the infrared waves except those that are absorbable by the HC and CO molecules.

Two optical detectors receive the infrared waves passing through the filters. There are two detectors for each cell: a detector for the wavelength that measures hydrocarbons and one that measures carbon monoxide. The detectors themselves then convert the remaining infrared waves to electrical signals. The difference between the absorption of the infrared waves within the reference and sample cells determines the strength of the electrical signals.

The signals then pass to the amplifier circuitry. These components amplify and convert the ac signals to direct current, which is used to operate the HC and CO meters. The meters read the concentrations of hydrocarbons and carbon monoxide in the metered amount of exhaust gas within the sample cell.

The two meters located on the analyzer shown in Fig. 3–5 also have dual scales. The HC meter, for example, indicates readings from 0 to 500 and from 0 to 2,000 ppm. The CO meter, on the other hand, has a range from 0 to 2.5 and from 0 to 10 percent. Which scale, on either meter, a mechanic reads depends on the positioning of the range button.

Other than the task of measuring emission levels to determine if a vehicle's exhaust emissions comply to legal standards, the infrared analyzer has two other very common usages. First, a mechanic can use this machine to perform emission tests at various engine speeds and conditions to quickly uncover a variety of engine, ignition, and fuel system malfunctions. Second, the analyzer provides the accuracy

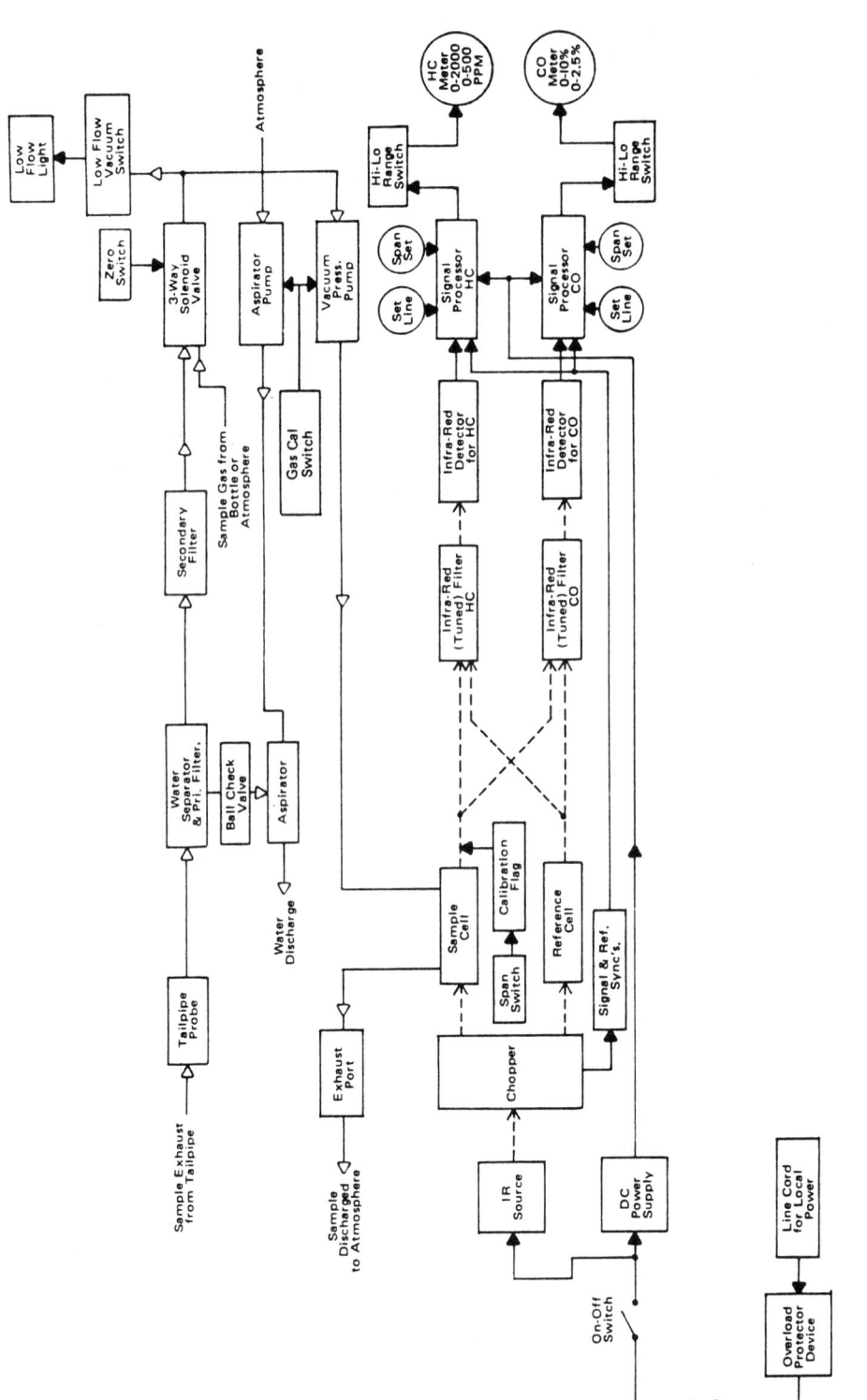

Figure 3-6. Block diagram showing the operation of an infrared analyzer (Courtesy of Sun Electric Corp.).

and range necessary for checking and adjusting most carburetors, except those found on many vehicles with catalytic converters.

Propane Enrichment Equipment

Many manufacturers now recommend the use of propane enrichment equipment when setting the idle mixture adjustment and curb-idle speed on their vehicles. This has become necessary with the advent of the catalytic converter, which normally permits an almost unmeasurable percentage of CO emissions from a vehicle's tail pipe. As a result, an infrared analyzer is not always accurate enough for a mixture adjustment. Therefore, a mechanic can set the idle mixture too rich, resulting in excessive emissions, and the analyzer may not indicate the problem.

In addition to the pollution problem, operating a catalytic converter-equipped vehicle with an overly rich mixture at idle can create several other problems. For example, over a period of time, a rich mixture reduces the effective life of the converter. Also, the same mixture causes the converter to produce a strong rotten-egg smell at the tail pipe. Finally, an excessively rich mixture can produce a rough idle or even engine stalling.

Propane enrichment equipment injects a controlled amount of propane into the engine while noting the effect on its speed (Fig. 3-7). Manufacturers recommend propane for this purpose because it is readily available in small pressurized containers and mixes well with the air/fuel mixture. The equipment shown in Fig. 3-7 consists of a propane cylinder, a main metering valve, propane metering valve, length of

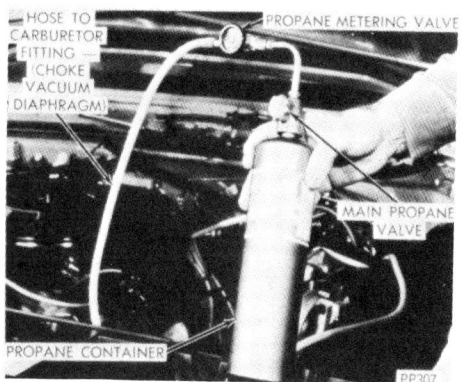

Figure 3-7. Many vehicles now require the use of this propane enrichment tool for correct adjustment of both idle mixture and speed (Courtesy of Chrysler Corp.).

hose, and the adapters to connect this hose to a carburetor or air cleaner fitting.

There is a direct three-way correlation between the carburetor's air/fuel ratio, the amount of propane injected, and the gain or loss in engine rpm. For instance, with a constant supply of pressurized propane to the engine adjusted to a given air/fuel ratio at idle, there should be a specified increase (gain) in engine speed. If the rpm gain is less than that specified by the manufacturer, the air/fuel mixture is too rich. On the other hand an rpm gain that is more than specifications indicates an air/fuel ratio that is too lean.

Vacuum Pumps

Hand-Operated Pumps

There are two common types of vacuum pumps in use for diagnosing emission control system components, electrically and hand-operated assemblies. A *hand-operated vacuum pump* is shown in Fig. 3-8. This device can test almost any type of vac-

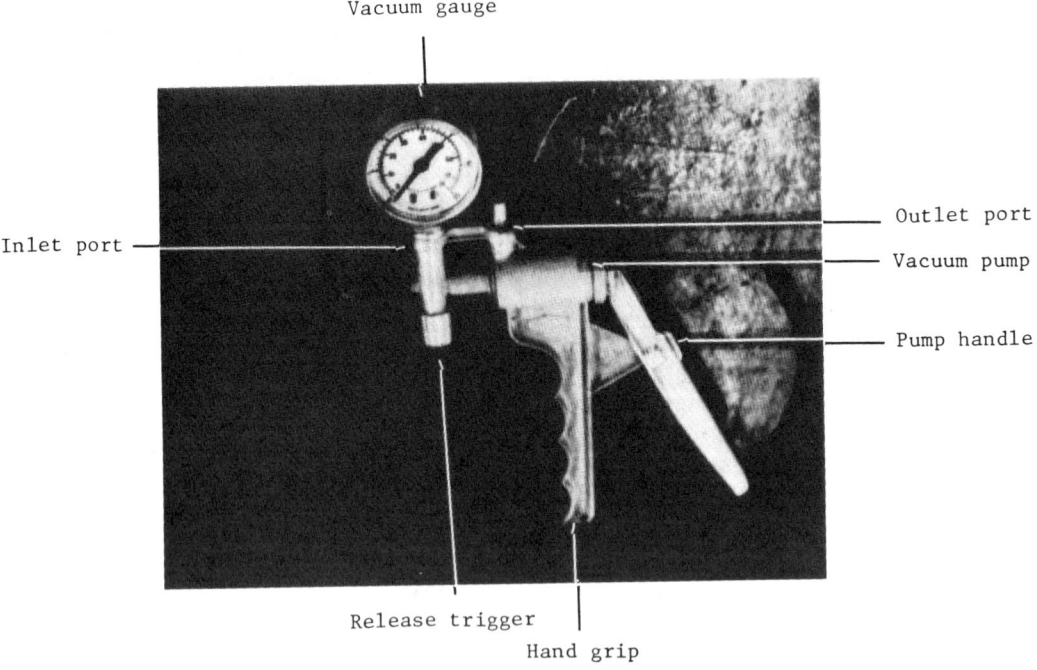

Figure 3-8. Hand-operated vacuum pump and gauge assembly.

uum-activated device such as a choke delay valve or vacuum advance diaphragm, with the unit on or off the vehicle. In addition, since the tool is hand operated and therefore requires no electrical power, it provides the mechanic with an easy and accurate method for checking vacuum devices not only in the shop but on a road call.

The portable pump assembly shown in Fig. 3-8 consists of a vacuum gauge, vacuum pump, inlet and outlet ports, release trigger, along with a hand grip and pump handle. The vacuum gauge connects externally, via a hose, to the test component and internally to the vacuum pump. The gauge scale has divisions ranging from 0 to 30 inches and 0 to 76 millimeters Hg.

The vacuum pump is inside the main housing. The handle mechanically activates this pumping mechanism; and it, in turn, evacuates the air from the component requiring a test, thus creating a vacuum.

The pump requires two valve-controlled ports, an inlet and outlet. The inlet port connects by a hose to the test component. The outlet port, on the other hand, provides an opening for the pump directly into the atmosphere.

The release trigger has but one important function: it vents all or part of any vacuum built up by the pump assembly into the atmosphere. In other words, by coordinating the movements of the pump handle and trigger, a mechanic can easily pump up a specific amount of vacuum in any unit.

The hand grip and handle merely provide a means by which the technician can hold and activate the pump. The palm of the operator's hand fits over the pump handle, with the fingers encircling the hand grip. This form of construction permits ease of control with one hand, leaving the other free to perform other tasks.

Electrically Operated Vacuum Pumps

Most distributor machines and some engine analyzers also have vacuum test devices on the consoles. In other words, these machines have built-in vacuum pumps, vacuum gauges, and the necessary controls to operate the pumps for the purpose of testing vacuum components. This type of vacuum pump assembly performs the same functions as the hand-operated unit, but it does require electrical power in order to function.

Summary

1. A vacuum gauge provides a quick analysis of an engine's mechanical condition through the vacuum effect inside the intake manifold.
2. The scale of a vacuum gauge has divisions ranging from 0 to 30 inches or 0 to 76 millimeters Hg.
3. The needle of a vacuum gauge indicates the difference in pressure between the intake manifold and the atmosphere.
4. All vacuum gauge readings are dependent on altitude.
5. A timing light is necessary in most cases to check ignition timing.
6. A mechanic uses a timing light to check basic ignition timing and the operation of the vacuum and centrifugal advance mechanisms.
7. A powered timing light tests basic ignition timing and quickly measures the number of degrees of total ignition advancement.
8. A powered timing light has two controls.
9. The powered timing light electronically moves the engine's timing mark in relationship to the movement of the indicator control on the unit.
10. A tachometer measures engine speed in revolutions per minute.
11. An infrared analyzer measures the amount of hydrocarbons and carbon monoxide in a vehicle's exhaust.
12. A fouled spark plug, defective spark plug wires, or burned valves increase the HC reading on the infrared analyzer.
13. The HC meter indicates HC levels in parts per million or grams per mile.
14. Carbon monoxide is a poisonous gas that results from incomplete fuel combustion.
15. A closed choke valve, restricted air cleaner, or extremely high carburetor choke level causes a high CO reading on the infrared analyzer.
16. A lean mixture produces the lowest CO reading but can cause a high HC reading on the infrared analyzer.
17. The infrared analyzer measures the amount of carbon monoxide in an exhaust sample, using a percentage figure.
18. The infrared analyzer measures the amount of HC and CO emissions from a vehicle's exhaust by comparing a sample of its content with the surrounding air.

19. Other than measuring emission levels, the infrared analyzer has two other very common uses.
20. Many manufacturers now recommend the use of propane enrichment equipment when setting the idle mixture adjustment and curb-idle speed on their vehicles.
21. An excessively rich mixture reduces the effective life of the catalytic converter and produces a rough idle.
22. Propane enrichment equipment injects a controlled amount of propane into the engine, while a notation is taken of its effect on engine speed.
23. There is a direct three-way correlation between the carburetor's air/fuel ratio, the amount of propane injected, and the gain or loss in engine rpm.
24. A hand-operated vacuum pump can test almost any type of vacuum-activated device on or off the vehicle.
25. A typical portable vacuum pump consists of a vacuum gauge, vacuum pump, inlet and outlet ports, release trigger, along with a hand grip and pump handle.
26. Most distributor machines and many engine analyzers have a vacuum test device on their console.

Review Questions

The questions listed below will assist you in determining how well you remember the material contained in this chapter. Read each question carefully before choosing the answer. If you cannot answer the question, review the section in the chapter that covers the question.

1. For every 1,000 feet above sea level, a vacuum gauge reads low by _____ inches Hg.
   a. 0
   b. 1
   c. 2
   d. 3
2. A _____ can accurately check both basic ignition timing as well as the operation of the advance units.
   a. powered timing light
   b. portable vacuum gauge
   c. portable infrared analyzer
   d. propane enrichment equipment
3. The device for measuring rpm is the _____.
   a. analyzer
   b. ammeter
   c. tachometer
   d. speedometer
4. The instrument for measuring HC and CO levels is the _____.
   a. timing light
   b. engine analyzer
   c. vacuum gauge
   d. infrared analyzer
5. The infrared analyzer measures _____ levels, using a percentage figure.
   a. HC
   b. CO
   c. CO_2
   d. NOx
6. Many manufacturers now recommend the use of _____ equipment when adjusting the idle mixture.
   a. propane
   b. butane

c. oxygen
d. CO_2

7. If an engine's air/fuel ratio is too rich, the effective life of the _____ is reduced.
   a. carburetor
   b. muffler
   c. converter
   d. tail pipe

8. The tool for testing vacuum devices is a vacuum _____.
   a. gauge
   b. pump
   c. tool
   d. advance mechanism

For the answers to these review questions, turn to the Appendix.

Chapter 4

Using Emission Control Test Equipment

Chapter 3 presented an overview of the function and design of the most common types of instruments used to measure vehicle exhaust emissions and test control devices. This equipment included the vacuum gauge, timing light, tachometer, infrared analyzer, propane enrichment equipment, and vacuum pump.

It is now time to turn our attention to the use of this equipment for testing purposes. The testing sequences presented in the following sections are general in nature with the purpose of familiarizing you with the use of these tools. This text will also discuss many other test procedures at the appropriate time.

Testing the Engine with a Vacuum Gauge

Basically, a vacuum gauge test of any engine provides the technician with information as to the mechanical condition of the power plant. In other words, any deviation from the normal action of the vacuum gauge needle indicates possible problems in an engine's rings, valves, head, or intake manifold gaskets.

To perform an accurate vacuum gauge test, follow this prescribed procedure:

1. Connect the vacuum gauge, using a sufficient length of hose, to a nonrestricted fitting on the intake manifold, where the gauge measures the total vacuum produced by the engine. The length of hose used (about 3 feet) must be long enough to dampen excessive vibrations out of the gauge needle. However, under certain conditions, it may be necessary to further dampen the pointer action by placing a small clamp around the hose in order to slightly restrict its opening.

2. Operate the engine until it reaches normal operating temperature.

3. Except when specified, perform all vacuum tests with the engine running at curb idle.

Vacuum Test Results and Indications

The following sections point out the results of vacuum tests performed on engines of varying mechanical conditions and what these results indicate:

1. A vacuum gauge needle that remains relatively constant between 16 to 21 inches Hg with no more than one-half-inch fluctuation indicates the engine to be in good mechanical condition (Fig. 4-1). Note: The gauge reading will be lower and/or unsteady if the engine is (a) brand new or recently overhauled and not as yet broken in, (b) a late-model engine with a cam providing a great deal of valve overlap, or (c) an engine with certain emission control equipment.

2. A low but steady gauge reading between about 12 to 16 inches Hg is a good indication that the engine has possible leakage around the piston rings, late ignition timing, or late valve timing (Fig. 4-2).

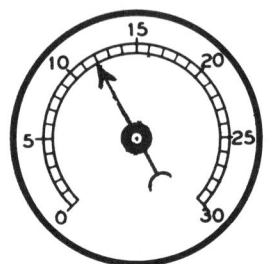

Figure 4-2. A low but steady gauge reading between about 12 to 16 inches Hg indicates either defective rings or late valve or ignition timing.

3. A gauge needle that oscillates slowly, then rapidly, between about 12 to 18 inches Hg provides a good indication that the ignition timing is too far advanced or the carburetor idle mixture is too lean (Fig. 4-3).

4. A gauge reading with an irregular pointer drop of 1 to 2 inches Hg is a sign of a possible sticky valve, carburetor that is out of adjustment, or an intermittent spark plug misfire (Fig. 4-4).

5. A gauge that shows a regular drop in vacuum from between 1 to 2 inches Hg is a symptom of a burned or leaky valve or a spark plug in one cylinder that is not firing at all (Fig. 4-5).

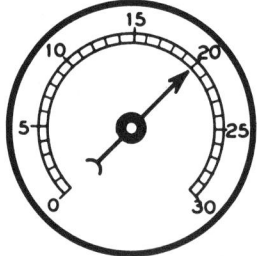

Figure 4-1. With the engine at idle, the gauge needle should hold steady between 16 to 21 inches Hg.

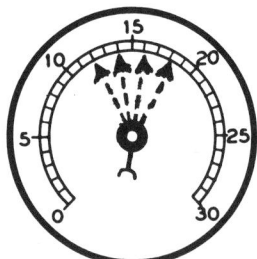

Figure 4-3. A slow oscillating reading between about 12 to 18 inches Hg is a good indication that the ignition timing is too far advanced or the carburetor mixture is too lean.

TESTING THE ENGINE WITH A VACUUM GAUGE

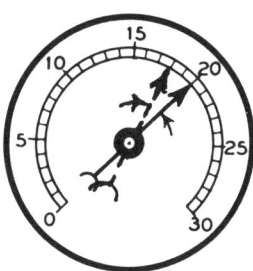

Figure 4–4. A stuck valve, carburetor out of adjustment, or an intermittent spark plug miss can cause an irregular needle drop of 1 to 2 inches Hg.

Figure 4–5. A regular needle drop of 1 to 2 inches Hg is a symptom of a defective valve or spark plug.

6. If the vacuum gauge reading is normal and steady at idle rpm but the pointer vibrates excessively at higher engine speeds, the valve springs are weak (Fig. 4–6).

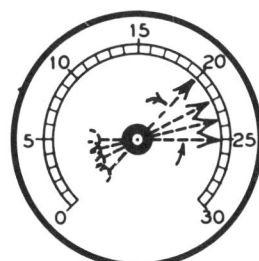

Figure 4–6. A vibrating pointer, at high speeds only, is a sign of weak valve springs.

7. If the gauge needle vibrates excessively at idle speed but steadies with increasing engine rpm, the valve guides probably are worn beyond tolerances (Fig. 4–7). Note: You can check valve guide condition by removing the valve covers and squirting motor oil at the top of the guides with the engine operating at idle rpm. If a large cloud of blue smoke is emitted from the tail pipe and the vacuum gauge pointer stabilizes, the valve guides are worn out.

8. If the needle of the vacuum gauge vibrates excessively at all engine speeds, the head gaskets are most likely leaking (Fig. 4–8).

9. If the gauge pointer fluctuates constantly from 3 to 9 inches Hg below nor-

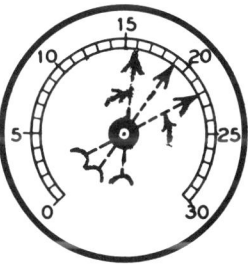

Figure 4–7. A vibrating needle, at idle rpm, points to the possibility of worn valve guides.

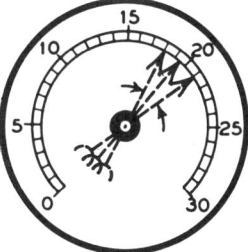

Figure 4–8. The head gasket has a leak if the gauge pointer vibrates excessively at all engine speeds.

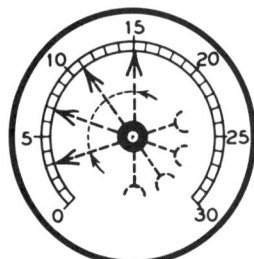

Figure 4–9. A gauge needle that fluctuates constantly from 3 to 9 inches Hg below normal indicates a vacuum leak in the intake system.

mal, there is a leak somewhere in the intake system (Fig. 4–9). The most common causes of a leak of this nature are defective intake manifold or carburetor mounting gaskets. You can test for vacuum leaks at these particular areas by squirting a noncombustible cleaning solvent along the gasket joints with the engine running at idle. If the vacuum reading now increases and the idle speed smoothes out, you have found the leak. Also, if the leak is large enough, you can see the vacuum pulling the solvent through the leak in the gasket joint.

Checking intake manifold gaskets with solvent will not always find the source of the leak because the lower edges of some gaskets on V-type and some in-line engines are not accessible. For example, the joints below the lower edges of the intake manifold and the cylinder heads of many V-type engines are inside the valve lifter valley, where a leak of this kind can occur. The best approach to confirm this form of leak is to check the exhaust for the presence of excessive smoke (a sign that manifold vacuum is pulling engine oil from the valley into the combustion chambers) and to eliminate all other possible causes of the gauge fluctuations.

Vacuum Gauge Testing of an Engine for Loss of Compression

You can also perform a vacuum test for loss of compression due to leakage around the pistons on any engine. This particular condition can result from stuck or worn piston rings, worn cylinder walls, or worn pistons. However, this type of test has no real value if the engine did not produce normal gauge readings on all the other vacuum tests listed earlier.

To perform this test follow this procedure:

1. Check the level and condition of the engine oil. The level must be full and the oil in good condition. Diluted or worn-out oil can create an incorrect reading, showing a loss of compression, where there is no real mechanical reason to cause such a leak.

2. Connect a vacuum gauge to the intake manifold.

3. Attach a tachometer, following the manufacturer's recommended procedures.

4. Start the engine and allow it to reach normal operating temperature.

5. From idle, accelerate the engine quickly until its rpm reaches 2,000. Next, quickly close the throttle valve. Note: If the carburetor has a dashpot that delays the closing of the throttle valve momentarily, you will have to first disconnect this device before performing this test or the results will not be valid.

6. Note the action of the vacuum gauge needle. As the throttle valve closes, the pointer will jump 5 or more inches Hg over the normal reading as long as the rings, cylinder walls, and pistons are in good condi-

tion. An increase in reading of less than 5 inches Hg is a good sign of a loss in engine compression.

7. Disconnect the hose from the intake manifold and remove the vacuum gauge and tachometer from the engine.

Checking Basic Ignition Timing and the Action of the Distributor Advance Units with a Timing Light

A standard timing light has two main functions. First, you can utilize this device to check basic ignition timing at curb-idle speed. Second, the timing light is a useful tool for testing the operation of both the vacuum- and centrifugal-operated advance mechanisms of the distributor.

To use a standard timing light to check basic ignition timing and the operation of the distributor advance units, follow these directions:

1. Following manufacturer's recommendations, connect both a standard timing light and a tachometer to the engine.

2. Remove and plug the hose to the vacuum advance mechanism. Note: Some distributors have both a vacuum advance and retard diaphragms. These units have individual hose connections at the distributor that require removal and plugging before performing the test sequence. The testing of these combination units will be discussed later.

3. Clean off the engine timing marks on both the crankshaft pulley or flywheel and timing flange. Then, go over the specified timing marks to brighten them with a piece of white chalk. Note: You can find the tim-

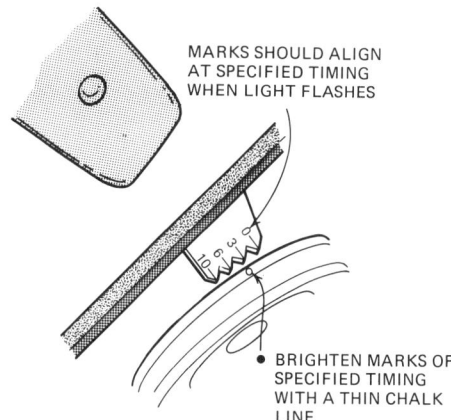

Figure 4–10. Checking basic ignition timing with a timing light.

ing specifications for a given engine either on the tune-up decal, located under the hood, or in the service manual for the vehicle.

4. Warm up the engine and operate it at the recommended rpm. Next, observe the initial (basic) timing (Fig. 4–10). As necessary, readjust the timing to manufacturer's specifications.

5. Raise and hold the engine rpm at 2,000.

6. While checking the timing with the light, unplug and connect the carburetor vacuum hose to the vacuum advance diaphragm fitting. The timing should immediately advance further than it already has from idle to 2,000 rpm (Fig. 4–11). If it does not, a problem exists either in the vacuum advance diaphragm or in the distributor itself.

7. Return the engine to the normal hot-idle setting, and recheck the basic timing with the light. The timing should retard back to the number of degrees indicated in step 4.

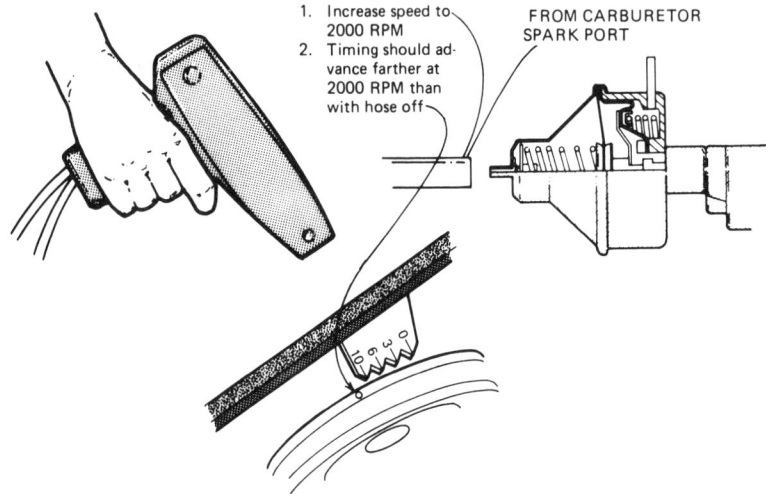

Figure 4-11. Checking the operation of the advance mechanisms with a timing light.

8. Disconnect and remove the timing light and tachometer.

Checking the Advance Curve of a Distributor with a Powered Timing Light

To measure the exact amount of ignition timing advancement provided by the vacuum and centrifugal advance units of a typical distributor, follow these instructions:

1. Connect the powered timing light and a tachometer to the engine, following the manufacturer's recommended procedures.

2. Start the engine and permit it to reach normal operating temperature.

3. At the specified idle rpm, turn the timing light switch on (Fig. 4-12) and check basic ignition timing in the same manner as with a standard light.

4. Raise and hold engine rpm to 2,000 or the speed recommended by the manufacturer.

5. With the timing light switch on and while viewing the timing marks, rotate the indicator control until the specified timing mark on the rotating pulley or flywheel aligns with the zero mark on the engine's timing flange.

6. Check the total amount of timing advancement by noting which degree, marked on the light's indicator control, lines up with the stationary mark on the control-unit body.

7. Compare this degree figure with specifications.

8. Return the engine to idle rpm and recheck basic ignition timing with the light. The timing should return to the degree figure from step 3.

ADJUSTING IDLE SPEED AND AIR/FUEL RATIO

Figure 4-12. To turn on this powered timing light, move the switch to the on position.

9. Disconnect and remove both the timing light and tachometer.

Adjusting Idle Speed and Air/Fuel Ratio with a Tachometer

For test purposes, there are many uses for the tachometer. As mentioned earlier, for example, a mechanic uses the tachometer, along with a vacuum gauge, to complete a series of mechanical engine tests and in conjunction with a timing light to check ignition timing. Later in this chapter we will discuss the utilization of a tachometer along with an infrared analyzer to measure exhaust emissions at various engine speeds. However, at this point, we will only explain the use of a tachometer for adjusting idle speed and air/fuel mixture.

Adjusting Idle Speed

To perform an idle speed adjustment on a typical engine, follow these steps:

1. Connect the tachometer to the engine, following the manufacturer's recommended procedures. This generally involves connecting the two meter leads to specific areas. One of the leads (usually with a red protective boot) attaches to the negative (−) terminal on the ignition coil (vehicles with negative-ground battery systems). In the case of electronic ignition systems, the lead connects directly to, or via an adapter, the tachometer (TACH) terminal of the coil or distributor assembly. The other lead connects to a good, clean engine ground.

2. Set the parking brake and block the drive wheels.

3. Start and operate the engine until it reaches normal operating temperature.

4. Make certain that after the engine warms up, the choke is wide open and the idle speed adjusting screw is off the fast-idle position on the cam.

5. Following the manufacturer's recommendations, disconnect the vapor hose to the carbon canister on vehicles with evaporation emission control systems, remove the air cleaner, turn the headlights and air conditioner on or off, and, if so equipped, place the automatic transmission in drive or neutral.

6. Adjust the rpm to the amount specified by the manufacturer, using the idle speed adjusting screw on the carburetor

linkage or the adjustment on the antidiesel solenoid.

7. Disconnect the tachometer leads from the ignition coil and ground and remove the meter.

Adjusting the Air/Fuel Mixture to Lean-Best Idle with a Tachometer

To adjust the engine's air/fuel ratio to lean-best idle with a tachometer follow this procedure:

1. Connect a tachometer to the engine, following manufacturer's instructions or those presented earlier in this chapter.

2. With the engine idling at normal temperature and at the specified rpm, turn the idle mixture adjustment screw in (clockwise) until the engine begins to slow down, as indicated by a drop in rpm by the tachometer needle.

3. Slowly adjust the screw outward (counterclockwise) until the engine operates at its highest steady rpm, as indicated by the tachometer needle.

4. As necessary, readjust the engine rpm to specifications.

5. On dual- or four-barrel carburetors with two idle mixture adjusting screws (Fig. 4–13), adjust each screw, one at a time, as outlined above. Note: These idle mixture adjustments can affect one another because the intake manifold or carburetor throttle flange may have balance passages between the two carburetor bores. Consequently, you may have to repeat the entire adjustment procedure over again on both mixture screws to achieve a balanced condition between the two idle systems.

6. Reset the idle speed to specifications

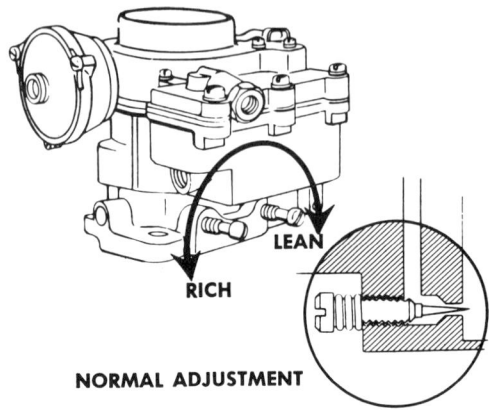

Figure 4–13. When setting the air/fuel mixture on an engine with a two- or four-barrel carburetor, both idle mixture screws require adjustment (Courtesy of United Delco).

as necessary. Then, disconnect the tachometer leads and remove the meter.

Adjusting the Air/Fuel Mixture Using the Lean-Drop Method

In some cases, the manufacturer suggests using the lean-drop method of adjusting air/fuel mixture as an alternative to utilizing the infrared analyzer. To adjust the mixture using this method:

1. Connect the tachometer to the engine and adjust the idle speed and mixture to lean-best idle.

2. Turn the *mixture adjustment screw* slowly in again until the idle speed drops down as specified by the manufacturer. Then, leave the screw at this setting. For instance, suppose the manufacturer's specifications call for 650 rpm at lean-best idle speed and a lean-drop setting of 630 rpm; you would turn the mixture screw of a single-barrel carburetor in (clockwise) until the rpm reduced 20 rpm or from 650 to 630.

On two- or four-barrel units, turn one mixture screw in to drop the idle speed 10 rpm and then adjust the other one in to lower the engine speed another 10 rpm. This action maintains idle system balance while accomplishing the specified lean drop. In both cases mentioned, the procedure reduced engine speed to 630 rpm.

3. Disconnect the tachometer leads and remove the meter.

Using the Infrared Analyzer

An infrared analyzer accurately measures the amount of HC and CO emissions from the internal combustion engine. However, before this machine can provide these accurate measurements, it requires two forms of calibration, initial and periodic. A technician initially calibrates the analyzer before placing it into service each morning and as often as necessary throughout the day to maintain its electronic calibration. In other words, if the analyzer has been shut off during the day, it requires electronic calibration.

Initial Electronic Calibration

To initially calibrate this analyzer (Fig. 4–14) perform the following procedure:

1. Plug the power cord into a suitable receptacle.

2. Depress in the zero-mode control button and permit the analyzer to warm up for at least 5 minutes.

3. Push in both the HC and CO low (LO) scale buttons.

4. Adjust the HC and CO zero knobs below both meters until their pointers read on the zero-set line.

5. Depress in and hold the span control button while adjusting the span knobs below the HC and CO meters, until their pointers both read on the span-set line.

6. Release the span button, and recheck the zero adjustment. Note: During the first 15 minutes of operation, many analyzers require slight calibration adjustments.

7. If the meters do not return to within one-pointer width of the zero-set line, repeat steps 2 through 6.

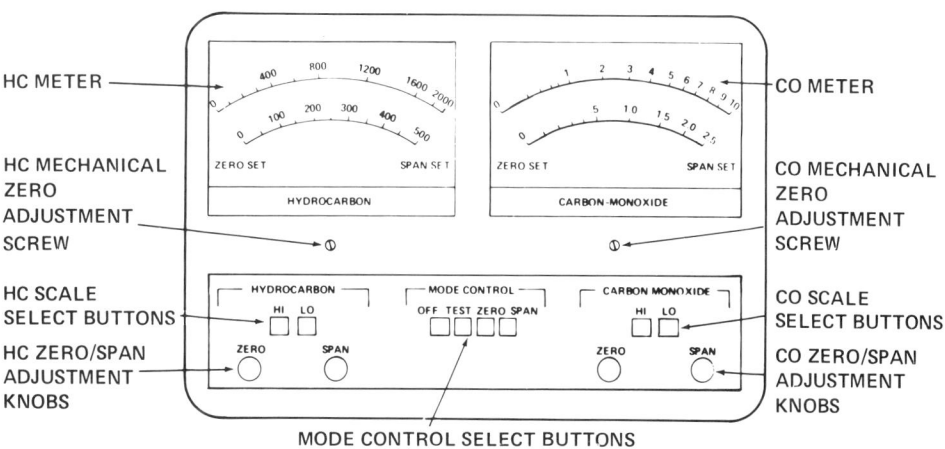

Figure 4–14. Meters and controls of a typical infrared analyzer.

8. Depress in the high (HI) buttons below each of the two meters. The HC and CO meters should indicate zero. If both meters do not return to within one-pointer width of the zero-set line, repeat steps 2 through 6.

Periodic Calibration

Certain states, such as California, have laws requiring periodic calibration of infrared analyzers. This periodic calibration verifies the accuracy of the machine in measuring concentrations of hydrocarbons and carbon monoxide.

Periodic calibration requires the use of a named gas and special equipment (Fig. 4–15). There are two kinds of calibration gas that technicians may use for this purpose, propane and N-hexane. However, many areas, such as California, require by law the use of propane gas for field calibration tests.

If N-hexane gas is used, it is not necessary to adjust its ppm concentration value; however, this is not true of propane. In other words, if a technician utilizes propane gas for calibrating the instrument, its ppm concentration requires a conversion into an N-hexane equivalent.

A specialist accomplishes this by multiplying the propane's correlation factor, appearing next to the calibration gas inlet on the analyzer, by the propane concentration stamped on the bottle. This value be-

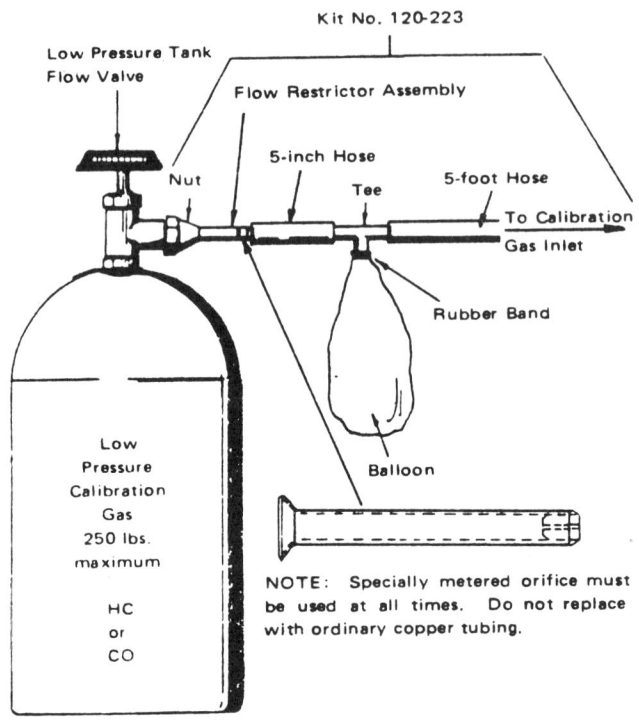

Figure 4–15. A named gas and special equipment is necessary to periodically check the accuracy of an infrared analyzer (Courtesy of Sun Electric Corp.).

comes the adjusted ppm concentration for the N-hexane equivalent of propane. For example, if the propane gas used for the calibration has a concentration of 3,054 ppm and its correlation factor is 0.520, then the N-hexane equivalent of the propane is 1,588 ppm (3,054 ppm × 0.520).

The CO concentration of the calibration gas is also marked on the bottle. But in this case, no adjustment of the propane's CO concentration value to that of N-hexane gas is necessary. In other words, whether you are using propane or N-hexane gas, no adjustment to any CO concentration value is necessary.

Preliminary Mechanical/Electrical Zero and Span Adjustments

To prepare the analyzer (Fig. 4–14) for the gas calibration, perform the mechanical/electrical zero and span adjustments as follows:

1. With the analyzer turned off, mechanically zero each meter pointer on the zero-set line by turning the set screw directly under both meters.

2. Connect the analyzer's power cord to proper electrical power source.

3. If the machine has an operating model/gas calibration (CAL) mode switch, set it in the operating mode position.

4. Depress in the mode control test button and allow the analyzer to warm up for at least 15 minutes. Note: If for any reason there is an interruption in electrical power from 1 to 5 seconds after the initial warm-up, allow an additional 10 minutes in this mode for the unit to restabilize.

5. After the warm-up period is complete, depress in the mode control zero button.

6. Depress in the low (LO) buttons under each of the two meters.

7. Zero set both meters in the LO scale mode by rotating the zero adjusting control knobs until each meter pointer is on its respective zero-set line.

8. Push in and hold the span mode control button. Next, adjust each of the span control knobs until each meter needle is on the span-set line.

9. Release the span control button; both meter needles should now return to the zero-set line. If the needles do not return to within one-pointer width of the zero-set line, repeat steps 5 through 9.

10. Depress in the high (HI) button under each of the two meters. Both of the meter needles should remain on the zero-set line. If they are not within one-pointer width of this set line, repeat steps 5 through 9.

Low and High Scale Comparison Checks

To perform these particular checks:

1. Push in the hydrocarbon LO button and turn the HC meter zero control knob until the HC meter indicates 400 ppm on its low scale.

2. Depress in the hydrocarbon HI button. The needle of the HC meter should still indicate 400 ppm, but this time on the high scale, and be within plus or minus one-half-pointer division.

3. Push in the hydrocarbon LO button and rotate the zero control knob until the HC needle is on the zero-set line.

4. Depress in the carbon monoxide LO button and turn the CO zero control knob

until the CO meter reads 2 percent on its low scale.

5. Push in the carbon monoxide HI button; now the CO needle should indicate 2 percent on the CO high scale and be within plus or minus one-half-pointer division.

6. Depress in the carbon monoxide LO button and turn the CO zero control knob until the CO needle reads on the zero-set line.

7. Push in both the hydrocarbon and carbon monoxide HI scale buttons. Both meter needles should now be on the zero-set line with no more than one-pointer width variation.

Gas Calibration Check

To gas calibrate the analyzer (Fig. 4-15), follow these steps:

1. Connect the test hose from the gas bottle to the inlet fitting labeled "calibration gas inlet" located on the rear of the unit.

2. If the analyzer has an operating mode/gas calibration mode switch, set it on the gas calibration (CAL) mode position.

3. Slowly open the low pressure tank valve or flow control valve on the calibration gas bottle until the in-line balloon begins to expand, showing a flow of gas into the analyzer. Caution: Do not fully inflate the balloon.

4. After the meter readings stabilize, in about 5 to 10 seconds, close the valve on the calibration bottle.

5. Read the indication on the HC meter appearing on the 0 to 2,000 scale. If the ppm reading of hydrocarbons is within ± 60 ppm of the N-hexane or the N-hexane equivalent concentration of propane, the HC meter section is within calibration. For instance, if the HC gas sample has a concentration of 1,588 ppm, the meter may indicate ± 60 ppm or from 1,528 to 1,648 and still be within specification limits.

6. Read the amount of carbon monoxide on the 1- to 10-percent scale of the CO meter. If this reading is within ± 0.3 percent of the certified percentage of the CO content in the calibration gas, the CO section of the analyzer is within calibration. For example, if the certified CO level is 7.56 percent, the meter may show ± 0.3 percent or from 7.26 to 7.86 percent carbon monoxide and still be in the specified tolerance limits.

7. If either the HC or CO section test results are not within tolerances, service is necessary on the analyzer. In this situation, contact a factory representative or a trained calibration technician.

8. Disconnect the test hose from the gas inlet.

9. Place the operating mode/gas calibration mode switch to the operating mode position. This position permits the analyzer to expel any calibration gas from its sample cell.

Infrared Analyzer Maintenance

Other than initial and periodic calibration, the infrared analyzer, like any other piece of equipment, requires some other types of maintenance. Because the infrared analyzer represents a sizable investment by the owner of a repair facility, performing the maintenance as recommended by its manufacturer protects the initial investment, reduces the analyzer's downtime, and keeps major repair costs to a minimum. The routine maintenance and inspections recommended by many analyzer manufacturers are to the filters, sampling hose, and probe.

Filter Replacement

The infrared analyzer illustrated in Fig. 4-14 has two filter assemblies, primary and secondary. These two filter units have a design that prevents particulate contaminants and water from entering the analyzing system.

The primary filter has an automatic water drain and therefore acts as the main water trap for the system. In other words, this device not only partially filters the exhaust gas but removes a large percentage of the water vapor found in the exhaust gases of the internal combustion engine. The trapping and removing of this water minimizes saturation of both particle filters.

The secondary filter has a manual-type bowl drain. This type of drain is necessary for removal of any moisture that may occasionally condense in the filter bowl.

The frequency or the required replacement period for analyzer filters does vary somewhat from one manufacturer to another. However, in the case of the analyzer shown in Fig. 4-14, the technician should service both filters along with the sample hose and probe whenever the low flow check filter warning light comes on during a test of a vehicle.

Primary Filter Service

To service the primary filter (Fig. 4-16) perform the following steps:

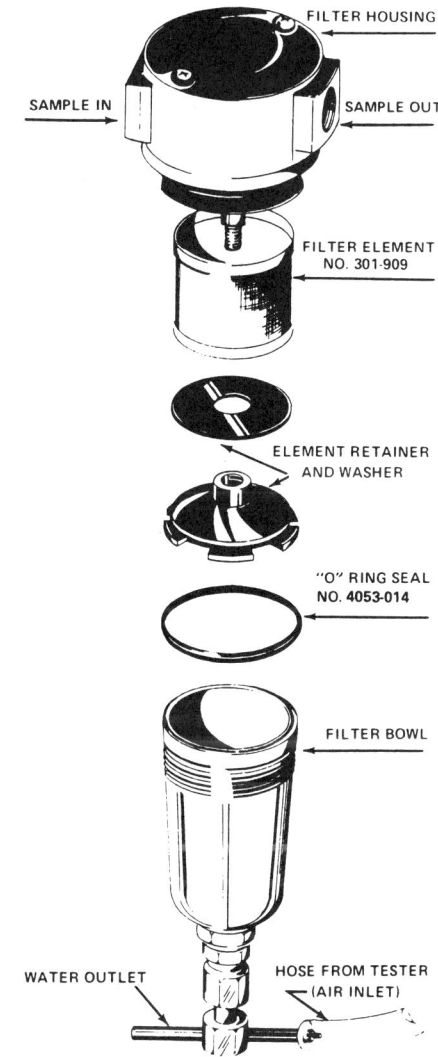

Figure 4-16. Servicing the primary filter assembly (Courtesy of Sun Electric Corp.).

1. Remove the inlet and outlet hoses from the T fitting located on the lower part of the filter bowl.

2. Unscrew the filter bowl from the housing by turning it counterclockwise. At this point, note the position of the O-ring seal on the bowl rim.

3. Turn the element retainer counterclockwise and remove the filter element.

4. Clean the bowl and filter element in a solution of detergent and water. Then, allow both units to air dry. If upon inspection the filter element does not come clean or shows signs of damage, replace it.

5. Put together the filter assembly in the reverse order, making sure the O-ring seal is in proper position in its groove at the lower edge of the bowl.

6. Reconnect the inlet hose to the side of the T fitting that has the smallest opening and the outlet hose to the other.

Secondary Filter Service

To service the secondary filter (Fig. 4-17), follow these instructions:

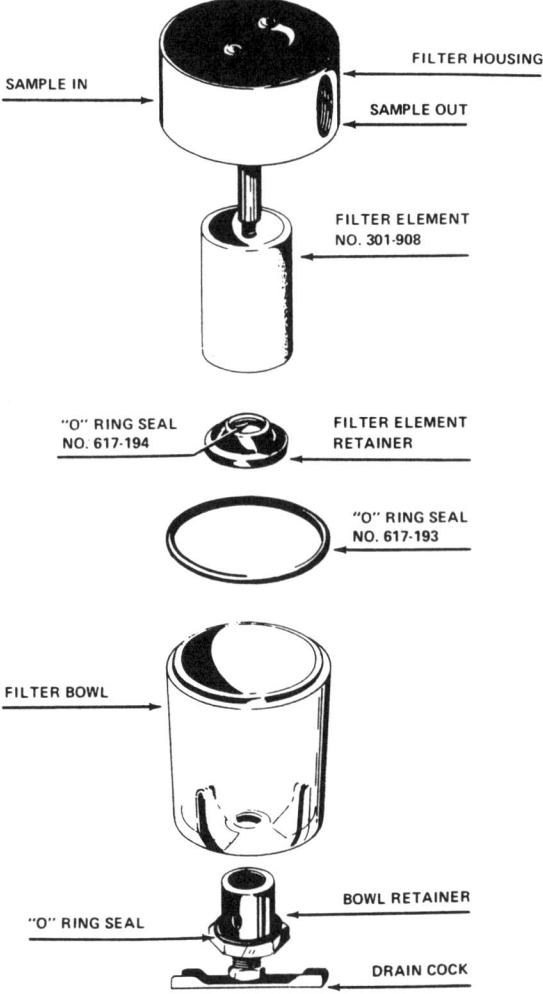

Figure 4-17. Servicing the secondary filter assembly (Courtesy of Sun Electric Corp.).

1. Remove the bowl retainer at the base of the bowl by rotating it counterclockwise. Next, remove the bowl.

2. Inspect both the O-ring seals on the retainer and bowl rim. Replace them if they are worn or deteriorated.

3. Remove the filter retainer by pulling it downward. Then, inspect the retainer's internal O-ring seal. Replace the seal if it is worn or deteriorated.

4. Remove the old filter element and replace it with a new one.

5. Wash the filter bowl in a solution of detergent and water and permit it to air dry.

6. Reassemble the secondary filter in the reverse order, making sure all the O-ring seals are in their proper positions and the drain cock is fully closed. Note: When servicing either or both of these filters, make sure that all the O-ring seals are in their proper positions and the units are put back together snugly to prevent the possibility of outside air leaking into the analyzer during a vehicle test. Any air leak reduces the accuracy of the analyzer's readings.

Extending Filter Life

The frequency of filter replacement depends largely upon the number of vehicles tested or the amount of usage the analyzer receives. However, there are two procedures that you can follow to extend the life span of the filters. First, do not start or warm up a vehicle's engine with the analyzer's test probe inserted into the tail pipe. Each time the engine starts, a large amount of loose deposits blow out of the tail pipe; these deposits do prematurely clog the filter elements.

Second, always purge the analyzer after use by permitting it to operate in fresh air away from the exhaust pipe. This action pulls any accumulated hydrocarbons from the hose.

Sampling Hose and Probe Service

The *flexible sampling hose* connects the probe to the infrared analyzer. This hose is quite expensive and therefore requires the following service and care to extend its useful life:

1. Visually inspect the sampling hose assembly daily for leaks, cuts, kinks, or damage caused by vehicles accidentally running over it.

2. Check the hose fitting connections for tightness and leaks.

3. Avoid driving over the hose with a vehicle because some sample hoses are fairly rigid and require replacement if collapsed.

4. Never insert the sample probe too far into the tail pipe. The sample hose itself may end up dangerously near the hot exhaust gases leaving the tail pipe. As a result, the hose may collapse or melt with the interior walls sticking together from the excessive heat.

5. Purge the sample hose with fresh air after each use. This action minimizes the buildup of hydrocarbons inside the hose. These hydrocarbons can later be released, while you are testing a vehicle, and result in an inaccurate reading.

To service the probe:

1. Daily visually inspect the probe for damage.

2. Examine the holes in the probe tip to make sure they are open. If the holes are clogged, use a paper clip or stiff wire to clear the holes of dirt or foreign material.

3. Check the tightness of the fitting connection to the base assembly.

Eliminating the Effects of Static Electricity

In certain areas, the buildup of static electricity on the plastic meter lens of the analyzer can result in inaccurate readings by affecting meter needle movement. There are two methods for eliminating the effects of static electricity, with a damp cloth or with antistatic spray.

The easiest method of removing static electricity is to dampen a clean shop towel and gently wipe off both meter lenses. This action minimizes the effects of static electricity while improving the appearance of the machine.

There are some commercial antistatic products on the market that are useful in reducing the effects of this problem. However, before using any such product, be sure to check with the instrument manufacturer to make sure the product will not harm the meter lens.

Testing Exhaust Emissions with an Infrared Analyzer

General Information

The testing of a vehicle's exhaust emissions with an infrared analyzer is a very simple task. It basically involves placing the analyzer's probe in a vehicle's tail pipe and waiting for the readings to appear on the meters.

Although this procedure sounds very simple, there are considerations you must take into account so that the exhaust sample represents a true picture of just what the vehicle's exhaust really contains. These considerations include the effects of exhaust and sampling system leaks, shallow probe insertion, dual exhaust, and the catalytic converter on HC and CO readings.

Effects of Exhaust System Leaks

The exhaust system of a motor vehicle has been and still is very notorious for rusting away and developing leaks. An exhaust leak can create two problems. First, an exhaust leak is dangerous because it can permit carbon monoxide to enter the inside of the vehicle. Remember, carbon monoxide is a very poisonous gas that can cause drowsiness, headaches, and even death if a person inhales sufficient quantities of this gas.

Second, if an exhaust leak becomes large enough, it produces an audible sound. This sound results from air being drawn into the exhaust between pressure pulses. These pressure pulses are the result of both positive and negative pressure fluctuations caused by the engine's various exhaust strokes.

Any air drawn into the exhaust system dilutes the volume of the exhaust sample pulled into the infrared analyzer. As a result, a lower than actual reading is shown on the meters (Fig. 4–18). Note in this diagram that the HC reading of 325 ppm represents the emissions without an exhaust leak; but with a leak, the meter shows approximately 125 ppm.

With an air leak, the CO meter will also indicate too low a reading. For instance, in Fig. 4–18, the CO meter should show, without a leak, about 1.2 percent. However, the meter actually indicates about 0.25 percent due to the effect of the air leak. As you can see, it is very important for accurate

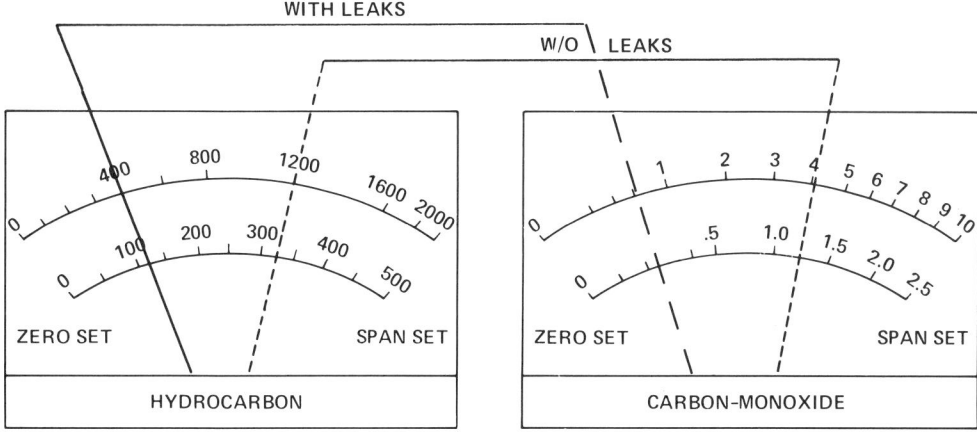

Figure 4-18. Effect on HC and CO readings by exhaust system leaks.

testing with an infrared analyzer that you visually check as well as listen for exhaust system leaks.

Effect of Sampling System Leaks

A pump within the analyzer draws an exhaust sample from a vehicle's tail pipe through an exhaust sample line. Along this sample line are many connections and components that can become loose or deteriorate. If any of these conditions occur, the result is similar to an exhaust system leak. That is, the analyzer's pump draws air into the sample line and dilutes the exhaust sample. This results in a lower than actual reading on both meters just as if there was a leak in the exhaust system.

The low-flow indicator on the infrared analyzer provides an easy method for checking for sample system leakage. To use the indicator to perform this check, simply block off the inlet hose at the end of the sample probe. If the sampling system is air tight, an indication of low flow should immediately occur. If there is no low-flow indication, make a systematic check, starting with the probe back to the analyzer itself until you locate and correct the source of the leak.

Effect of Shallow Analyzer Probe Insertion

The depth of insertion of the analyzer probe can also affect the readings of the machine. Positive and negative pulses are present in the tail pipe, which means that air can move some distance into this pipe during the negative exhaust pulses.

To minimize the effect of this air flow, you should place the probe about 8 to 12 inches into the tail pipe (Fig. 4-19). At this distance, there is very little chance of air passing into the probe.

However, on some vehicles, the probe will not go very far into the tail pipe because of a screen or other obstruction. For this situation, the analyzer manufacturer supplies an adapter that fits over the tail pipe, making it longer.

Shop exhaust recovery systems also create possible air leak problems. To reduce the problem, make a small slit into the re-

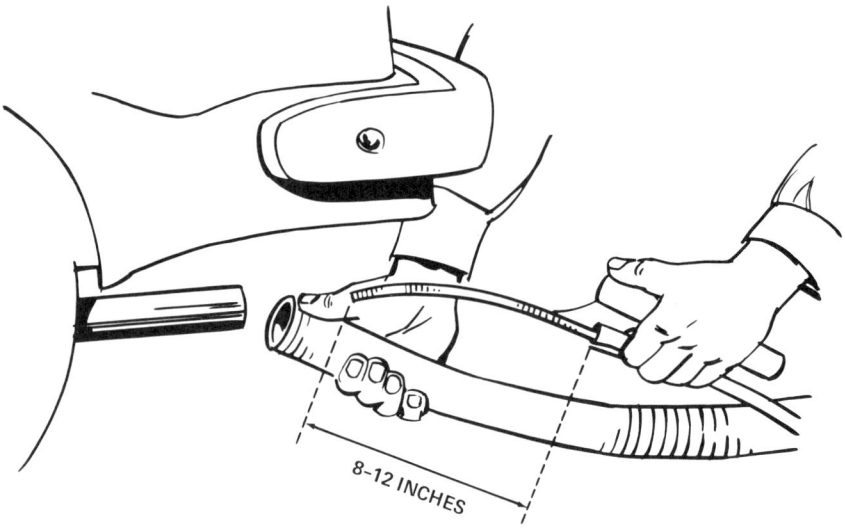

Figure 4-19. Proper probe insertion into a vehicle's tail pipe.

covery hose and insert the probe into the tail pipe through the slit opening.

Effect of Dual-Exhaust Systems

Many vehicles do have dual-exhaust systems. This can present a problem for measuring exhaust emission levels because it is possible that the left and right tail pipe emission levels will not be the same.

To eliminate possible problems, it is always a good idea on any vehicle with dual exhaust to probe both exhaust pipes and check the emission levels. If the sample taken from either tail pipe shows excessive emissions or there is a large difference in the two readings, there is some type of problem that deserves further attention.

If for whatever reason, you are only going to measure the emissions from one tail pipe, always insert the analyzer probe into the pipe *without* the heat-riser valve. If the vehicle's exhaust system has no heat-riser valve, insert the probe in either tail pipe in order to perform the emission test.

Effect of the Catalytic Converter and/or Air Pump

Currently, automobiles and many other gasoline-operated motor vehicles have a variety of emission control devices installed on them. For example, a vehicle may have a catalytic converter, catalytic converter/air pump combination, or an air pump by itself, to name a few of the devices. All of the above-named devices are known as post emission control devices. In other words, they modify or change the actual HC and CO emissions that result from combustion but do so outside of the combustion chamber. Consequently, if these devices are functioning properly, the infrared analyzer will show lower emission levels than those actually produced during the combustion process.

Effects of the Catalytic Converter

On a properly tuned vehicle with a catalytic converter, HC readings will be somewhat less than on a vehicle without this device. Furthermore, the CO levels from a converter-equipped vehicle will be close to zero at idle. If the vehicle has a defect that slightly increases HC emissions, the analyzer indicates these changes; but the readings will be lower in magnitude due to the action of the converter itself.

However, this is not the case with a defect that would normally increase CO levels. For instance, if a vehicle without a catalytic converter has its carburetor idle circuit set too rich, the infrared analyzer shows high CO emissions. This is not true in the case of the converter-equipped vehicle because the unit reduces excessive amounts of CO emissions to nearly zero at idle. Therefore, the analyzer will not show a maladjusted carburetor that causes high CO levels on vehicles with a catalytic converter.

Effects of a Catalytic Converter and Air Injection System Combination

On a vehicle with both a catalytic converter and an air injection system, the infrared analyzer shows little or no increase in hydrocarbons even when shorting out a spark plug. The converter, in combination with the air injection system, does an excellent job of eliminating HC and CO emissions even when they are excessive. As a result, these units do hide the effects of HC and CO levels as well as many other engine or system malfunctions.

Effect of an Air Injection System

The air injection system reduces HC and CO emissions by forcing a quantity of air into the hot exhaust gases. Consequently, the infrared analyzer should indicate overall lower HC and CO readings on vehicles with air injection over those without this system. The catalytic converter and air injection systems will be discussed in later chapters.

Infrared Test Procedures— Vehicles without Catalytic Converters

To check a vehicle's exhaust for HC and CO content with an infrared analyzer, follow this procedure:

1. Warm up and initially calibrate the infrared analyzer as mentioned earlier in this chapter.

2. Connect a tachometer to the engine, following manufacturer's instructions.

3. Start the engine and permit it to reach normal operating temperature.

4. Depress in both the HC and CO HI meter scale buttons on the analyzer.

5. Insert the analyzer's sampling hose 8 to 12 inches into the vehicle's tail pipe (Fig. 4-19).

6. Depress in the analyzer's test selector button on the mode control.

7. With the air cleaner in place, operate the engine at idle or the rpm specified by the manufacturer.

8. Note and record the readings on both the HC and CO meters. Note: If the HC meter reads less than 500 ppm and/or the CO meter reads less than 2.5 percent, depress in both meter LO scale buttons. This provides for more convenient scale reading.

9. Increase the engine speed to 2,500 and hold it at this rpm.

10. Note and record the HC and CO levels. Both should be as low as or lower than those taken at idle. Note: When checking some electronic fuel-injected vehicles without a load, only the idle readings will be accurate. Testing of these engine types at 2,500 rpm requires the use of a dynamometer to load the engine.

11. Return the engine to its normal hot-engine idle. After a momentary increase on deceleration, both readings should return to the levels recorded in step 8.

12. After completing these tests, depress in the analyzer's zero button on the mode control. This action, during intervals between testing procedures, maintains the analyzer ready for use without the need for further warm-up and calibration along with increasing its filter life.

13. To turn the analyzer completely off, depress the off button on its mode control.

14. Remove the sample probe from the tail pipe and the tachometer from the engine.

Infrared Test Procedures—Vehicles with Catalytic Converters

With few exceptions the procedure just outlined for testing exhaust emissions with the infrared analyzer applies also to a vehicle with a catalytic converter. One of the notable differences will be in the test results; the HC and CO levels should be much lower. The reason for this is that the converter itself almost eliminates these emissions; consequently, the analyzer will barely indicate the remaining concentrations. This is especially true with the CO levels. It is for this reason that many manufacturers no longer provide CO percentages with their tune-up specifications. However, if the vehicle has a problem that causes it to be a gross polluter, the analyzer will, of course, indicate the levels of increased emissions which the converter cannot reduce.

Another difference in the procedure is the area into which you insert the test probe. In many cases, the probe still fits into the tail pipe. However, in a few instances, manufacturers do provide a plug in front of the converter to provide access for the infrared probe (Fig. 4–20). This access permits the analyzer to sample the exhaust gases before they enter the catalytic converter.

This is especially necessary if the technician is to use the infrared analyzer to make carburetor air/fuel adjustments. Without the test plug, the mechanic can use the machine to check a vehicle's emissions against federal or state regulations,

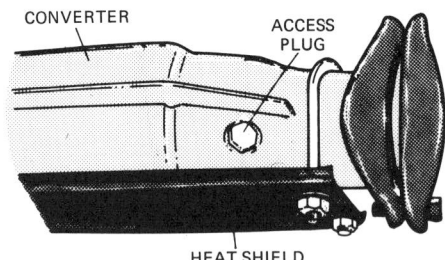

Figure 4–20. Some converters have a plug access for the analyzer probe so that the unit can sample the exhaust gases before they enter the converter.

and the analyzer will indicate if the levels are within limits. But in many cases, the CO emissions, due to converter action, are so low at idle that the analyzer is no longer an accurate tool for performing carburetor adjustments. In these situations, the manufacturers do recommend the use of propane enrichment adjustments, which this text will cover later on.

The key to when you should use an infrared analyzer to adjust a carburetor is found in the manufacturer's specifications for the particular vehicle being repaired. These specifications are either on a decal located under the hood or are in the appropriate service manual. In either case, these specifications or instructions will inform you as to the exact procedure to follow to correctly set idle mixture and speed along with the CO level. By following these instructions, you should have no difficulty in obtaining the proper idle mixture that provides a relatively smooth idle and low emissions.

Infrared Test Results and Indications—Idle Speed

1. If at idle the HC and CO readings are within specifications, the engine along with the ignition, fuel, and emission control systems are functioning normally. The HC readings should be in the range of about 50 to 300 ppm while CO levels should fall between 0.3 to 3 percent for an emission-controlled engine. However, these specifications should only serve as a guide, since there are federal or state standards as to the allowable emissions a given vehicle can produce and still comply with the law.

2. If the analyzer shows a high HC reading but a normal CO level, there is a possible problem in the ignition system (Fig. 4-21). In this situation, the HC readings are usually greater than 1,000 ppm while the CO level ranges between 0.3 to 3 percent. In addition, the engine will usually have a rough idle along with a steady or intermittent spark plug misfire.

3. If the analyzer indicates a higher than normal HC emission with a CO reading within specified limits, the ignition timing may be too far advanced (Fig. 4-22). In this case, the HC reading will be in the range of 300 to 900 ppm, depending upon the vehicle, while the CO level should still remain in the normal range. Other symptoms of overadvanced ignition timing include possible rough engine idle, pinging, or detonation.

4. If the analyzer shows a high HC reading but a normal CO level, the engine may have a mechanical problem or it has a missing, stuck open, or incorrect thermostat. Other symptoms of a mechanical engine problem are the possibility of one or more misfiring cylinders, mechanical noises, poor performance, and excessive smoke in the exhaust.

There are several other indications of a thermostat problem. These include delayed engine warm-up, poor heater operation, and a reduction in fuel economy.

5. If the HC reading is higher than normal and possibly wavering, usually accompanied by a low CO level, the engine most likely has a vacuum leak. Depending on just how bad the leak is, the HC reading will vary and possibly waver from above normal to full-scale deflection, and the CO level may be less than 0.8 percent. Note: For this particular problem, less than 0.8 percent is the cutoff figure for a vehicle without a catalytic converter. However, a similar problem on a converter-equipped

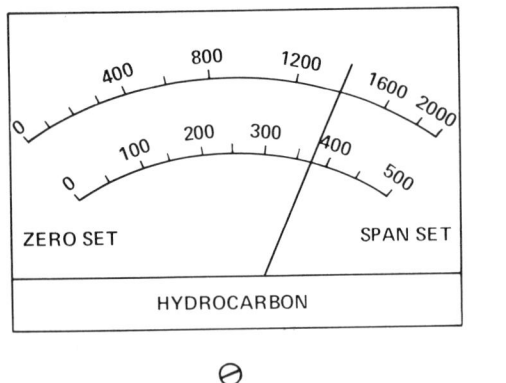
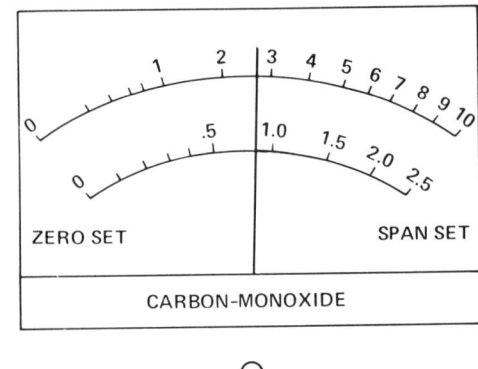

Figure 4-21. A high HC reading with an average CO level is a symptom of an ignition system malfunction.

vehicle could result in a very insignificant or a zero CO reading.

There are other symptoms of an engine vacuum leak. For example, the engine may experience a rough idle or misfire that may be steady or intermittent, depending on the severity of the leak. In addition, there may be a hissing sound caused by air passing into some portion of the intake system.

6. If the analyzer shows a high HC reading and a very low CO level, the carburetor may also be set too lean, causing a lean misfire condition (Fig. 4-23). In response to this problem, the analyzer indicates an HC reading that is higher than normal with a possible wavering pointer, accompanied by a very low CO level (less than 0.3 percent). Note: For this particular problem, the less than 0.3 percent cutoff point is for a vehicle without a catalytic converter. The same malfunction on a converter-equipped vehicle usually results in a very insignificant or a zero CO reading.

There is also another symptom for a

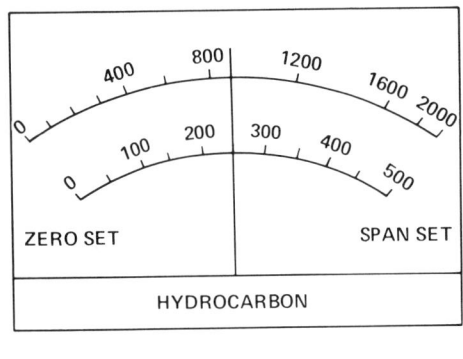
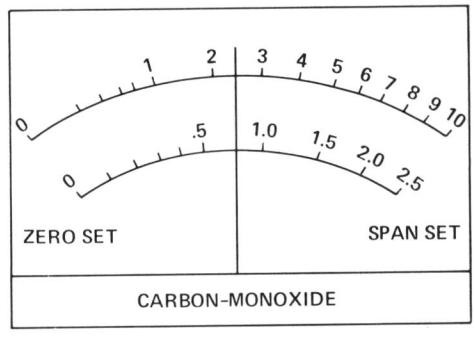

Figure 4-22. If the timing is too far advanced, the HC emissions will be excessive.

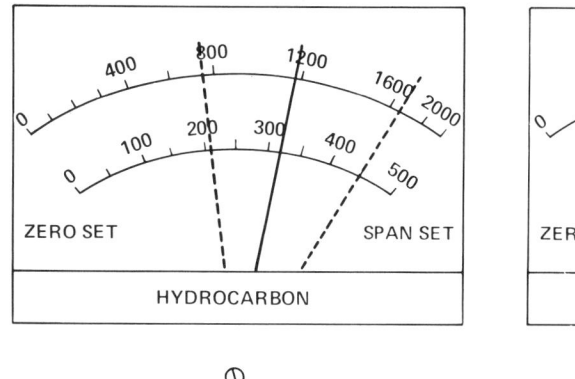
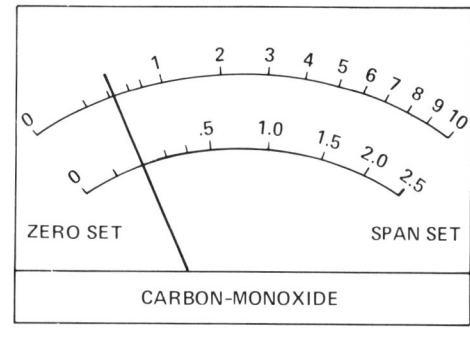

Figure 4–23. Typical HC and CO readings caused by a lean misfire.

lean misfire. That is, the engine has a rough idle that may be steady or intermittent, depending upon just how lean the air/fuel mixture is.

7. A HC analyzer reading that is higher than normal and possibly wavering, accompanied by an acceptable CO level, is a very good indication of improper exhaust gas recirculation (EGR) system operation (Fig. 4–24). Along with these readings, the engine will have a rough idle or a misfire that may be steady or intermittent, depending upon how severe the EGR malfunction is.

There is another symptom for improper EGR system operation, namely, a reduction in intake manifold vacuum. Normally, this vacuum should be in the range of 16 to 21 inches Hg. A reading of less than this on a vacuum gauge, along with the above-mentioned analyzer readings, points to a defect in the EGR system. The troubleshooting procedure for this system will be covered in a later chapter.

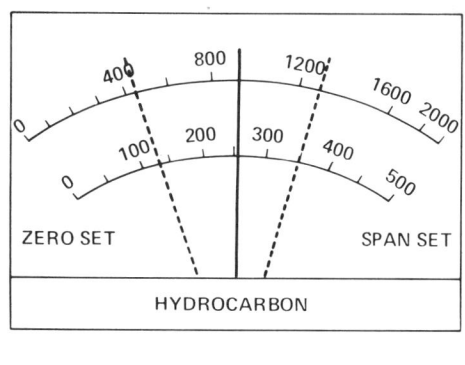
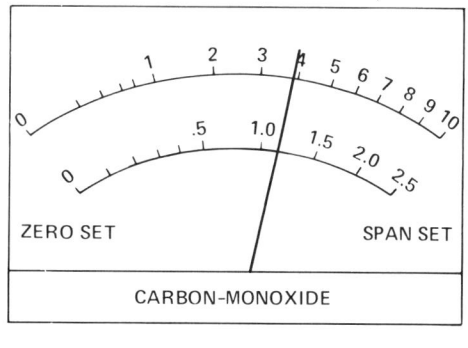

Figure 4–24. Typical analyzer readings that result from an EGR system malfunction.

8. If the analyzer shows a HC reading within specifications but the CO level is slightly high, either the air cleaner element is dirty, the choke valve is stuck partially closed, the carburetor is out of adjustment, the idle speed is not to specifications, or the positive crankcase ventilation (PCV) valve has a restriction. In order to locate the exact cause of the high CO level, it will be necessary to check all of these items, one at a time, until the cause of the problem is located.

9. If both the HC and CO readings are high, the problem can be that air/fuel ratio is extremely rich, there is an internal carburetor malfunction, the PCV system is inoperative, the heat-riser valve is stuck open, the air injection system is inoperative or disconnected, the thermostatic air cleaner preheat door is sticking, or the catalytic converter is defective.

There are other symptoms of an extremely rich air/fuel ratio or an internal carburetor problem. These include:

a. Very smooth idle. A smooth idle on an emission-controlled engine is a good indication of a slightly rich air/fuel mixture from either a maladjusted carburetor or an internal carburetor problem.

b. Black smoke. Black smoke emitting from a vehicle's tail pipe is a very good indication of a grossly rich mixture or an internal carburetor malfunction. The driver may also complain about a rough idle or the engine loading up with excessive amounts of fuel.

c. Decreased fuel economy. If the problem is severe enough, the vehicle will use excessive amounts of fuel.

d. Presence of an odor. The majority of vehicles with catalytic converters will emit a rotten-egg odor during idle engine operation if for any reason the air/fuel ratio is too rich.

If the air/fuel ratio or the carburetor itself was not the cause of the high CO and HC readings, you will have to check the remaining items mentioned previously, one at a time, until the defect is found. This text will explain how to check the air injection system, thermostatic air cleaner, and the catalytic converter in later chapters.

Infrared Test Results and Indications—2,500 rpm

If at 2,500 rpm the HC and CO levels are at or slightly below those noted at engine idle, the engine, along with the fuel, ignition, and emission control systems, is functioning properly. However, if either or both readings are above specifications, check all the items listed for the idle speed test as well as the following items:

1. If the analyzer shows a high HC but a low CO reading, there may be a fouled plug, defective spark plug cable, extremely lean air/fuel mixture, or a floating exhaust valve.

2. If the HC level is high while the CO reading is normal or slightly low, there may be a vacuum leak, sticking valve, or arcing, bouncing, or misaligned ignition contact points.

3. If the test shows a normal HC reading but a high CO level, the choke valve may be partially closed, the air cleaner element has a restriction, or the carburetor has a defective high-speed or power circuit.

Causes of and Diagnosing Excessive HC Emissions

A HC reading above about 300 ppm is usually a good indication that excessive amounts of unburned fuel are now passing out of the exhaust. This unburned fuel emission can result from a malfunction or defect within any of seven engine systems: the ignition, timing advance, intake, PCV, carburetion, EGR, and compression.

Ignition System Malfunctions

The following items represent the most common possible causes of high HC emissions, a steady or intermittent misfire, along with a rough engine idle due to a malfunction within the ignition system.

1. An open spark plug wire. A spark plug wire or wire connector can, for a variety of reasons, develop enough resistance or become open-circuited so that the ignition coil is incapable of providing sufficient voltage to jump the air gap at the spark plug. As a result, there is a misfire of one or more spark plugs.

2. A bridged or fouled spark plug. If the gap between the center and the ground electrodes becomes bridged due to oil or carbon deposits, an electrical arc cannot jump across this air gap. Consequently, the air/fuel mixture in the effected combustion chamber(s) will not ignite and burn.

3. A spark plug failure. Conditions such as burned internal resistor or cracked insulator can cause a spark plug not to fire. This also prevents the air/fuel mixture in the affected cylinder from igniting.

4. A loose spark plug wire connection. A loose connection or corroded connector, in a similar manner as a wire with high resistance, can create a misfire under engine load conditions or, if severe enough, a continuous misfire.

5. The breakdown of spark plug wire insulation. This problem usually results from high underhood temperatures that cause the wire itself to become hard and brittle or sometimes to even crack open. When the insulation becomes defective, the coil's high-voltage surge follows the path of least resistance to the nearest ground. This can be a valve cover, wire loom, or any other object capable of providing good electrical pathway to ground.

6. Crossed spark plug wires. Crossed spark plug wires, ones attached to the wrong plugs, create a misfire in at least two cylinders because the attached spark plugs fire at the incorrect time.

Diagnosing Ignition System Malfunctions

There are two general methods used by technicians to locate malfunctions within the ignition system: the oscilloscope check or visual inspection. The fastest and usually the most accurate way of locating the majority of ignition system malfunctions is through the use of an ignition oscilloscope, which is part of an engine analyzer. The oscilloscope itself shows a graphlike picture or pattern of what occurs during the entire sequence of ignition system operation. By comparing the pattern of a malfunctioning system with one of a normal system, the mechanic quickly pinpoints the cause of a problem.

However, the oscilloscope may not always indicate some defects within the system such as crossed spark plug wires. In this case, it is necessary to visually check all of these components to make sure they are connected to the proper spark plugs.

Ignition Timing System Malfunctions

The following defects are some of the most common encountered by mechanics within the ignition timing system. These problems can also cause excessive HC emissions.

1. Basic ignition timing too far advanced. If someone advances basic ignition timing too far beyond manufacturer specifications at curb idle, there will be incomplete burning of the air/fuel mixture. This results in HC readings from between about 300 to 900 ppm.

2. Full manifold vacuum applied to advance diaphragm. The vast majority of vehicles now use a ported vacuum signal to activate the vacuum advance unit of the distributor. At engine idle, this type of signal is nearly zero. Therefore, there is no timing advance provided by the vacuum advance diaphragm. However, if a source of manifold vacuum reaches the advance unit instead of ported vacuum, the end result is overadvanced ignition timing and high HC emissions at idle.

3. Failure of the centrifugal advance mechanism. Although somewhat unusual, a problem within the centrifugal mechanism can occur. If, for example, one centrifugal advance spring is weak or broken, the timing can overadvance at idle.

There are several other indications of another form of centrifugal advance problem. If, for instance, this device fails to advance the timing the proper amount or not at all, there will be a noticeable lack of engine power and poor fuel economy.

4. Binding vacuum advance mechanism. If a portion of this mechanism binds whenever vacuum is applied to the diaphragm, this can prevent the retarding of ignition timing as the carburetor throttle valve closes, and the vacuum signal drops to zero. As a result, there is overadvanced ignition timing at idle because the vacuum advance diaphragm spring does not return the breaker plate to the retarded position.

5. Failure of vacuum advance controls. Many different types of devices are now necessary to control the operation of the vacuum advance for different vehicle running conditions. If one of these devices fails to function or someone installs it incorrectly, a full manifold vacuum signal may reach the advance unit or become trapped in it during engine idle. This, of course, results in overadvanced ignition timing.

Diagnosing Timing Advance System Malfunctions

There are three basic pieces of equipment used by the industry to check for malfunctions in the ignition timing system. These include the standard timing light, power timing light, and a distributor tester. As mentioned earlier, a standard timing light is necessary in most cases to check and adjust basic ignition timing.

A powered timing light, on the other hand, is a valuable tool for checking several timing-related factors. For example, the technician uses this device not only for checking and adjusting basic ignition timing but for testing problems within the vacuum and centrifugal advance mechanisms as well. Most important of all, a mechanic can perform these tasks without the need of removing the distributor from the engine.

Figure 4-25. Typical distributor tester.

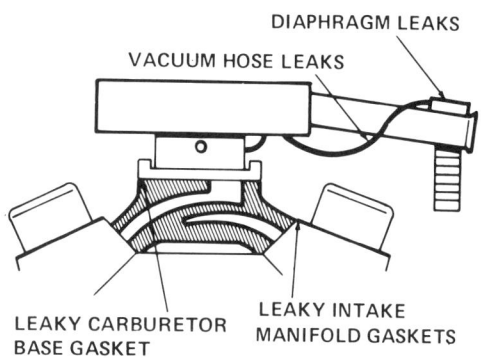

Figure 4-26. Points where air can leak into the engine's intake manifold.

However, many mechanics prefer to use a distributor tester (Fig. 4-25), for diagnosing problems within the vacuum and centrifugal advance mechanisms. This particular machine accurately tests the operation of both of these units at various distributor rpm. The only drawback to the machine's use is that the distributor requires removal from the engine before the operator can test its operation.

Intake System Malfunctions

The main malfunction within the intake system that creates excessive HC emissions is a vacuum leak. The following discussion covers the main causes and locations of vacuum leaks in this system:

1. Broken or disconnected vacuum hoses. Vacuum hoses do become brittle when exposed to high underhood temperatures. When a hose becomes hard and brittle, it breaks very easily due to vehicle or engine vibrations. In addition, due to the large number of vacuum hoses on an emission-controlled vehicle, it is easy for one to be knocked off or left off during any type of engine maintenance.

In either case, the result is that additional air passes into the intake manifold (Fig. 4-26). This air thins out the already lean mixture, resulting in a misfire condition.

2. A ruptured diaphragm in a vacuum-activated unit. A ruptured diaphragm in any component permits air movement into the intake manifold below the carburetor. This air has the same effect, as mentioned above, of thinning out the lean air/fuel mixture at idle and causing a misfire.

3. Leaking intake manifold or carburetor base gasket. If a leak develops between the mating surfaces of the intake manifold and cylinder head or between the carburetor base and intake manifold, additional air will also pass into the engine below the carburetor (Fig. 4-26). This forms an excessively lean air/fuel ratio that results in a misfiring problem.

Diagnosing Malfunctions (Vacuum Leaks) in the Intake System

Mechanics use one of two similar methods to locate the source of a vacuum leak in the intake system. For many years, technicians have utilized noncombustible cleaning solvent along with a vacuum gauge to locate vacuum leaks. This procedure has already been outlined in an earlier section.

A vacuum gauge and a bottle of propane can also be used for this purpose. The bottle of propane must have attached to it a control valve and a length of hose. The valve controls the amount of gas emitted from the bottle while the hose transports the gas to probable leak areas.

To locate a vacuum (air) leak with this equipment:

1. Attach a vacuum gauge and tachometer to the engine.

2. Start the engine and permit it to reach normal operating temperatures.

3. Crack the propane valve open slightly and direct the end of the propane hose around all the potential vacuum leak points as shown in Fig. 4–26.

4. When the propane reaches a source of an air leak, the engine idle will smooth out and its rpm will usually increase, as indicated by both the tachometer and vacuum gauge.

Caution: Propane gas is very explosive; consequently, only use it in a well-ventilated area and never close to flames or other ignition sources.

Malfunctioning or Incorrect PCV Valve

If the PCV valve stays open too far (high-flow position) or if someone installs a valve with an excessive flow rate, the engine's air/fuel mixture leaks out. This causes a lean misfire. Chapter 6 will explain the various methods used to test for proper PCV valve operation.

Incorrect Carburetor Idle Mixture or Speed Adjustments

If the carburetor mixture and speed adjustments are set so that the engine operates too lean, a lean misfire condition can result. In this case it is first important to ensure that there are no other air leaks causing the misfire before attempting to perform carburetor mixture and speed adjustments.

Several methods for making these adjustments have already been explained. However, Chapter 10 will also discuss the use of the infrared analyzer and propane enrichment equipment for carburetor adjusting on vehicles with emission-controlled engines.

EGR System Malfunctions

Listed below are some of the most likely causes of excessive idle HC emissions due to problems within the EGR system.

1. A leaking EGR valve gasket. A gasket is necessary to seal between the EGR valve itself and the intake manifold or appropriate fitting. This gasket performs three functions:

   a. It stops exhaust leaks on the exhaust side of the EGR valve.

   b. The gasket prevents any air leaks on the intake manifold side of the valve.

   c. It provides a seal between the intake and exhaust passages of the EGR valve.

If a leak develops due to warpage or a blown gasket, exhaust gases can flow from the exhaust to the intake passage (Fig. 4–27). Without the EGR valve controlling the flow, there will be excessive exhaust

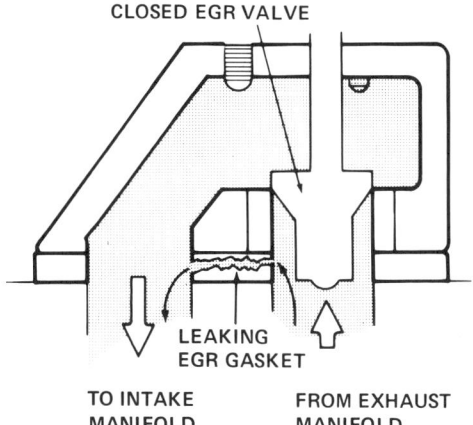

Figure 4-27. Exhaust gas leaking through a defective EGR valve gasket.

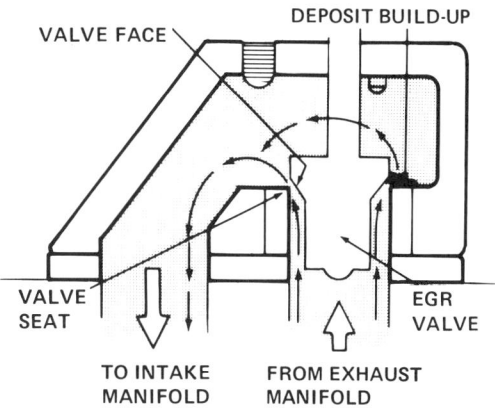

Figure 4-28. Exhaust gas leaking around a partly open EGR valve.

gas dilution of the idle mixture, causing a misfire.

2. EGR valve not fully closed. If an EGR valve does not close tightly due to a buildup of foreign material on the valve face or its seat, exhaust gases flow from the exhaust system into the intake manifold at all times (Fig. 4-28). At idle, these gases dilute the already lean air/fuel mixture found on a properly adjusted vehicle. The resulting exhaust gas dilution creates a rough idle and the possibility of a misfire condition.

There are several methods suggested by manufacturers to check the operation of the EGR valve. Chapter 16 will explain how to test, service, or replace typical EGR valves.

3. Improper vacuum hose routing and connections. Pre-1976 EGR valves do not receive a full manifold vacuum signal at idle. Instead, they use a ported signal. However, many 1976 and newer vehicles with exhaust pressure sensors do utilize direct manifold vacuum for EGR valve operation.

If for any reason a vacuum signal from the wrong source reaches the EGR valve, it will open at idle. This results in excessive exhaust gas dilution of the mixture at idle and misfiring.

There are a large variety of EGR valve vacuum control systems on the market. For this reason, a good EGR system service manual is necessary for checking vacuum hose routing, proper vacuum sources, and correct procedures for operational checks.

Malfunctions in the Engine's Compression System

The following are some of the mechanical malfunctions most likely found in an engine's compression system that can cause excessive HC emissions.

1. Burned exhaust valve. If an exhaust valve burns out, it permits the piston on its compression stroke to push part of the air/fuel mixture out of the cylinder in an unburned state. As a result, there are excessive HC emissions from the vehicle's exhaust.

2. Improper camshaft application. If an improperly ground or incorrect camshaft is installed in an emission-controlled engine, it produces excessive HC emissions. Basically, this results from the engine's inability to properly contain fuel combustion during its operating cycle.

3. Worn piston rings and/or valve guides. If an engine has worn piston rings and/or valve guides, oil is drawn into its combustion chamber and burned. Since engine oil is usually a petroleum product, it creates excessive HC emissions as it burns within the cylinders.

Diagnosing Compression System Malfunctions

There are several methods available for checking valve, piston ring, or in some cases, valve guide condition. Earlier, we presented a procedure for testing these components with a vacuum gauge. A compression or leakage tester are other valuable tools for checking the condition of the piston rings and valves.

In the case of a suspected incorrect camshaft with no markings, it is almost impossible for a technician to diagnose the problem by a visual inspection of the shaft. The only way to pinpoint this problem is to accurately check the engine's complete valve timing, using the recommended procedure provided by the engine manufacturer in the appropriate service manual.

Causes and Diagnosis of Excessive CO Emissions

Listed below are many of the main causes for excessive CO emissions along with the methods used to diagnose them.

1. Restricted air cleaner element. An air filter element that has trapped sufficient dirt and other materials can restrict the air flow into the engine. This can create or contribute to excessively high CO emissions. There are two ways to check for a restricted air cleaner element: a visual inspection or an infrared test. To visually inspect the element:

   a. Loosen and remove the air cleaner attaching hardware. Then, remove the cover and element.

   b. With a droplight positioned inside the element, visually check for light passing through the unit as you rotate the element completely around the stationary light (Fig. 4–29).

   c. If the light does not pass through at least 50 percent of the filtering material, replace the element.

To check the filter element for restrictions with an infrared analyzer:

   a. Perform the idle test on the engine with the infrared analyzer as outlined earlier. Note the results.

   b. Stop the engine and remove the air cleaner element as outlined above. Next, reinstall the cover and its attaching hardware.

   c. Repeat the infrared idle test. If the CO level is now within specifications, the air filter element is dirty and requires replacement.

2. Improper choke valve operation. A choke valve that sticks or will not open completely restricts the air flow. This causes excessive fuel flow from various carburetor circuits into the engine. This increases CO emissions and lowers fuel economy.

The easiest way to ascertain if the choke valve is operating properly is a visual check. To perform this check:

   a. Run the engine until it reaches normal operating temperature.

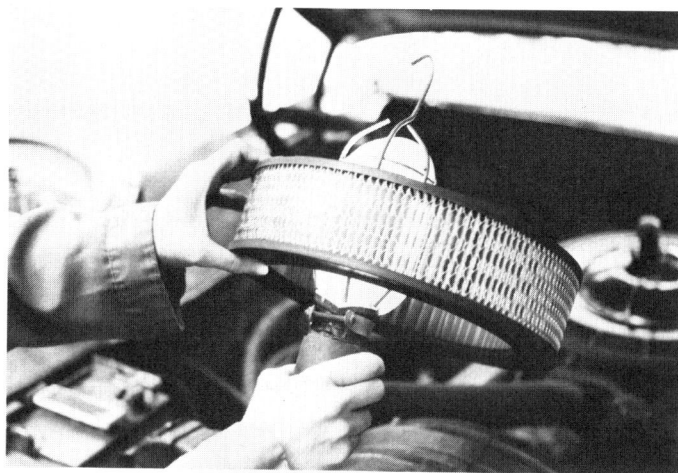

Figure 4-29. Checking an air filter element for restriction with a droplight.

b. Loosen and remove the air cleaner cover attaching hardware. Then, remove the cover and, if necessary, the filter element.

c. Check the position of the choke valve. With the engine at normal operating temperature, the valve should be wide open. If not, check the valve for binding in the air horn.

3. Engine oil diluted with fuel. High CO emissions occur if the engine oil becomes diluted by gasoline. In this case, fuel vapors pass out of the crankcase to the combustion chambers by means of the PCV system. These vapors enter the intake manifold below the carburetor and add additional fuel to the calibrated mixture supplied by the carburetor. As a result, the engine produces higher than normal CO emissions.

There are many causes of engine oil dilution. For example, some likely problem areas to investigate are improper choke valve operation, fuel-saturated carburetor floats, considerable amounts of intercity driving in cold weather, or a defective or plugged PCV valve. Chapter 6 will explain the various methods of testing the PCV system for serviceability.

4. Air injection system failure. If the air injection system is inoperative or not functioning properly, the engine will produce higher than normal amounts of both CO and HC emissions. Chapters 11 and 12 will describe this system and the methods of testing and servicing it.

5. Improper air/fuel mixture or speed adjustments. The improper adjustment of either the idle mixture adjusting screws and/or the throttle valve (idle speed) screw is probably the largest single contributing factor to high CO emissions.

We have already discussed a few of the methods for making these critical adjustments. However, Chapter 10 will also cover the procedures for using the infrared analyzer and propane enrichment equipment for these same adjustments. These two pieces of equipment are extremely necessary for obtaining the proper air/fuel ratio for low CO emissions at engine idle.

6. Other carburetor problems. There are other internal carburetor problems that can create high CO emissions. For ex-

ample, a float saturated with fuel or set too high causes a high fuel level in the float bowl. This permits fuel to flow from the main discharge nozzles during engine idle. In addition, a porous casting or leaking internal seals or gaskets can bring about uncalibrated fuel discharge into the engine, resulting in high CO emissions.

There is one other good symptom of an internal carburetor malfunction. If, at the correct idle speed, you can turn in the idle mixture adjusting screw(s) until lightly seated without any effect on engine operation or the engine's idle smooths out, there is a definite internal problem in the carburetor.

During this procedure, the CO needle on the infrared analyzer should begin to drop toward a lower percentage. However, with an internal carburetor problem, the CO pointer will never drop down to the specified amount even with the mixture screw(s) lightly seated.

There is only one of two corrective measures for an internal carburetor malfunction. That is, the carburetor requires removal for overhaul or for replacement with a new or remanufactured unit.

7. Catalytic converter failure. If the catalytic converter is defective, the engine produces high CO and HC emissions at all times. Chapters 17 and 18 will describe this device and how to test and, in some cases, service it.

Other Test Procedures with the Infrared Analyzer— Accelerator Pump

There are many other special tests you can perform with an infrared analyzer. These include the testing of the carburetor's accelerator pump and power circuit. We will also cover other procedures at appropriate times throughout the text.

To test the accelerator pump within the carburetor:

1. Start the engine and permit it to operate at its normal hot-idle rpm.

2. Warm up and calibrate the infrared analyzer. Next, insert its probe into the vehicle's tail pipe.

3. While observing the CO meter, quickly open and release the throttle.

4. The engine should increase in speed without hesitation, and the CO meter should show an increase in the percentage of carbon monoxide.

Results and Indications of the Accelerator Pump Test

If the CO reading decreases more than 0.5 percent before increasing or it does not increase at all, the carburetor requires either an accelerator pump, linkage adjustment, or repair of the pump mechanism.

Carburetor Power Circuit Test with the Infrared Analyzer

To test the operation of the carburetor's power valve or power circuit:

1. Connect a tachometer to the engine.

2. Start the engine and permit it to reach normal operating temperature.

3. Warm up and initially calibrate the infrared analyzer. Then, insert its probe into the vehicle's tail pipe.

4. While operating the engine at 2,500 rpm, quickly snap the throttle wide open and then release it.

5. Note the reading on the CO meter. The meter should indicate a slight but quick increase in CO emissions. This shows that the power valve or circuit is functioning properly.

Results and Indications of the Power Circuit Test

If the reading of the CO meter does not increase, the power valve or circuit is inoperative and requires service. As previously stated, accelerator pump operation also causes an increase in the CO reading; but this increase will be much higher than the one created by the action of the power circuit and occur at a lower throttle valve opening point.

Carburetor Adjustments with the Infrared Analyzer and Propane Enrichment Equipment

The checking and possible subsequent adjustment of a carburetor's air/fuel ratio with either an infrared analyzer or propane enrichment equipment is necessary to make sure a vehicle is not emitting excessive amounts of carbon monoxide and hydrocarbons. Because either one of these procedures involves, in most cases, the adjustment of other carburetor-assist devices, which are usually part of the exhaust emission modified engine, this text will cover both of these procedures in detail in Chapter 10.

Using a Portable Vacuum Pump

There are a large number of vacuum-operated emission control devices found on modern motor vehicles. The number, type, and design of these devices depend largely upon the particular manufacturer, the engine size, type, plus the year model of the vehicle. Therefore, when testing a particular vacuum-operated device, it will be necessary, in most cases, to use factory instructions and specifications.

For this reason, this section only presents one sample test procedure using the portable vacuum pump along with a separate vacuum gauge to familiarize you with the instrument. However, other test procedures using the vacuum pump will be discussed in other chapters.

Testing a Choke-Delay Valve with a Vacuum Pump

To check a typical choke-delay valve, as found on some late-model Ford engines, follow these instructions:

1. Remove the valve from the vacuum line between the intake manifold and the choke vacuum diaphragm or piston.

2. Connect the hand-operated vacuum pump to the manifold side of the valve and a vacuum gauge to the other end (Fig. 4-30).

3. With the vacuum pump, apply 10 inches Hg to the valve. Then, note the number of seconds required for the second vacuum gauge to indicate the 10-inch Hg reading.

4. Compare this time delay in seconds to factory specifications. Replace the valve

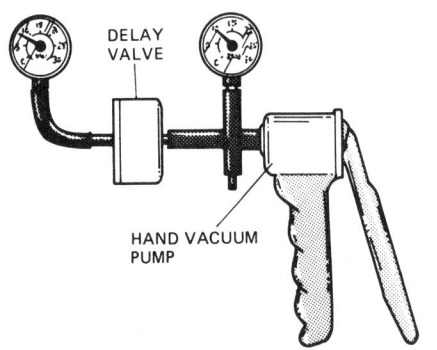

Figure 4-30. Testing a typical external-type choke-delay valve.

if the delay period is not within specified tolerances.

Summary

1. A vacuum test provides the technician with information as to the mechanical condition of the engine.
2. To perform an accurate vacuum test, follow the procedure outlined in this chapter.
3. Compare the readings taken during the vacuum gauge test with those within the results and indications section of this chapter.
4. A vacuum gauge is also a useful tool for measuring a loss of engine compression due to any form of leakage around the pistons.
5. To utilize a standard timing light to check basic ignition timing and the operation of the distributor advance units, follow those directions presented in this chapter.
6. Use a powered timing light to measure the exact amount of ignition timing advance, following the instructions provided in this text.
7. Use the tachometer to adjust the carburetor's idle mixture and engine rpm by following the procedures outlined in this chapter.
8. Adjust the carburetor's air/fuel mixture to lean-best idle or lean drop the ratio by following the instructions presented in this text.
9. An infrared analyzer requires initial calibration before placing it into service each morning.
10. Initially calibrate an infrared analyzer using the manufacturer's instructions or those outlined in this chapter.
11. Certain states have laws that require periodic calibration of the infrared analyzer.
12. Periodic calibration requires the use of a named gas and special equipment.
13. Perform a periodic calibration on an infrared analyzer, following manufacturer's instructions or those presented in this chapter.
14. Other than initial and periodic calibration, an infrared analyzer requires other types of maintenance.
15. Analyzer filters prevent particulate contaminates and water from entering the analyzing system.
16. Service an analyzer's filters using manufacturer's instructions or those outlined in this text.
17. Analyzer filter life depends upon the number of vehicles tested or the amount of usage the machine receives.
18. Inspect and service the analyzer's sampling hose and probe by following manufacturer's instructions or those presented in this chapter.
19. Eliminate the effects of static electricity on analyzer meters using the procedures outlined in this text.

20. An exhaust leak can permit carbon monoxide to enter the inside of a vehicle and air to enter the sampling hose of the infrared analyzer.
21. With any form of air leak into the analyzer's sampling system, both the HC and CO meters will indicate lower readings than acceptable.
22. The low-flow indicator on the analyzer provides an easy method for checking the sampling system for leakage.
23. The depth of insertion of the analyzer probe can affect the HC and CO readings.
24. Insert an analyzer's probe 8 to 12 inches into a vehicle's tail pipe so outside air cannot enter the sampling system.
25. It is always a good idea on any vehicle with dual exhaust to probe both tail pipes and check the emission levels.
26. If taking only one reading from a vehicle with a dual-exhaust system, always probe the tail pipe from the side without the heat-riser valve.
27. The catalytic converter, or a combination converter and air pump, or the air injection system by itself reduces HC and CO emission samples taken by the infrared analyzer.
28. To check a vehicle's exhaust for HC and CO content with an infrared analyzer, follow the procedure outlined in this text or manufacturer's instructions.
29. Some vehicle manufacturers provide a plug in the front of the catalytic converter to permit an access for the analyzer probe.
30. Without the converter access plug, the infrared analyzer may not be accurate enough for performing carburetor adjustments.
31. The key to when you should use the infrared analyzer to adjust a carburetor is found in the manufacturer's specifications.
32. Compare the readings taken during the infrared test with those provided in the results and indications section of this chapter.
33. If the analyzer shows a high HC reading but a normal CO level, there is most likely a problem within the ignition system.
34. A high HC reading but a normal CO level can also mean that the engine may have a mechanical problem or a missing or stuck open thermostat.
35. If the HC reading is higher than normal and possibly wavering with a low CO level, the engine has a vacuum leak.
36. If the analyzer indicates a high HC reading and a very low CO level, the carburetor may be set too lean.
37. A HC reading that is higher than normal and possibly wavering with an acceptable CO level are good indications of a malfunction within the EGR system.
38. If the CO level is slightly high but the HC reading is normal, either the air cleaner element is dirty, the choke valve is stuck partially closed, the carburetor is out of adjustment, the idle speed is not to specifications, or the PCV valve has a restriction.
39. If both the HC and CO readings are high, the air/fuel ratio is too rich, there is internal carburetor malfunction, the PCV system is inoperative, the heat-riser valve is stuck open, the air injection system is inoperative or discon-

nected, the thermostatic air cleaner's preheat door is sticking, or the catalytic converter is defective.
40. A high HC reading can result from a malfunction or defect within either the ignition, timing advance, intake, PCV, carburetion, EGR, or compression systems.
41. Locate ignition system problems using an oscilloscope or visual check.
42. There are three basic pieces of equipment utilized to locate malfunctions in the ignition timing system: the standard timing light, powered timing light, and the distributor tester.
43. Either use noncombustible cleaning solvent or propane to locate the source of an engine vacuum leak.
44. The vacuum gauge, compression or leakage tester are valuable tools for checking the mechanical condition of an engine.
45. A restricted air filter element, improper choke valve operation, engine oil diluted with fuel, air injection system failure, improper air/fuel ratio or speed adjustments, internal carburetor problems, or a defective catalytic converter can all cause high CO emissions.
46. Check the operation of the carburetor's accelerator pump and power circuit with an infrared analyzer.
47. An infrared analyzer, propane enrichment equipment, tachometer, and a vacuum gauge are necessary pieces of equipment to properly set a carburetor's idle mixture and speed adjustments.
48. A portable vacuum pump can test almost any vacuum-activated device.

Review Questions

The questions listed below will assist you in determining how well you remember the material contained in this chapter. Read each question carefully before choosing the answer. If you cannot answer the question, review the section in the chapter that covers the material.

1. The vacuum gauge reading at idle of an engine in good mechanical condition should be between _____ inches Hg.
   a. 16 to 21
   b. 15 to 20
   c. 14 to 19
   d. 13 to 18

2. Use propane or _____ to check the engine for a vacuum leak.
   a. water
   b. oil
   c. gasoline
   d. solvent

3. Use a _____ to check basic ignition timing.
   a. distributor tester
   b. timing light
   c. engine analyzer
   d. infrared analyzer

4. Use a (an) _____ to adjust idle rpm.
   a. analyzer
   b. ohmmeter
   c. tachometer
   d. ammeter

5. The infrared analyzer requires _____ type(s) of calibration.
   a. three
   b. zero

c. one
d. two
6. For periodic calibration, a named gas such as _____ must be used.
   a. propane
   b. hydrogen
   c. oxygen
   d. helium
7. During periodic calibration, the infrared analyzer should warm up at least _____ minutes.
   a. 15
   b. 10
   c. 5
   d. 0
8. Infrared analyzers have filters that prevent particulate contaminants and _____ from entering the analyzing system.
   a. carbon monoxide
   b. hydrocarbons
   c. water
   d. gasoline
9. Service the infrared analyzer filters whenever the _____ light comes on.
   a. HI-flow
   b. LO-flow
   c. flow
   d. purge
10. _____ can cause inaccurate infrared analyzer readings.
    a. Improper warm-up
    b. Initial calibration
    c. Periodic calibration
    d. Static electricity
11. Always insert the infrared analyzer probe _____ inches into the tail pipe.
    a. 1 to 4
    b. 4 to 8
    c. 8 to 12
    d. 12 to 16
12. An air leak into the infrared analyzer's sampling system _____ the HC and CO readings.
    a. zeroes
    b. raises
    c. lowers
    d. none of these
13. The _____ reduces CO levels to nearly zero at idle.
    a. catalytic converter
    b. air pump
    c. PCV valve
    d. tail pipe
14. A high HC reading with a normal CO level indicates a(an) _____ problem.
    a. valve
    b. ignition
    c. vacuum
    d. carburetor
15. If the HC reading is high but the CO level is low, the problem may be a(an) _____ leak.
    a. carburetor
    b. converter
    c. vacuum
    d. exhaust
16. HC emissions above _____ ppm are excessive for an average emission-controlled engine.
    a. 900
    b. 700
    c. 500
    d. 300
17. A ported vacuum signal provides ____

_____ inches of vacuum to the distributor at idle.
   a. 0
   b. 2
   c. 4
   d. 6

18. Use a visual check or a (an) _____ _____ to locate ignition system malfunctions.
   a. vacuum gauge
   b. ignition oscilloscope
   c. vacuum pump
   d. timing light

19. A vacuum leak causes a _____ misfire.
   a. lean
   b. rich
   c. plug
   d. manifold

20. A compression or leakage tester and a (an) _____ are valuable tools for checking the mechanical condition of the engine.
   a. infrared analyzer
   b. engine analyzer
   c. ignition oscilloscope
   d. vacuum gauge

21. Visually check or use a (an) _____ _____ to determine if the air filter element is restricted.
   a. infrared analyzer
   b. engine analyzer
   c. ignition oscilloscope
   d. timing light

22. The single most contributing factor to high CO emissions at idle is a maladjusted _____.
   a. filter
   b. throttle
   c. carburetor
   d. distributor

For the answers to these review questions, turn to the Appendix.

Chapter 5

Crankcase Emission Control Systems

The first device, factory installed on vehicles in 1961 in an attempt to reduce HC emissions, was the positive crankcase ventilation (PCV) system. The basic function of this system was and still is to prevent HC emissions from escaping from an engine's crankcase into the atmosphere. These emissions, as previously stated, amounted to about 20 percent of total HC pollution from the motor vehicle.

Need for an Engine Ventilation System

The problem of ventilating an engine's crankcase has existed since the beginning of the automobile. This problem occurs because there is no piston ring, new or old, that can provide a perfect seal between the piston and cylinder. Consequently, some unburned air/fuel mixture and other by-products of combustion escape past the rings on the engine's compression and power strokes. These resulting gases are known as *crankcase vapors* or more commonly *blow-by gases* (Fig. 5–1).

Blow-by gases have three detrimental effects if permitted to remain within the engine. First, without some form of control device, the hydrocarbons, not burned during the combustion process, will enter the crankcase. Later, these harmful pollutants would pass from the crankcase and enter the atmosphere.

The second detrimental effect is to the engine itself because the blow-by gases contain many harmful ingredients. For example, the hydrocarbons, which can condense back to liquid form, mix with the oil to reduce its viscosity and lubricating properties. In addition, moisture from the combustion process also condenses and makes its way into the crankcase along with the unburned fuel, soot, and dust to form sludge. Finally, this water can also

CRANKCASE EMISSION CONTROL SYSTEMS

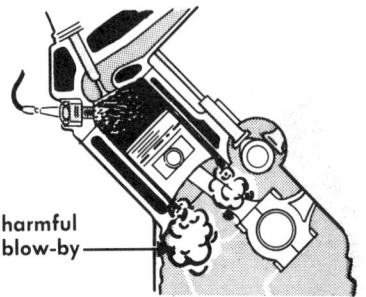

Figure 5-1. Blow-by are gases forced past the piston rings (Courtesy of Union Carbide Corp.).

combine with the unburned hydrocarbons, portions of its additives, and sulfur from the original crude oil to form carbonic acid, sulfuric acid, hydrobromic acid, nitric acid, and hydrochloric acid. These particular acids are responsible for engine wear brought about by the etching, corrosion, and rusting of its internal components.

The final negative effect of blow-by gases is an increase in crankcase pressure. The presence of blow-by gas, along with the movement of the crankshaft, rods, and pistons, creates a pressure buildup within the crankcase. The resulting pressure can eventually build up to a point where the engine seals or gaskets will no longer contain it. Therefore, oil leakage past these units can result.

It should be obvious that every internal combustion engine must have some form of ventilation system so it can breathe. Breathing within an engine's crankcase is the direct result of permitting a charge of fresh air to enter the crankcase and mix with the blow-by gases. Then, the ventilation system removes this combined mixture from the crankcase area. However, the ventilation system on modern engines cannot permit these contaminants to pass out of the power plant and into the atmosphere, where they would create air pollution.

Early Ventilation Systems

Before the introduction of the PCV system, manufacturers used a road-draft tube system to take care of the breathing requirements (ventilation) of most engines (Fig. 5-2). This system consisted of a road-draft tube along with an open oil filler or breather cap. The draft tube fitted into the upper crankcase or a rocker-arm cover and projected down the side of the engine until it terminated near the base of the oil pan.

Installed over an opening in the oil filler tube or rocker-arm cover was a filler or breather cap. This cap was open, which simply means the cap permitted atmospheric air to pass through it and enter the crankcase itself. However, the cap had a filtering element to prevent dust and other contaminants in the air from entering the engine's crankcase.

Operation of the Road-Draft Ventilation System

The operation of this system was relatively simple. With the vehicle in motion, air moved past the opening in the base of

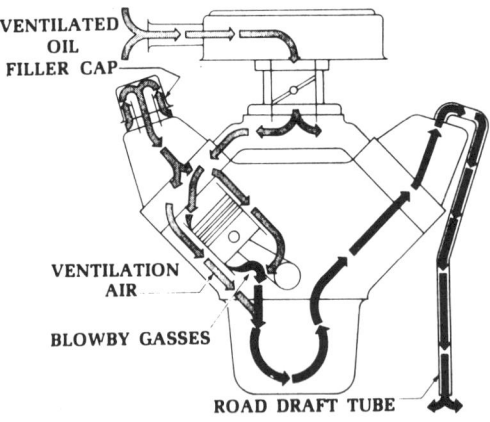

Figure 5-2. Design and operation of the road-draft tube ventilation system.

the road-draft tube, thus creating a low pressure area or vacuum (Fig. 5-2). The vacuum caused atmospheric pressure outside the engine to force a quantity of fresh air through the breather cap. This air passed through the crankcase and mixed with the blow-by gases. Then the combined mixture flowed out the end of the road-draft tube and into the airstream. Although this action ventilated the crankcase, it also contributed to the pollution problem.

In addition to contributing to air pollution, the road-draft system had two other major shortcomings. First, although this system was very simple in its operation, it was not very efficient at low vehicle speeds. Unless the vehicle was moving at 20 to 25 mph, the air velocity around the road-draft tube was insufficient to create enough of a vacuum to draw the vapors and air from the crankcase. As a result, the formation of sludge increased within the crankcase.

Second, at high vehicle speeds, there was so much air flow that the resulting crankcase vacuum pulled some oil from this area along with the vapors. This action created excessive oil usage by the engine at high vehicle speeds.

Positive Crankcase Ventilation Systems

Manufacturers eliminated the drawbacks to the road-draft system with the introduction of controlled crankcase ventilation. Controlled or positive crankcase ventilation relies on intake manifold vacuum to move air into the crankcase and the combined mixture from this area into the intake manifold. This results in a positive movement of air through the crankcase whenever the engine is running. Then, the manifold returns the crankcase vapors to the combustion chambers, where they are burned with the air/fuel mixture.

Types of PCV Systems

Since the introduction of the PCV system, manufacturers have developed four basic types: Type 1 (open system), Type 2 (restricted system), Type 3 (tube-to-air cleaner system), and Type 4 (closed system). The first three systems are known as the open type because the crankcase has some form of opening into the atmosphere

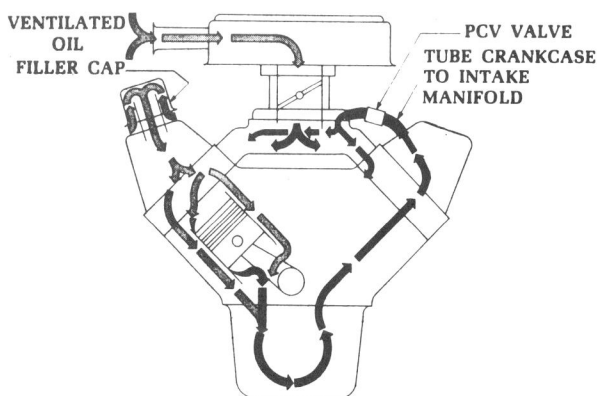

Figure 5-3. Design and operation of a Type 1 open PCV system.

through an unrestricted or partially restricted oil filler or breather cap.

Type 1—Open System

The first factory-installed PCV device used for pollution control that eliminated the road-draft tube completely as well as its inherent problems was the Type 1 open system (Fig. 5-3). In this particular system, the manufacturer still installed an open-type breather cap into the upper portion of the crankcase. In the illustration, this cap is shown in place over an opening in a rocker-arm cover. The cap, along with its filtering element, performed the same function as the cap utilized with the road-draft system.

However, in this system, the manufacturer installed a PCV valve into the crankcase in place of the road-draft tube. In the diagram of the engine shown Fig. 5-3, the valve fits into a hose connected between the rocker-arm cover opposite from the breather cap location and the intake manifold.

The PCV valve can be nothing more than a metered orifice or flow jet. However, the majority of these systems use a valve similar to the one illustrated in Fig. 5-4. This particular valve consists of a housing, valve plunger, and calibrated spring.

The valve housing has two port openings. One opening connects directly into or via a hose to the engine crankcase. The other port fits inside a PCV hose attached to the intake manifold.

The plunger valve operates in a special bore inside the housing. The function of the plunger is to meter the flow of air and blow-by gases through the system during the various phases of engine operation: idle, heavy acceleration, and cruise.

The action of the plunger itself is under the control of intake manifold vacuum, the tension of the calibrated spring, and in some situations, the pressure within the crankcase. For instance, during periods of high vacuum, such as during engine idle and deceleration, the vacuum overcomes the spring's tension and the plunger bottoms in the manifold end of the valve housing (Fig. 5-4b). Because of the design of the valve face and seat, this postion restricts, but does not completely stop, the flow of crankcase vapors through the valve.

When there is a zero-vacuum condition within the intake manifold, such as when the engine is not operating or is running momentarily at wide-open throttle, the calibrated spring moves the plunger so that it bottoms on the crankcase end of the valve (Fig. 5-4c). In this valve position, no fumes from the crankcase can move through the valve. However, if during heavy acceleration blow-by pressure reaches a given amount, some valve designs will open slightly to permit some flow of gases through the valve.

At any engine rpm between idle and wide-open throttle, the valve plunger assumes a position as determined by the resulting intake manifold vacuum operating against the tension of the calibrated spring

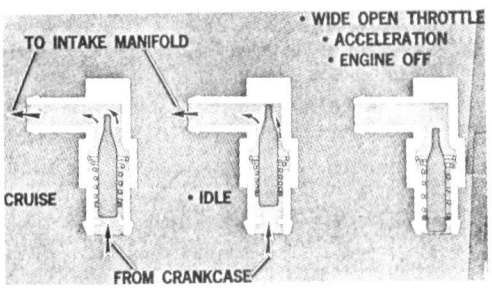

Figure 5-4. Design and operation of a typical PCV valve (Courtesy of Chrysler Corp.).

(Fig. 5-4a). In other words, when spring tension equals the effect of engine vacuum, the plunger assumes a given position in its housing bore. This position then determines the flow rate during that particular phase of engine operation. Consequently, the flow rate through the valve will be lower at idle but higher during vehicle cruise conditions.

In the event of an engine backfire, the valve plunger moves to the same position as it did during heavy acceleration. In other words, the plunger seats against the crankcase side of the housing. This action is due to gas pressure from the intake manifold and the action of the spring on the plunger. As a result, the closed valve stops any fire attempting to travel through the valve to the crankcase, where it could possibly ignite the volatile blow-by gases.

Type 1 System Operation

When an engine with an open Type 1 PCV system (Fig. 5-3) is running at idle rpm, the valve permits about 1 to 3 cubic feet (0.03 to 0.09 cubic meter) of air per minute to enter the oil breather (filler) cap. This clean air flows through the crankcase, picking up the hydrocarbons, moisture, and other forms of gaseous combustion by-products. Then these vapors move through the PCV valve, as described earlier, and enter the PCV hose leading to the intake manifold, where they mix with the air/fuel charge. Finally, the entire charge enters the various combustion chambers, where it burns.

During moderate acceleration and cruising speeds of the engine, the PCV valve opens further than at engine idle. As a result, it permits a higher air flow rate through the breather cap of approximately 3 to 6 cubic feet (0.09 to 0.17 cubic meter) per minute. After passing through the valve and hose, this air flow, along with the increasing amounts of crankcase vapors, absorbs into the air/fuel mixture within the intake manifold. The additional air flow, at this time, does not affect engine performance, as the entire charge moves into the combustion chambers.

Under heavy acceleration engine vacuum decreases considerably while crankcase vapor pressure builds up. As mentioned earlier, under PCV valve operation, if a zero vacuum exists within the intake manifold, the valve will close; while under moderate acceleration, it assumes a position based upon spring tension and manifold vacuum. Consequently, under a heavy acceleration condition, the valve's closed or nearly closed position does not allow the increasing amounts of crankcase vapors to pass into the intake manifold. As a result, the increasing pressure in the crankcase forces some of the vapors out of the open cap and into the atmosphere. Crankcase vapors will also pass through the vented cap and into the air if the PCV valve, hose, or intake manifold fitting become clogged. For these reasons, a Type 1 system only partially controls crankcase vapors and is therefore only about 75 percent efficient.

Type 2— Restricted System

The Type 2 system is in some respects quite similar to most Type 1 devices in that it meters crankcase gases to the intake manifold through a variable orifice valve (Fig. 5-5). However, there are some design differences in the oil filler cap and PCV valve that make this system restricted and permit the crankcase to operate under a given amount of vacuum.

The oil filler (breather) cap in a Type 2

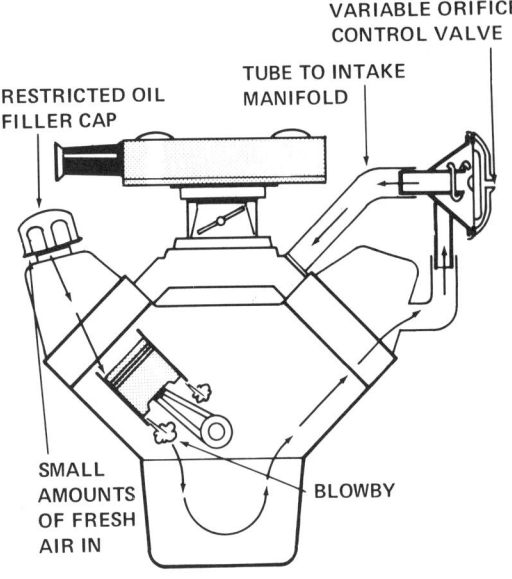

Figure 5-5. Design and operation of a Type 2 restricted system.

system is not fully open, as an atmospheric vent for the crankcase. This particular cap has a small hole—equivalent to about 0.060 inch—that provides the restricted air flow necessary to maintain a partial vacuum in the crankcase. However, the cap does have a filtering element to prevent contaminants from entering the crankcase with the air.

The Type 2 PCV valve consists of a housing and spring-loaded diaphragm (Fig. 5-6). This particular valve housing has three openings. The lower port opening fits either into a rubber grommet or hose fastened into the crankcase. With this arrangement, crankcase vapors enter the valve through this hole.

The right-hand opening of the valve connects via a PCV hose to the intake manifold. Therefore, with the engine operating, intake manifold vacuum will act on one side of the diaphragm.

On the left-hand side of the valve housing, opposite to the manifold port, is an atmospheric vent or breather hole. This opening vents the small chamber on the other side of the diaphragm to the atmosphere at all times.

Stretched across the inside of the housing, between the breather hole chamber and the vacuum chamber, is a flexible diaphragm. The diaphragm moves slightly back and forth within the metal housing due to its interaction between vacuum and spring tension on one side and atmospheric pressure on the other.

Mounted in the center of the diaphragm on the vacuum chamber side is a modulator ball. As the diaphragm moves, this ball controls a passage leading from the crankcase opening to the intake manifold port. For example, in Fig 5-6, the modulator ball is in its full right-hand (closed) position, thus blocking the passage from the crankcase to the intake manifold. This would normally prevent any crankcase fumes from passing into the intake manifold.

However, the ball has an idle groove. This groove allows a given amount of crankcase vapors to pass into the manifold at engine idle when the modulator ball is on its seat.

Type 2 System Operation

The Type 2 PCV valve varies its modulator ball opening in relation to the type of pressure within the crankcase. For instance, when an engine in good mechanical condition is running at idle, it produces high intake manifold vacuum and reasonably small amounts of blow-by gas. In this situation, the engine vacuum acts on the right side of the diaphragm; and the diaphragm moves toward the spring, increasing its tension (see Fig. 5-6).

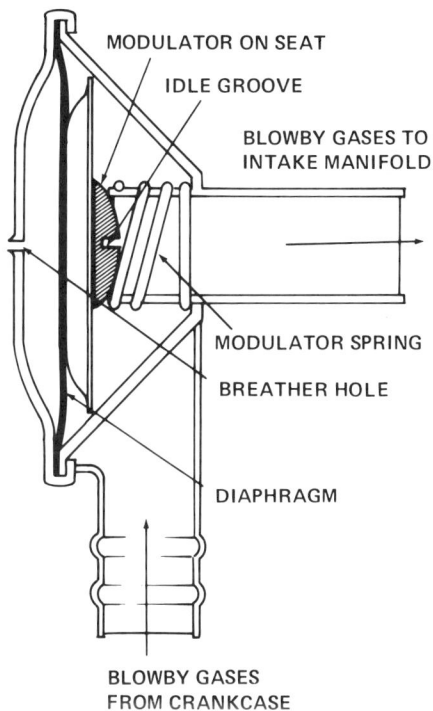
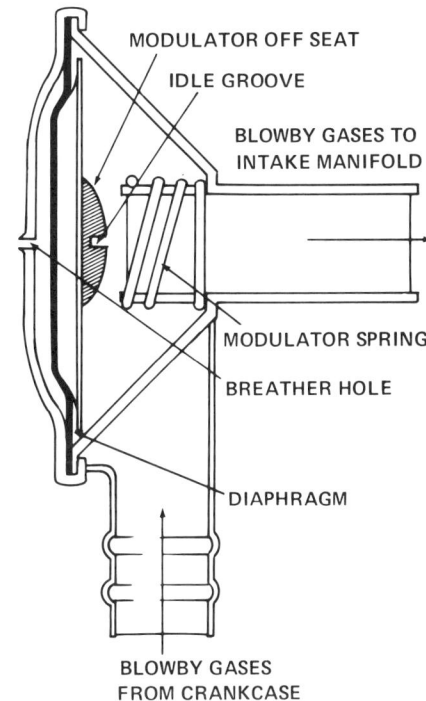

Figure 5–6. Design and operation of a Type 2 PCV valve at idle.

Figure 5–7. Operation of a Type 2 PCV valve in its maximum flow position.

With the modulator ball now on its seat, there is a vacuum acting on the crankcase via the idle groove in the ball. With a vacuum on the crankcase, atmospheric pressure forces small amounts of fresh air through the orifice in the breather cap and into the engine. This air mixes with the crankcase vapors and assists in carrying the contaminants into the intake manifold via the idle groove on the ball.

Since only a restricted amount of air enters through the breather cap, about 3 cubic feet (0.09 cubic meter) per minute, there is a partial vacuum (negative pressure) within the crankcase. This tends to hold the spring-loaded diaphragm in the right-hand (closed) position with its modulator ball seated.

As engine speed increases above idle, crankcase vapor pressure builds up, decreasing the amount of crankcase vacuum (Fig. 5–7). In this situation, the diaphragm begins to move to the left due to the tension of the calibrated spring, an increasing amount of crankcase pressure, and possible reduction in manifold vacuum. The resulting diaphragm movement forces the modulator ball off of its seat and permits more of the blow-by gases to pass through the valve and into the intake manifold. However, the amount of ventilating air permitted to flow with the blow-by gas remains limited at this time due to the restriction within the breather cap.

The important thing to remember about the Type 2 PCV valve is that the type of crankcase pressure, negative or positive, determines its position and the resulting vapor flow rate. When, for example, a negative pressure (vacuum) exists within the crankcase due to a reduced production of blow-by gases at idle, the valve closes. However, as the engine develops increased amounts of blow-by gases at high engine speeds, there tends to be a positive pressure in the crankcase. This positive pressure, the tension of the spring, and possible reduction in manifold vacuum forces the modulator ball off its seat. This permits more of the engine vacuum to act on the crankcase in an attempt to reestablish its vacuum. As a result, a larger volume of blow-by gases pass through the valve and into the intake manifold. Therefore, the flow rate of the valve depends on the amount of blow-by gases generated by the engine at any given time.

Moreover, in order for the valve to respond to the type of crankcase pressure, it is necessary to limit the flow of ventilating air by means of the restricted cap. If the flow of air has no limitations due to a possible leak in a valve cover gasket or other sources, the PCV valve tends to open fully. This allows ventilated air to pass through the system up to the maximum capacity of the valve, causing excessive lean-out of the air/fuel mixture. Therefore, for the proper operation of the Type 2 system, the breather cap must be the specified restricted type, and the engine must have few or no air leaks into the crankcase.

Since an engine with a Type 2 device requires almost perfect sealing and a restricted cap in order for the system to function, this form of emission control device prevents, under normal circumstances, any blow-by gases from entering the atmosphere. In other words, very little if any gas can escape through the restricted cap into the outside air even during heavy acceleration or very high engine speeds. Instead, under these conditions, the PCV valve opens further to permit the additional volume of blow-by gas to pass into the intake manifold, which distributes them along with the air/fuel mixture to the various combustion chambers.

Type 3—Tube-to-Air Cleaner System

In the early 1960s a few domestic automobile manufacturers, Ford and American Motors, used a Type 3 PCV system. Also, some foreign manufacturers have used this system in one form or other.

A typical open Type 3 system consists of an open cap and a tube or hose; the system has no PCV valve (Fig. 5-8). The oil filler (breather) cap has about the same design and serves the same function as the one

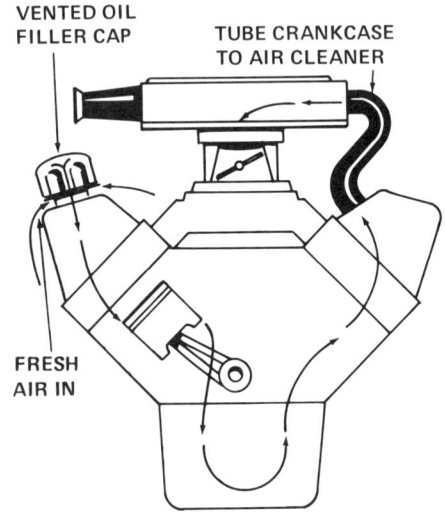

Figure 5-8. Design and operation of a Type 3 PCV system.

used in both the road-draft tube and Type 1 systems.

The tube or hose connects between the engine's crankcase and the carburetor air cleaner. Its function, when the engine is operating, is to carry the blow-by gases and ventilating air between these two locations.

Type 3 System Operation

When the engine is running, there is a pressure drop at the end of the tube or hose connected to the air cleaner due to the air flow through the filter itself. With a reduction in air pressure at this location, atmospheric pressure forces air through the oil filler cap. This air, as in the other two systems described, mixes with the blow-by gases and assists in carrying contaminants through the tube or hose into the air cleaner. At this point, the combined mixture joins the incoming air, passes through the carburetor, and moves into the intake manifold.

Design Variations of Type 3 Systems

There are design variations in Type 3 systems dealing with the location of the tube or hose and the means of limiting the flow of ventilating air and oil through the system. For example, some foreign-built vehicles do not have an open cap. In this case, the Type 3 system only provides an escape path for the blow-by gases, and it has no provision for introducing ventilating air to the engine's crankcase. This particular variation of the Type 3 device is known as a *sealed system*.

On some early six-cylinder Ford engines with this system, the crankcase required an oil separator. This component attached to the crankcase outlet to reduce the amount of oil drawn through the PCV hose to the air cleaner to a minimum.

Type 3 Systems Enrich the Air/Fuel Mixture

A Type 3 system tends to enrich the carburetor's air/fuel mixture when the engine is operating. Since blow-by gases contain a great deal of unburned hydrocarbons mixed with air, the addition of this mixture to the upstream side of the carburetor causes a second charge of fuel to mix with the gases as they flow through the carburetor. As a result, the mixture is too rich. Consequently, carburetors used with a Type 3 system require adjustment or recalibration to compensate for this problem.

Benefits of the Open and Restricted PCV Systems

The open and restricted PCV systems provide three positive benefits. First, they promote longer engine life by removing most of the harmful vapors from the crankcase. Second, the systems reduce if not eliminate HC emissions into the atmosphere. Third, an open or restricted PCV system does increase fuel economy by recirculating most, if not all, of the unburned hydrocarbons in the blow-by gases back to the intake manifold.

Type 4—Closed PCV System

The Type 4 closed system has been in successful use for many years. It first became required equipment on new automobiles in California beginning in 1964. The device became standard factory-installed equipment nationwide by 1968 and is still used today on all new domestic-built automobiles, light trucks, and similar imported vehicles.

The main reason for utilizing a closed Type 4 system is to achieve 100 percent control of crankcase emissions into the atmosphere. Remember, except for the Type 2 restricted systems, an open PCV device permits some blow-by gases to escape into the air during heavy acceleration. In addition, fumes can also slip out from the engine if it is in poor mechanical condition or if the open PCV system becomes inoperative. However, this is not the case with a Type 4 closed system.

Figure 5-9 illustrates a typical Type 4 closed PCV system. Although the system design varies somewhat between manufacturers, it usually consists basically of components from both the Type 1 and 3 open devices. For example, a variable or fixed orifice valve connects via a hose between the intake manifold and the engine's crankcase as in the Type 1 system. In addition, the manufacturer installs some form of sealing device around the dipstick to prevent any air from leaking into the crankcase after its installation. This also prevents any leakage of crankcase fumes out of the engine from this area.

Moreover, the Type 4 system does away with the open-type breather cap. In its place, a sealed or closed cap stops any outside air from entering the crankcase and any of its fumes from leaving the engine from this point.

However, many engines, such as the one illustrated in Fig. 5-9, use a modified cap that does have one port opening which indexes with the crankcase. This opening accommodates a second PCV hose, which fits between it and the air cleaner. This hose installation is similar to the one used in the Type 3 device.

But in a closed PCV system, the hose from the air cleaner to the crankcase serves two functions. First, as in a Type 3 system, the hose carries excess blow-by vapors to the air cleaner. Second, this same hose transports the ventilating air necessary to purge the crankcase of contaminants.

To perform this latter function, the hose connection on the air cleaner can be either on the clean or dirty side of its filter element. When, for example, the hose connects to the dirty side of the element, a PCV air filter is necessary to trap foreign particles in the air that would otherwise enter the crankcase. Some manufacturers install

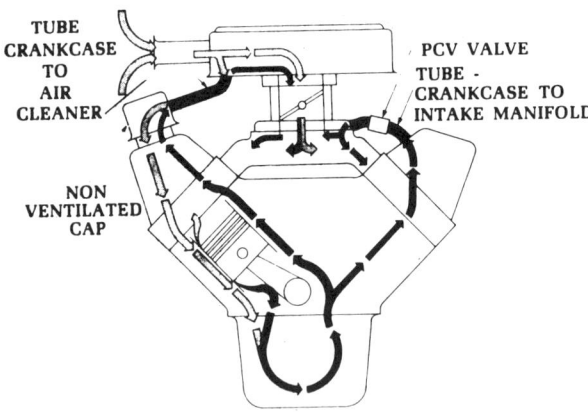

Figure 5-9. Design and operation of a closed Type 4 PCV system.

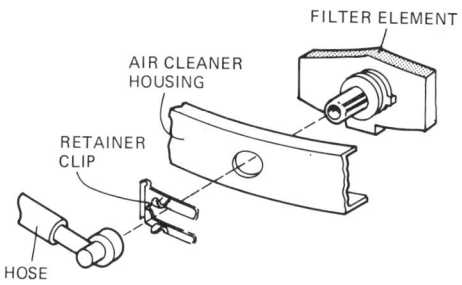

Figure 5-10. PCV inlet filter installed inside the air cleaner housing.

this filter inside the air cleaner housing, where the PCV inlet hose connects (Fig. 5-10). However, in other applications, the filter may also be inside the closed oil filler cap (Fig. 5-11) or in the inlet hose when it connects directly to a valve cover fitting.

When the hose connects to the clean side of the air cleaner element, it acts as the PCV filter. But in this situation, a flame arrester, such as the one shown in Fig. 5-12, is necessary. This particular device in nothing more than a fine wire screen installed over the opening in the PCV inlet hose fitting within the air cleaner housing.

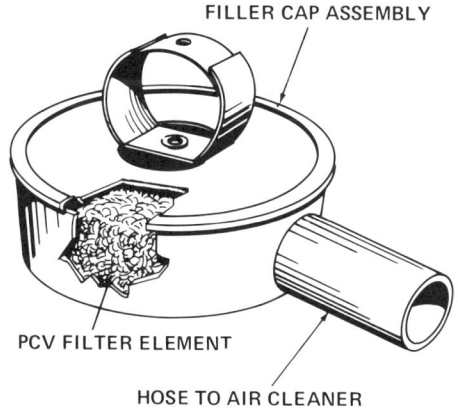

Figure 5-11. PCV filter located inside an oil filler cap.

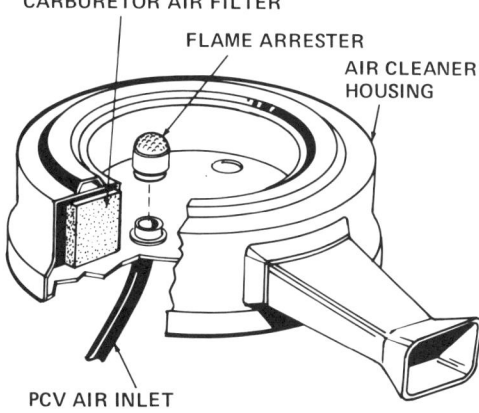

Figure 5-12. Flame arrester. Used when the PCV inlet hose attaches to the clean side of the air cleaner.

However, a similar device may also be located where the PCV hose attaches to the valve cover. In either case, this component prevents a crankcase explosion by extinguishing any flame created by an engine backfire. Finally, on a system where the hose is on the dirty side of the filter element, a flame arrester is not necessary since the element itself performs this same function.

Type 4 Closed System Operation

During all other phases of engine operation except for heavy acceleration, the closed PCV system (Fig. 5-9) functions in much the same way as the open Type 1 device. The only real difference is that the fresh air that enters the crankcase comes through the inlet hose connected to the air cleaner. Then, this air flows through the crankcase, PCV valve, and into the intake manifold.

If the Type 1 portion of this system becomes clogged or the engine is operating

Figure 5-13. During heavy acceleration, excess vapors vent to the air cleaner (Courtesy of Chrysler Corp.).

under a heavy acceleration condition, the additional blow-by vapors reverse their direction (Fig. 5-13). In other words, the gases flow back up the air inlet PCV hose and into the air cleaner. From there, the vapors mix with the incoming air and flow into the carburetor. Due to this arrangement, any buildup of crankcase vapors cannot pass into the atmosphere from the system.

Benefits of the Type 4 System

When in operation, the closed Type 4 PCV system provides three benefits. First, the system promotes longer engine life by completely ventilating the engine to prevent the buildup of sludge and other harmful materials detrimental to the engine. Second, it completely eliminates crankcase vapors, such as hydrocarbons, from entering and polluting the atmosphere. Finally, the closed system increases fuel economy even more than the open types by recirculating all the unburned hydrocarbons within the blow-by gases back to the intake manifold.

Retrofit Crankcase Emission Systems

Since the early 1960s, manufacturers have developed and produced all four types of retrofit crankcase emission control devices. A *retrofit device* is one designed for installation on a vehicle that never had a PCV system or one with either a Type 1 or 3 system.

Retrofit devices are necessary so that a vehicle can comply with a given state law. For example, for some time now, California has required that all automobiles and light trucks built between 1955 and 1967 have either a Type 2 or 4 system in order to be registered in that state.

This created somewhat of a problem for aftermarket manufacturers because vehicles built between these years came from the factory either with no device at all or some form of open system. Therefore, in order to supply devices to meet all these vehicle configurations, manufacturers produced all four types of PCV systems.

A Type 1 system was necessary for a vehicle that came from the factory with a Type 3 device. In other words, by installing the Type 1 retrofit system along with the original Type 3 unit, the combination became a Type 4 device.

As mentioned before, many vehicles had a factory-installed Type 1 system. In order to convert it to a Type 4 closed device, a mechanic installed the retrofit Type 3 unit that together with the original equipment completed the required conversion.

Vehicles with a road-draft ventilation system require either a Type 2 or 4 system in order to comply with the state law. A Type 2 system is legal because it permits very little if any blow-by vapors into the atmosphere, even during heavy acceleration. However, this device is not always available in all areas; thus, many of the vehicles requiring a complete system must have a Type 4 retrofit device installed.

Summary

1. The purpose of a PCV system is to prevent HC emissions from escaping into the atmosphere from the crankcase.
2. Blow-by gases result primarily from leakage of combustion chamber gases past the piston rings.
3. Blow-by gases contain hydrocarbons that pollute the atmosphere if permitted to leave the crankcase.
4. Blow-by gases are detrimental to the engine because they contain many harmful ingredients.
5. Blow-by gases increase crankcase pressure.
6. Internal engine breathing involves the introduction of fresh air into the crankcase.
7. The first form of crankcase ventilation system was the road-draft type.
8. The road-draft tube system consists of a draft tube and an open cap.
9. The road-draft tube system contributed to air pollution and was not very effective at vehicle speeds below 20 to 25 mph.
10. A PCV system relies on intake manifold vacuum to move air into the crankcase and blow-by gases from this area into the intake manifold.
11. The first form of PCV system, factory installed, was the Type 1 open system.
12. A Type 1 PCV system consists of an open cap, valve, and hose.
13. The majority of Type 1 and 4 PCV systems use a variable orifice valve that is under the control of the intake manifold vacuum, the tension of a calibrated spring, and in some situations, the pressure within the crankcase.
14. A typical PCV valve consists of a housing, valve plunger, and calibrated spring.
15. Along with controlling the flow rate of the blow-by gases, a PCV valve also prevents an engine backfire from igniting the volatile blow-by gases.
16. A Type 1 system is only about 75 percent efficient.
17. A Type 2 system is known as a restricted type because of the design of its breather cap and PCV valve.
18. A Type 2 PCV valve consists of a housing, a diaphragm, and a spring.
19. In a Type 2 system, the type of crankcase pressure determines the position of the PCV valve and the resulting vapor flow rate.
20. For proper operation of the Type 2 system, the breather cap must be the restricted type and the engine must have few or no air leaks into the crankcase.
21. A Type 2 PCV system prevents any blow-by gases from entering the atmosphere.
22. A typical Type 3 system consists of an open cap and a hose connected between the crankcase and the air cleaner.
23. Some Type 3 systems are the closed type.
24. A Type 3 system tends to enrich the carburetor's air/fuel mixture.

25. An open, restricted, or closed PCV system provides three benefits.
26. A Type 4 system achieves 100 percent control of crankcase emissions into the atmosphere.
27. A Type 4 system consists of components from both Type 1 and 3 open devices.
28. A Type 4 system utilizes a sealed oil filler cap.
29. In a Type 4 system, the hose to the air cleaner can be either on the clean or dirty side of the element.
30. When the hose connects to the clean side of the element, a flame arrester is necessary.
31. When a Type 4 system is in operation, the crankcase receives its fresh air through the inlet hose connected to the air cleaner.
32. Since the early 1960s, manufacturers have developed and produced all four types of retrofit crankcase devices.
33. A retrofit device is necessary so that a vehicle can comply with a given state law regarding crankcase emission control devices.

Review Questions

The questions listed below will assist you in determining how well you remember the material contained in this chapter. Read each question carefully before choosing the answer. If you cannot answer the question, review the section in the chapter that covers the material.

1. Primarily, blow-by gases contain _____ _____ that can pollute the atmosphere.
   a. carbon monoxide
   b. hydrocarbons
   c. carbon dioxide
   d. water
2. The first type of ventilation system used on automobiles was the _____ _____ device.
   a. road-draft tube
   b. Type 1
   c. Type 2
   d. Type 3
3. An open-type PCV system utilizes a (an) _____ type oil filler cap.
   a. restricted
   b. closed
   c. open
   d. none of these
4. At idle, a Type 1 PCV valve permits _____ cubic feet of air to enter the crankcase.
   a. 0 to 3
   b. 1 to 3
   c. 3 to 5
   d. 5 to 7
5. A Type 1 system is only about _____ percent efficient.
   a. 45
   b. 55
   c. 65
   d. 75
6. A Type 2 breather cap has a small opening in it equivalent to about _____ inch.
   a. 0.060
   b. 0.080
   c. 0.100
   d. 0.120
7. Crankcase _____ is one of the fac-

tors determining the position of a Type 2 PCV valve.
   a. flow
   b. displacement
   c. pressure
   d. volume
8. A Type _____ system tends to enrich the air/fuel ratio.
   a. 4
   b. 3
   c. 2
   d. 1
9. A Type 4 system uses a (an) _____ oil filler cap.
   a. flow-type
   b. restricted
   c. open
   d. closed
10. Under heavy acceleration of an engine with a Type 4 PCV system, excess fumes _____.
   a. enter a hose leading to the air cleaner.
   b. pass through the PCV valve.
   c. pass out of the breather cap.
   d. pour out of the road-draft tube.

For the answers to these review questions, turn to the Appendix.

Chapter 6

Testing and Servicing PCV Systems

For some time after their initial introduction, PCV systems, in general, did not receive proper maintenance for several reasons. First, vehicle owners and many mechanics would not accept the fact that air pollution was really a problem, intensified by the number of motor vehicles in use. Therefore, any device, such as a PCV system, installed on an automobile would not have any affect on air quality.

In addition, like many other newly installed devices, technicians for the most part did not at first understand how the PCV system operated and how to service it. Remember, before the advent of this device, vehicles had a road-draft tube ventilation system that was quite simple in design and operated well for long periods of time with little or no maintenance.

However, the PCV system does require routine service or it will not function properly. The system requires maintenance basically due to the nature of the materials flowing through it, namely, crankcase vapors. These gases tend to form sludge and sediment that plug up components within the system.

Symptoms of Improper PCV System Operation

Whenever a PCV system of any design is not functioning properly, the engine does not receive adequate ventilation or release its blow-by gases. A malfunctioning PCV system can cause:

1. Oil contamination with a reduction in additive life. This results in sludge formation that reduces oil circulation along with increasing acid etching and rusting of critical highly polished, internal engine parts.

2. Increased oil consumption.

3. Blow-by vapors escaping from the

dipstick tube, oil filler cap, valve cover gasket areas, and other openings to the crankcase. These gases contribute to the problem of air pollution.

4. Objectionable and sickening oil fume odors inside the vehicle.

5. Rough engine idle speed and stall tendencies caused by uneven or overrich carburetor operation at idle and low speed.

In the event of a rough idle, an attempt should never be made to compensate for the condition by disconnecting the PCV system or by making carburetor adjustments without checking this form of ventilation system. The removal of PCV components from the engine adversely affects its ventilation and reduces fuel economy. Without a serviceable PCV system, the engine does not have correct crankcase breathing, which shortens engine life due to formation of sludge and loss of oil through engine seals.

Also, keep in mind that on non-PCV engines, the hydrocarbons escaping from the crankcase are lost to the atmosphere. However, on vehicles equipped with a PCV system, these vapors burn in the combustion chambers. This improves the fuel economy of the vehicle.

Maintenance of a PCV System

On the average, most vehicle manufacturers require some form of service on the PCV system at a given mileage or time interval. For example, many domestic automobile manufacturers recommend an inspection of the system, along with replacement of the PCV valve and inlet filter, at 12,000 miles or every 12 months. Also, some manufacturers have requirements for the cleaning of the filter in either the open or closed oil filler cap.

When performing preventive maintenance or service work on a PCV system, always consult the vehicle manufacturer's instructions and specifications. While some general service procedures may be formulated, it is very important that the manufacturer's instructions and service schedules be followed.

PCV System Inspection

The first form of PCV system service is always a complete inspection of all of its components. Listed below are the necessary inspection areas for most PCV systems in use:

1. On all types of systems, check all hoses for cracks and clogging, making sure they are properly connected in the system (Fig. 6-1). Replace, clean, or properly connect the hoses as necessary. Note: To clean a PCV hose, wash it in a cleaning solution such as mineral spirits and use a bore brush to scrub out the interior walls of the hose. Never soak a hose in a strong solvent such as carburetor cleaner because such a solution causes the hose to swell.

2. On a Type 1 and 4 system, remove the PCV valve. Then, shake the valve. You should hear a clicking noise, indicating the valve plunger is free within its housing bore.

3. On Type 1, 2, or 3 open or restricted systems, inspect the condition of the breather or oil filler cap. If the cap's filter is dirty, wash it in mineral spirits and permit it to air dry. Note: Never use com-

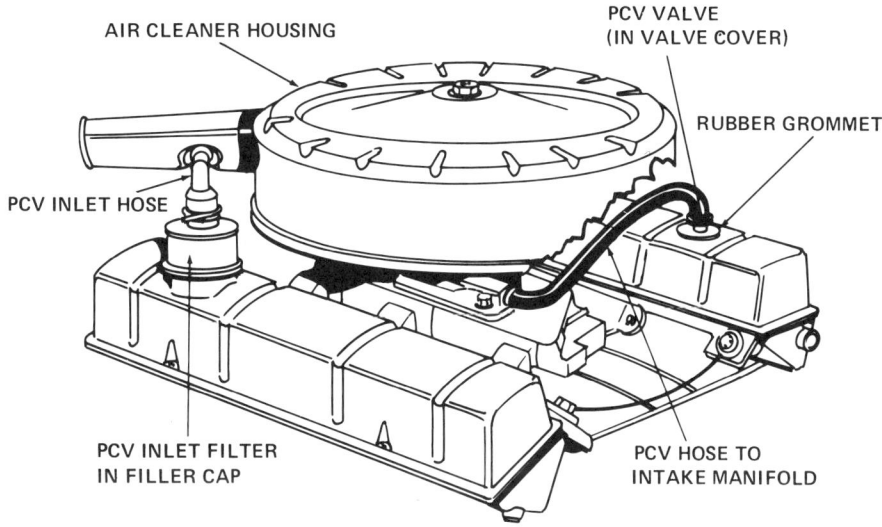

Figure 6-1. Inspection points within a typical PCV system.

pressed air to blow dry the cap because the procedure may dislodge the filtering material. Also, on a Type 2 retrofit installation, always follow the manufacturer's recommendations when servicing the breather cap.

4. In a Type 4 closed system with the inlet hose attached to the clean side of the air cleaner element, check the flame arrester to make sure it is open. If the unit is dirty, wash it in clean mineral spirits and permit the arrester to air dry. In addition, check the filter element itself for restrictions because blow-by vapors on high mileage engines tend to clog this unit with oil.

5. On a Type 4 system with the inlet hose attached to the dirty side of the filter element, check the inlet PCV filter within the air cleaner housing, in the hose itself, or inside the oil filler cap. The PCV filter in the air cleaner housing cannot be cleaned; therefore, it requires replacement when dirty or at the given change interval. However, the filter contained in the oil filler cap is serviceable by a cleaning and reoiling process. These procedures will be explained below.

PCV System Testing

Vacuum Test

You can test the operation of a Type 1, 2, or 4 PCV system by performing a vacuum test with a piece of stiff paper. To perform this test:

1. Start the engine and permit it to idle at the normal operating temperature.

2. Remove the PCV valve or its hose from the rocker-arm cover or the crankcase. If the valve is not plugged, you should hear a hissing noise as air passes through the valve.

3. Place your finger over the inlet side of the valve or the end of the hose. You should feel the strong pulling effect of engine vacuum (Fig. 6-2).

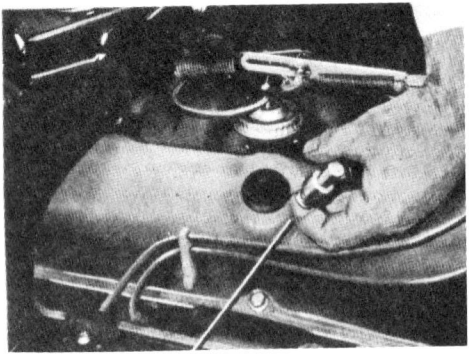

Figure 6-2. Testing the operation of a PCV valve with your finger (Courtesy of Chrysler Corp.).

4. Reinstall the PCV valve or its hose into the valve cover or crankcase.

5. Remove either the inlet hose or the oil filler cap on Type 4 systems or the breather cap from a Type 1 or 2 system.

6. Hold a piece of stiff paper over the opening from which you removed the hose or cap. After waiting about a minute for crankcase pressure to reduce, a vacuum should hold the paper against the surface of the opening with considerable force (Fig. 6-3).

7. Remove the paper and reinstall either the inlet hose, filler, or breather cap in the rocker-arm cover or crankcase.

Results and Indications of the Vacuum Test

1. If the PCV system passed all the above-mentioned tests, its operation is satisfactory.

2. If the system failed checks 2, 3, or 6, replace the PCV valve and recheck the system. Do no attempt to clean the old valve unless it is the type that can be completely disassembled because there is no adequate method to clean and inspect a modern sealed-type valve.

3. If the system fails to pass check 6 with a new valve, inspect the PCV hose and its carburetor passage for obstructions. Replace or wash a plugged hose or clean out restricted passages utilizing the procedures outlined below.

Figure 6-3. Checking the operation of a typical PCV system with a piece of stiff paper (Courtesy of Chrysler Corp.).

Checking the System with a PCV Tester

Some technicians prefer to use a PCV tester, similar to the one illustrated in Fig. 6-4, to check the operation of Type 1, 2, or 4 systems. Essentially, this tester indicates whether a pressure instead of a vacuum exists within an engine's crankcase. For example, if during the test the colored ball moves into the good zone (as illustrated in the drawing), there is a slight vacuum in the crankcase indicating the system is operating normally. However, if when testing the device the ball moves quickly into the repair side of the tester, it is showing a pressure instead of a vacuum. In this latter situation, the system requires some form of service.

To perform a PCV system check with this special tester:

1. Perform the first four preliminary steps of checking a Type 1, 2, or 4 system according to the vacuum testing procedures outlined above.

2. Position the PCV tester as shown in Fig. 6-4 over the opening in the rocker-arm cover or crankcase from which you removed the inlet hose, filler cap, or breather cap.

3. Shut off the engine and reinstall the inlet hose, filler cap, or breather cap.

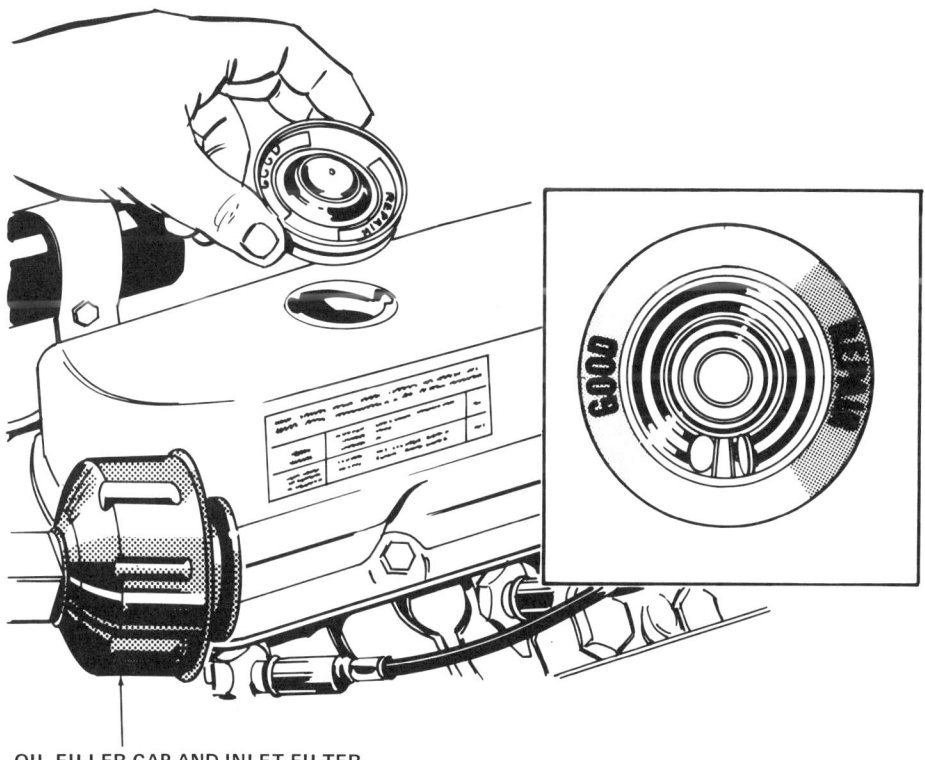

OIL FILLER CAP AND INLET FILTER

Figure 6-4. Checking a PCV system with a special tester.

TESTING AND SERVICING PCV SYSTEMS

Results and Indications

1. If the colored ball within the tester moved immediately into the good zone, the PCV system is operating properly.

2. If the system failed checks 2 or 3 of the vacuum test or the colored ball moved rapidly into the repair zone of the tester, replace the PCV valve and recheck the system. Do not attempt to clean the old valve unless it is the type that can be completely disassembled because there is no adequate method to properly clean and check a modern sealed-type valve.

3. If the system with a new valve again fails to pass check 2 of this procedure, inspect the PCV hose and its carburetor passage for obstructions. Replace or wash a plugged hose and clean out the carburetor passages as outlined below.

Testing PCV Systems with a Tachometer

You should use the tachometer method of testing either a Type 1, 2, or 4 PCV system when a special tester is not available or if the results of the vacuum test were not conclusive. To perform this particular test:

1. Connect a tachometer to the engine, following the manufacturer's recommended procedures.

2. Start the engine and permit it to run at its normal hot curb-idle speed.

3. Note and record the engine speed at this point.

4. Disconnect the PCV valve or its hose from the rocker-arm cover or crankcase. Then, completely restrict the air flow through this portion of the system by placing your finger over the opening in the valve or hose.

5. Note and record the engine rpm with the PCV system's air flow stopped.

Results and Indications

1. An engine speed drop from 40 to 80 rpm, with the system first operating with normal air flow and then with restricted air flow, indicates a properly operating PCV device.

2. An rpm drop of less than 40 or no speed reduction indicates that a valve or hose is partially to fully clogged. However, a drop of less than 40 rpm may also mean that the system has an incorrect PCV valve.

Testing a PCV System with an Infrared Analyzer

You can check the operation of a Type 1, 2, or 4 PCV system with an infrared analyzer. To perform this check, follow this procedure:

1. Start the engine and permit it to operate at its normal hot curb idle.

2. As outlined in an earlier chapter, warm up and initially calibrate the infrared analyzer. Then, insert the probe into the vehicle's tail pipe.

3. With the engine idling, note the CO reading on the analyzer.

4. Remove the PCV valve or its hose from the rocker-arm cover or crankcase.

5. Observe the CO reading on the analyzer; it should drop to a leaner value than the one noted in step 3.

6. Plug the end of the PCV valve or hose with your finger and again note the CO level on the meter. It should now be greater (richer) than the one taken in step 5.

Results and Indications

If there are no changes in the CO readings during the test, there is a malfunction within the PCV system. As necessary, replace the PCV valve or clean the hose or carburetor passages. Then retest the system.

Replacing a PCV Hose

All PCV hoses, either the inlet or the hose from the valve to the intake manifold, require replacement if cracked, severely plugged, or damaged. When replacing any of these hoses, only use the special hose made specifically for PCV systems. Do not use ordinary heater hose because it cannot withstand the effects of blow-by gases.

To replace a typical PCV hose:

1. If so equipped, loosen and slide back the hose clamps.

2. Twist each end of the hose slightly and remove it from its fittings using a steady pulling action.

3. Using the old hose as a guide, cut a new section from a roll of PCV hose stock to the correct length.

4. Slide the clamps over the new hose and install the hose over the ends of both fittings. Then, tighten the clamps securely.

Cleaning the PCV Passages in the Carburetor

To properly clean out the PCV passages located within a typical carburetor, follow these steps:

1. Following the manufacturer's instructions, remove the carburetor from the intake manifold. However, it is not necessary to disassemble the unit in order to clean out the PCV passages.

2. By hand, carefully turn a drill through the passages to dislodge the solid particles. As necessary, use a drill bit smaller than the diameter of the passage openings so that you do not remove any metal. Then, blow the passages clean with low-pressure compressed air.

3. Reinstall the carburetor and recheck the operation of the PCV system.

Replacing a PCV Valve

To replace a typical PCV valve (Fig. 6-5) follow these procedures:

1. Disconnect the lower vent hose from the rocker-arm cover or crankcase and the upper hose from the carburetor or intake manifold fitting.

2. If so equipped, loosen and remove the clamps that secure the ends of both hoses to the PCV valve.

3. While grasping the valve, remove each vent hose using a twisting and pulling motion.

4. Blow out each hose with compressed air to be certain there are no obstructions.

5. Install a new PCV valve after checking its part number against manufacturer's specifications. Note: On some Ford engines, it is necessary to install a specified high-flow PCV valve if the engine has been operated a given number of miles.

This high-flow valve is necessary because the air/fuel mixture tends to become too rich after the engine has operated so many miles. Consequently, the engine requires a different valve with a higher flow rate in place of the standard unit. This re-

TESTING AND SERVICING PCV SYSTEMS

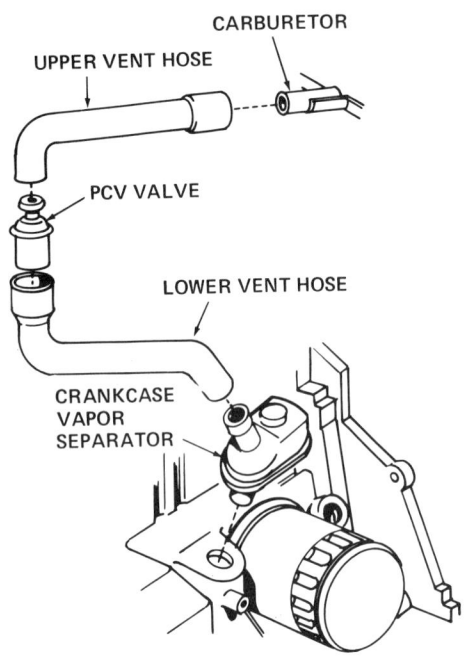

Figure 6-5. Replacing a typical PCV valve.

placement valve leans out the mixture enough to keep emissions down to within specified limits. Finally, the engine decal provides the maintenance interval at which you must change the PCV valve along with the part number of the required high-flow valve.

6. Insert the crankcase end of the new PCV valve into the lower vent hose and the intake manifold end into the upper vent hose. Next, if so equipped, install both hose clamps and tighten them securely.

7. Install the free end of the upper vent hose into its carburetor or intake manifold fitting.

8. Reinstall the free end of the lower vent hose over its fitting in either the rocker-arm cover or crankcase.

9. Test the system for proper operation.

Quite a number of PCV valve installations have only one hose (see Fig. 6-1). In these situations, the valve itself seats in a rubber grommet located in the rocker-arm cover or, in some cases, the intake manifold.

To replace a valve of this type:

1. Disconnect the PCV hose from the valve cover or intake manifold.

2. As necessary, loosen and remove the hose clamp around the PCV valve.

3. Remove the hose from the valve using a twisting and pulling action.

4. Blow compressed air through the hose to make sure that there are no obstructions.

5. Remove the PCV valve from its grommet in either the intake manifold or rocker-arm cover (Fig. 6-1).

6. After checking the new valve's part number against manufacturer's specifications, insert the correct end of the valve into the rubber grommet. For easier installation, coat the inside edge of the grommet with silicone lubricant.

7. Slide the PCV hose over the open end of the valve. If the hose has a clamp, install it over the hose and tighten securely.

8. Install the free end of the PCV hose over its fitting in either the intake manifold or rocker-arm cover.

9. Test the PCV system to make certain it is functioning properly.

Replacing a PCV Inlet Filter in the Air Cleaner Housing

A polyurethane-type PCV inlet filter, like the one illustrated in Fig. 6-6, is not serviceable by cleaning. Therefore, when this filter is dirty or requires replacement

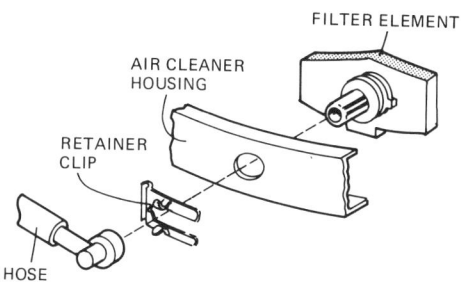

Figure 6-6. Replacing the PCV inlet filter within the air cleaner housing.

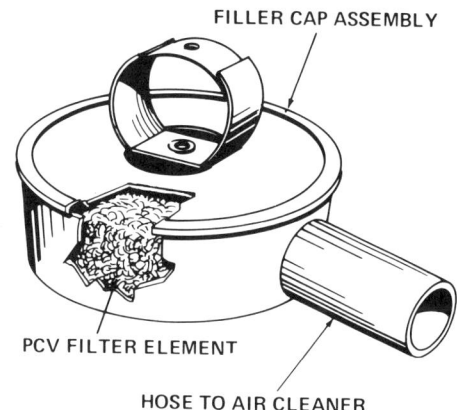

Figure 6-7. The inlet filter in the filler cap is serviceable by washing it in mineral spirits.

during preventive maintenance, change it following this process:

1. Remove the air cleaner cover.

2. Unsnap or remove the hose from the filter fitting on the air cleaner housing.

3. Slide the retainer clip from the filter's nipple and remove the filter element from the air cleaner housing.

4. Insert a new element and then install the new retainer clip, supplied with the filter kit. If the kit does not have a new retainer, clean and reinstall the old clip.

5. Reconnect or snap the hose onto the filter fitting on the air cleaning housing.

6. Reinstall the air cleaner cover.

Cleaning and Servicing the Filter Inside the Oil Filler Cap

To clean a typical PCV inlet filter inside an oil filler cap, follow these instructions (Fig. 6-7):

1. Remove the cap from the rocker-arm cover.

2. Loosen and remove the inlet hose clamp and slide the cap off the hose using a twisting and pulling action.

3. Soak the cap completely in a solvent such as mineral spirits.

4. Allow the cap to drain and air dry. Do not use compressed air to blow dry this type of filter because the air blast may dislodge part of the element.

5. Inspect the cap and filter assembly. If the cap shows any signs of damage or is still fully to partially clogged, replace the cap.

6. If the cap is still serviceable after cleaning and drying, invert its inlet fitting from the air cleaner (Fig. 6-8). Next, fill

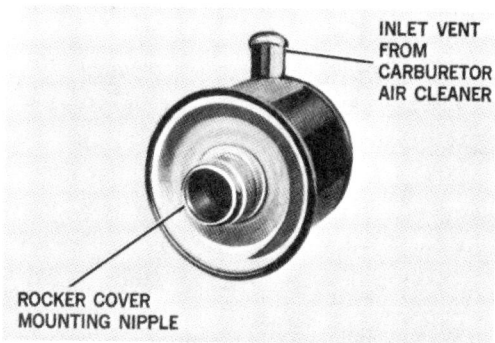

Figure 6-8. Servicing the element inside a filler cap with oil (Courtesy of Chrysler Corp.).

this opening with SAE 30 oil, until the oil runs out of the mounting nipple.

7. Reconnect the cap inlet fitting to the hose leading to the air cleaner.

8. Reinstall the cap into the rocker-arm cover.

Summary

1. For some time after their initial introduction, PCV systems did not receive proper maintenance for several reasons.
2. The PCV system requires routine service, or it will not function correctly.
3. Whenever a PCV system is not functioning satisfactorily, the engine does not receive adequate ventilation or release of its blow-by gases.
4. There are five symptoms of improper PCV system operation.
5. Never disconnect the PCV system to compensate for a rough engine idle.
6. Most vehicle manufacturers require some form of service on the PCV system at a given time or mileage interval.
7. When performing preventive maintenance or service work on a PCV system, always consult the manufacturer's instructions and specifications.
8. The first form of PCV system service is always a complete inspection of all of its components.
9. When inspecting a PCV system, check the hoses, valve, cap, and inlet filter.
10. On a Type 1, 2, or 4 system, check the operation of the device by performing a vacuum test with a piece of stiff paper.
11. Some mechanics prefer to use a PCV tester to check the operation of a Type 1, 2, or 4 system.
12. Use the procedure outlined in this text or the manufacturer's instructions when checking a PCV system with a special tester.
13. Use a tachometer to test a Type 1, 2, or 4 system whenever a special tester is not available or the results of the vacuum test were not conclusive.
14. Perform a tachometer test on a PCV system using the procedures outlined this text.
15. You can test the operation of a Type 1, 2, or 4 PCV system with an infrared analyzer following manufacturer's instructions or those outlined in this chapter.
16. Replace any PCV hose if cracked, severely plugged, or damaged using the steps presented in this text.
17. Clean out the PCV passages in a typical carburetor utilizing the process outlined in this chapter.
18. Replace a PCV valve using manufacturer's recommendations or those steps outlined in this text.
19. Some Ford engines require a high-flow valve in order to lean out the air/fuel mixture.
20. Replace the PCV inlet filter in the air cleaner housing following the manufacturer's instructions or those presented in this chapter.
21. Clean and service a PCV inlet filter inside an oil filler cap using the procedures outlined in this text.

Review Questions

The questions listed below will assist you in determining how well you remember the material contained in this chapter. Read each question carefully before

REVIEW QUESTIONS

choosing the answer. If you cannot answer the question, review the section in the chapter that covers the material.

1. Many manufacturers recommend service on the PCV system once a year or every _____ miles.
   a. 10,000
   b. 12,000
   c. 14,000
   d. 16,000

2. If the PCV system is inoperative, the crankcase will have a buildup of _____.
   a. sludge
   b. oil
   c. air
   d. vacuum

3. The first form of PCV maintenance is a complete _____.
   a. filter change
   b. hose replacement
   c. system inspection
   d. valve cleaning

4. You can perform a vacuum test on a PCV system with a (an) _____.
   a. piece of stiff paper
   b. infrared analyzer
   c. vacuum gauge
   d. engine analyzer

5. You should never attempt to clean a PCV _____.
   a. cap
   b. hose
   c. valve
   d. none of these

6. During a tachometer test of a PCV system, a less than 40-rpm drop can mean _____.
   a. a dirty inlet filter
   b. a plugged valve
   c. a plugged hose
   d. the wrong valve was installed into the system

7. Some PCV valves seat in a rubber grommet in the intake manifold or the _____.
   a. air cleaner
   b. lower crankcase
   c. rocker-arm cover
   d. engine breather

8. Some _____ engines require a high flow rate PCV valve at a given mileage interval.
   a. AMC
   b. Ford
   c. Chrysler
   d. Buick

For the answers to these review questions, turn to the Appendix.

Chapter 7

Evaporation Emission Control Systems

Function

Due to California's stringent emission control laws, beginning in 1970, all new automobiles sold in that state had to have an evaporation emission control (EEC) system. However, federal requirements for the installation of EEC devices did not begin until 1971. Thus, all domestic vehicles, regardless of where they were sold, had to have this system after that date.

An EEC system (Fig. 7-1) effectively reduces the escape into the atmosphere of gasoline vapors from a vehicle's fuel tank, carburetor vents, and on some newer-model vehicles even from the throat of the carburetor and intake manifold. These particular vapors amounted to about 20 percent of the total HC emissions from a typical motor vehicle. Since these vapors are essentially raw hydrocarbons, their release into the atmosphere contributes to the formation of photochemical smog. Consequently, the trapping of the vapors by an EEC system and then directing them to the engine for burning reduces their potential as air pollutants and provides a slight increase in the vehicle's fuel economy.

Types of EEC Systems

Since 1970, automobile manufacturers have produced two basic types of EEC devices—one using crankcase storage and the other utilizing carbon canister storage. All 1970–71 Chrysler- and some Ford-built vehicles, for instance, employed the crankcase as the vapor storage area. In these particular systems, fuel vapors from the tank and carburetor accumulated within the engine's crankcase when the vehicle was not in operation. Later, the PCV system moved the vapors from the crankcase

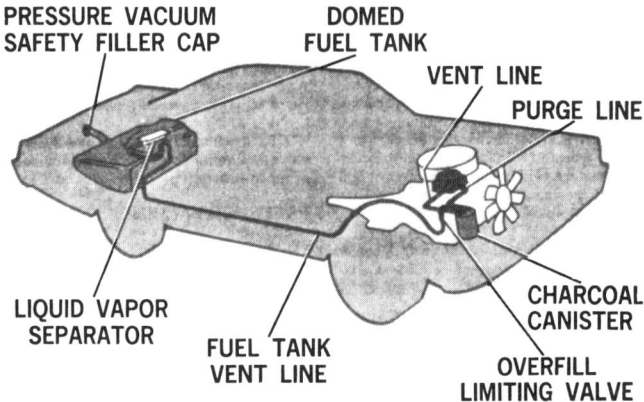

Figure 7-1. Typical EEC system (Courtesy of Chrysler Corp.).

into the engine during its various operating phases.

All domestic automobiles, starting with the 1972 model, use the popular carbon canister as their storage device. Because this type of system is the most common one in use, we will concentrate on its design and operating principles.

However, each vehicle manufacturer modifies the canister storage system somewhat to fit a given vehicle configuration. Therefore, we will only present an overview of the components found in a typical system and describe some of the variations in design the reader may encounter in the field.

EEC System Design

Filler Caps

A typical EEC system consists of a special filler cap, a pressure-vacuum control mechanism, an overfill-limiting device, a liquid-vapor separator, a number of rollover devices, fuel filter separator, carburetor vent valve, one or more carbon canisters, a carbon element, purge control valve, and a thermovalve (Fig. 7-2). The fuel tank of any EEC system requires a special filler cap that is different from the one used on a non-EEC-equipped vehicle. The cap is a sealed type that prevents any liquid fuel spillage due to either gasoline surging within the tank or heat expansion along with the escape of fuel vapors into the atmosphere.

The fuel cap may also contain a combination pressure and vacuum valve (Fig. 7-3) or just a vacuum valve. The combination valve protects the tank from physical damage in the event of a system malfunction or problem in a vent line that causes either excessive pressure or vacuum. Excessive pressure can develop within the tank due to heat expansion of its volume of fuel. Vacuum, on the other hand, develops above the level of the fuel in the tank as a result of the action of the fuel pump as it transfers fuel to the carburetor. This will not occur if the tank has the proper ventilation.

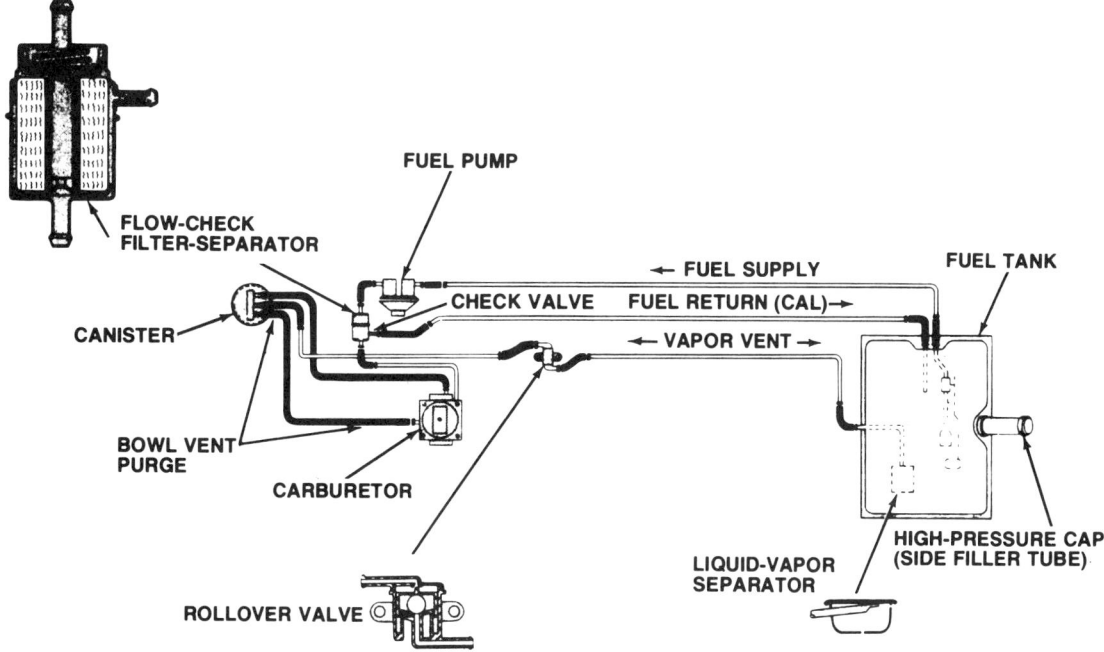

Figure 7-2. Schematic showing the components of an EEC system (Courtesy of Chrysler Corp.).

Operation of a Cap's Pressure-Vacuum Valve

Figure 7-4 illustrates the action of a pressure valve when the fuel tank has excessive pressure. In this situation, the vacuum valve has closed, but the pressure valve is open. This action occurs when there is an excessive pressure that, if not relieved, would damage the tank itself. With the pressure valve open, pressure can pass around the valve and out of the cap's opening. However, when the pressure in the tank drops to that of the atmosphere, the pressure valve closes.

Figure 7-5 shows the operation of the vacuum valve of an EEC cap whenever the tank requires venting to the atmosphere. This is necessary to prevent a vacuum lock that can prevent fuel flow or damage to the tank itself. When low pressure builds up in the tank, the vacuum valve moves open while the pressure valve remains closed. This valve position allows atmospheric air to enter the cap's air vent, pass around the

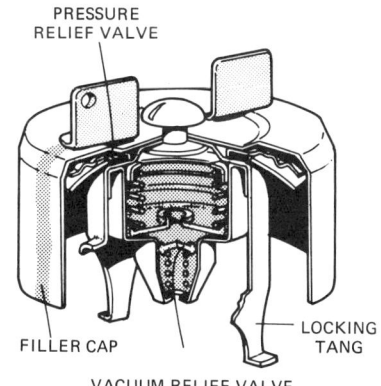

Figure 7-3. Combination pressure and vacuum valve in a typical EEC filler cap.

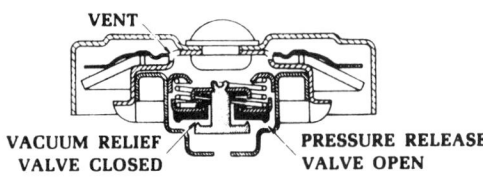

Figure 7-4. Operation of the pressure valve in an EEC system filler cap.

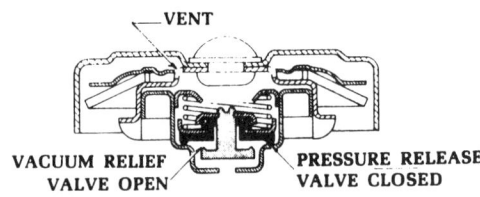

Figure 7-5. Operation of the vacuum valve within an EEC system filler cap.

open vacuum valve, and enter the fuel tank. This action equalizes the pressure in the tank to that of the atmosphere. As soon as the pressure is equal, the vacuum valve quickly closes.

Some late-model EEC filler caps have a design that prevents fuel vapors from leaving the cap under all conditions but does have a vacuum relief valve (Fig. 7-6).

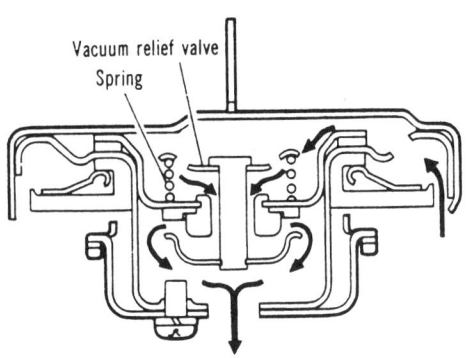

Figure 7-6. EEC filler cap with a vacuum relief valve (Courtesy of Chrysler Corp.).

Since the cap has no pressure relief valve, its mounting seals prevent any liquid fuel or vapors from escaping the fuel tank and entering the atmosphere.

However, the incorporated vacuum valve does protect the fuel tank if a malfunction occurs in the system. For example, if the pressure in the tank drops below a specified negative pressure, the valve will open in a manner similar to the unit in the combination assembly previously described. When the valve is in the open position, atmospheric air enters the cap's air vent, passes around the open valve, and enters the fuel tank. This action breaks up a vacuum by equalizing tank pressure to that of the atmosphere. Then, the valve quickly closes.

A EEC fuel cap, on some late-model vehicles, is secured with a two-step latching device and has an extended skirt (Fig. 7-7). The cap itself has two pairs of tangs arranged like those on a radiator cap. These permit the gas-station attendant to break the tank-to-cap seal in order to relieve tank pressure, without the complete separation of the cap from the filler tube. Then, in order to remove the cap, another 90° turn of the unit is necessary. Both of these features prevent tank pressure from forcing liquid fuel out of the tank if someone removes the cap too quickly.

One of the most important things to remember here is that if an EEC cap requires replacement, always use an exact duplicate. Never install, for example, a cap without the pressure-vacuum valve into a system that requires one. If you do, a vacuum lock can develop in the fuel supply system, or the fuel tank may sustain damage due to pressure from fuel expansion or a vacuum from fuel pump action.

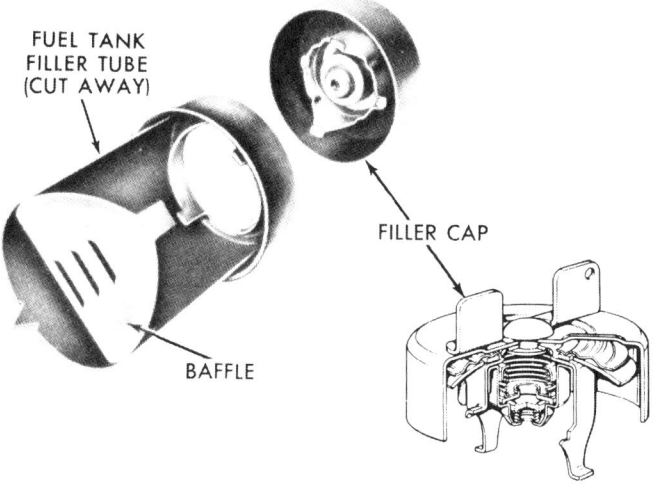

Figure 7-7. Fuel filler cap with a two-step latching mechanism (Courtesy of Chrysler Corp.).

Overfill-Limiting Devices—
Two-Way Valve

The EEC system usually has some form of overfill-limiting device. This component prevents the normal complete filling of the fuel tank, thus providing an internal expansion space into which liquid fuel or fuel vapors can safely expand on a hot day. This space amounts to approximately 10 to 12 percent of the total fuel tank capacity.

An overfill-limiting device may be in the form of either a two-way valve, an expansion tank, or a specially designed filler tube. Figure 7-8 is a schematic of an overfill, two-way limiting valve. This valve may be within or on the charcoal canister or in the vapor line between this unit and the fuel tank. In either case, this particular valve serves three functions.

First, the overfill-limiting portion of the valve prevents the complete filling of the tank. The manufacturer accomplishes this by only permitting a small amount of vapor pressure, caused by the rising level of the fuel in the tank during the refueling process, to pass through an orifice to the charcoal canister. This action causes fuel to back up in the filler tube before the tank is actually full. As a result, the service nozzle shuts off automatically.

The second function of this limiter valve, is to stop excess pressure from developing inside the tank. To accomplish this goal, the unit has a spring-loaded pressure valve. This valve opens whenever the fuel tank's internal pressure reaches a predetermined amount. With the valve open, the excess pressure passes around the valve and vents into the canister.

The final purpose of this particular overfill limiter is to prevent a vacuum from forming above the fuel in the tank. To perform this function, the limiter incorporates a vacuum valve similar to the one installed in some EEC caps. This valve serves the same function in that it opens whenever

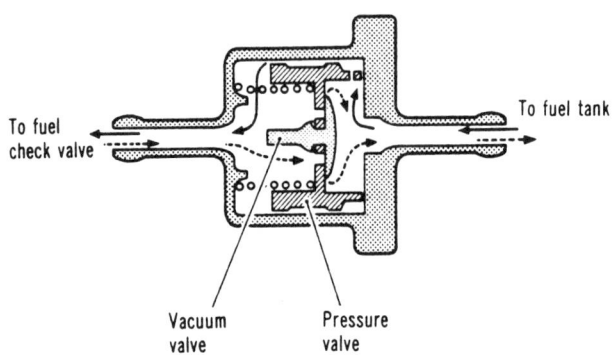

Figure 7-8. Overfill, two-way valve (Courtesy of Chrysler Corp.).

the pressure in the fuel tank drops below a preset value.

When the EEC system utilizes an overfill-limiting valve, its fuel tank has a raised section that appears much like an inverted dishpan on its otherwise flat surface. This additional space serves as a vapor collection area.

Another type of overfill-limiting device is in the form of a small tank mounted on the inside upper surface of the fuel tank. This smaller tank takes up about one-tenth of the main tank's volume or about the space of two gallons of gasoline.

This unit is sometimes known as an expansion tank (Fig. 7-9) and has a series of small holes machined into it. The size of these openings is such that it would normally require 10 to 15 minutes to fill the expansion tank with gasoline during the refueling process. Consequently, there remains about a two-gallon space-free area above the level of the fuel in the main tank, when its gauge reads full. This space takes care of any fuel or vapor expansion within the main tank in the event the vehicle sits in the sun for a prolonged period. The space also serves as a vapor collection area for the EEC system.

Some manufacturers install the fuel tank filler tube in such a manner that it acts as an overfill-limiting device. In this case, the filler tube extends a given distance below the top of the tank, thus preventing it from being filled 100 percent. This action provides an adequate space, between 10 to 12 percent of the tank's capacity, at its top to permit room for the

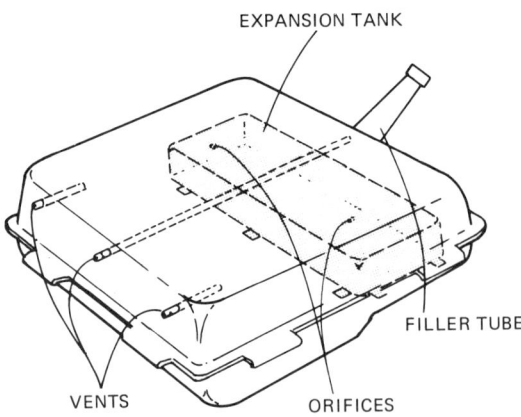

Figure 7-9. Location of a typical expansion tank.

fuel to expand and provide a vapor collection area (Fig. 7–10).

It is not unusual for an automobile manufacturer to utilize several methods to prevent overfilling of the fuel tank (Fig. 7–11). For example, some vehicles have an overfill-limiting valve, thermo expansion area, and a filler tube that extends down below the top of the tank. Due to the action of the overfill-limiting valve, fuel that continues to pump into the tank, above the lower end of the filler tube, returns to its upper end via an external leveling tube. As a result, the filler tube has a higher fuel level than the main tank that shuts off the automatic fuel-servicing nozzle.

Liquid-Vapor Separators

Open-Foam Type

All EEC systems require some form of liquid-vapor fuel separator. This device stops any liquid fuel from reaching the charcoal canister, which would overload its normal capacity. There are three common types of separator units in use: the open-cell foam, standpipe, and float. Figure 7–12 illustrates a foam-type liquid-vapor separator. This unit usually mounts at the top center of the tank so that the internal expansion area at the very top provides an adequate breathing space for the separator.

This separator consists of a quantity of open-cell foam that is inside a small container with two openings. The opening at the bottom of the unit permits the entrance of fuel vapors from the tank's expansion area. The top opening has a restrictor orifice and connects to the vapor line leading to the storage canister.

With this type of device, only fuel vapors can pass through the open-cell foam as they move from the tank, through the orifice, and into the vapor line to the canister. Any liquid droplets of fuel cannot pass through the foam material and, therefore, do not enter the vapor line. The liquid, in this case, just returns to the fuel tank.

Standpipe Liquid-Vapor Separators

Figure 7–13 is a schematic of a standpipe-style separator. This device fits above the fuel tank and has lines of different lengths running to each corner of the tank. In addition, the unit has a vent to the storage area of the charcoal canister that is at the highest point in the separator. This vent, in some cases, may also have an orifice to minimize the possibility of liquid fuel transfer to the canister.

The multiple vapor lines from the separator into the tank serve two functions. First, they constantly act as vents for the fuel tank as the engine consumes gasoline. Second, at least one of the lines provides a liquid fuel drain-back into the tank of any fuel droplets accumulated within the separator.

Any time fuel vapors condense in this type of separator from the standpipes, the resulting liquid drains back into the tank

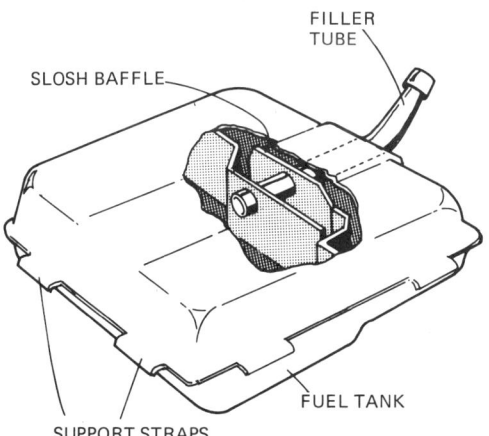

Figure 7–10. Location of the filler tube can prevent the complete filling of the fuel tank.

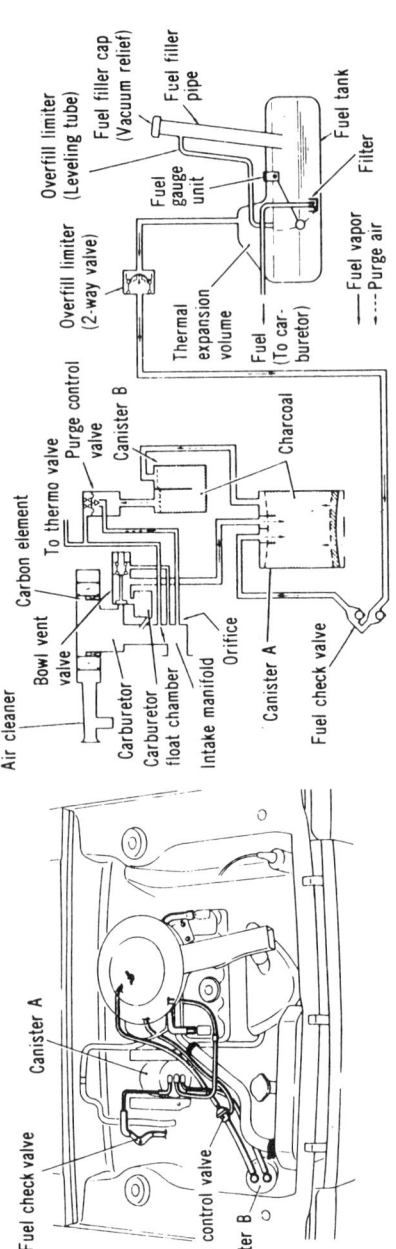

Figure 7–11. Some EEC systems utilize more than one method to prevent overfilling the fuel tank (Courtesy of Chrysler Corp.).

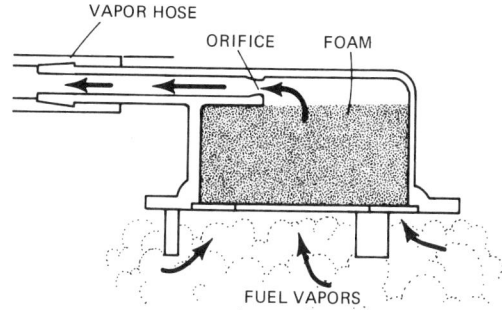

Figure 7-12. Open-cell foam-type liquid-vapor separator.

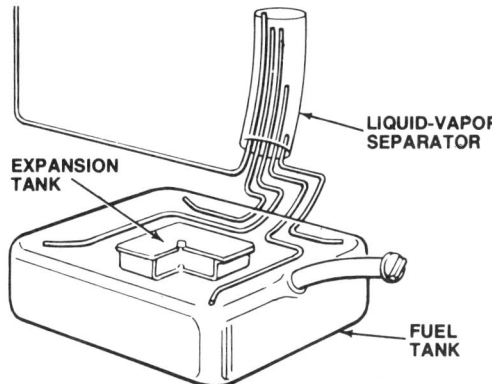

Figure 7-13. Schematic of a standpipe style of liquid-vapor separator.

through the shortest standpipe or a drain opening in one of the other vent lines. Also, if a driver parks a vehicle with a standpipe separator on an incline, any collected fuel that does not drain back to the tank will remain in the unit until the vehicle operates once again on a level road. This eliminates the flooding of the charcoal canister with liquid fuel. Therefore, the standpipe separator constantly acts as a baffle to prevent fuel from ever entering the canister regardless of the amount of fuel-vapor condensation or vehicle attitude.

There is a later-style liquid-vapor separator that functions essentially the same as the standpipe type. The main difference is the mounting positon of the units, due to changes in vehicle design. The newer type mounts in a horizontal position instead of vertical as shown in Fig. 7-13.

Float-Type Liquid-Vapor Separators

Figure 7-14 illustrates the third common type of separator, the float type. On some models this particular device is built into the tank, while on others it mounts directly on top of the tank itself. But regardless of its mounting location, the unit basically consists of a sealed float that functions inside a housing that has two openings. The opening in the base of the housing permits the entrance of fuel va-pors into the unit from the tank, below the float. The second opening has a fitting that connects to a vent hose and line leading to the canister.

The float itself is a sealed unit that rises in the housing whenever liquid fuel enters the separator. On one end of the float is a needle valve that contacts a seat that indexes over the orificed outlet opening.

If sufficient fuel enters the separator housing, the float rises until it reaches its uppermost position, where the valve shuts off the orificed opening. With the valve seated, neither fuel vapors nor liquid fuel can transfer to the canister. The valve remains closed until the fuel drains back into the tank, and the float lowers in the hous-

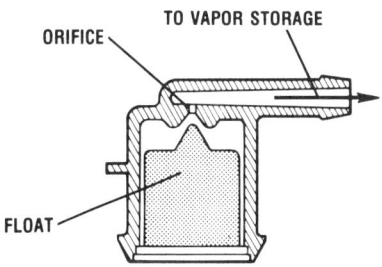

Figure 7-14. Design of a float-type liquid-vapor separator.

ing. At this point, vapors can once again pass through the opening and flow to the canister.

Rollover Leakage Protection Devices

Beginning in 1976, domestic automobile manufacturers began to install rollover leakage protection devices. In other words, each vehicle was equipped with one or more devices to prevent fires caused by liquid fuel leaks if the vehicle rolled over during an accident. Manufacturers install some of these units into the EEC system.

Figure 7–15 is a schematic of a Chrysler EEC system showing several of these devices, including a rollover valve, a flow check fuel filter separator, and a redesigned fuel cap. Almost all rollover devices operate on the check-valve principle. That is, they usually permit vapor flow in one direction but prevent liquid gasoline that accumulates in the same line, during a rollover accident, from flowing in the other direction. The fuel filler cap is the only exception to this rule in that it may have valving which controls tank pressure and vacuum.

A rollover valve may be positioned at the midpoint in the vapor line or incorporated into the liquid-vapor separator valve on top of the fuel tank. The fuel check valve shown in Fig. 7–16 connects into the vapor line. This valve contains two balls; under normal operating conditions, the balls assume the positions as show in the illustration. This permits gasoline vapors to move from the tank to the canister.

However, if the vehicle rolls over, one or both of the balls close the entire passage-

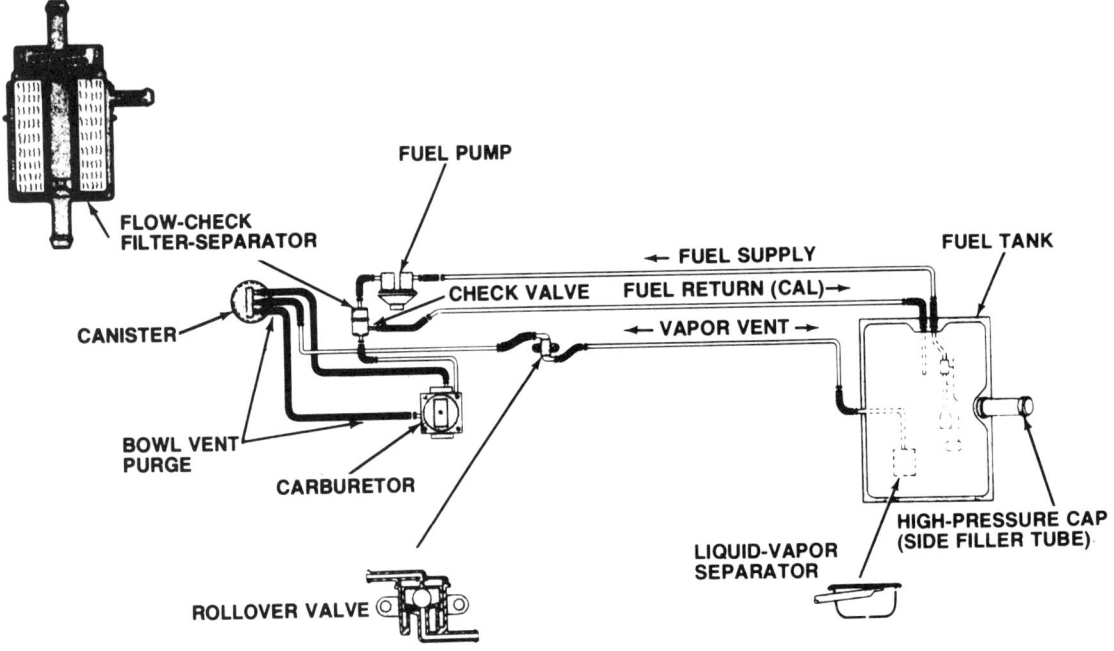

Figure 7–15. Schematic showing typical rollover protection devices incorporated into EEC system components (Courtesy of Chrysler Corp.).

EEC SYSTEM DESIGN   123

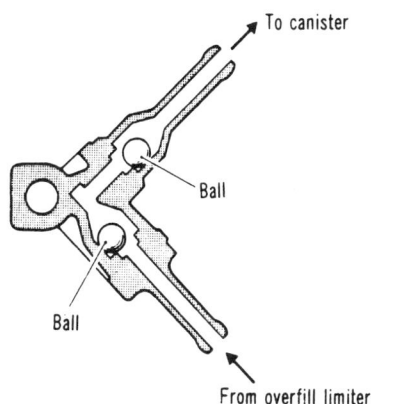

Figure 7-16. Construction of a rollover check valve (Courtesy of Chrysler Corp.).

way, thus preventing the fuel in the tank from reaching the canister. Otherwise the fuel would soon fill the canister and begin overflowing onto the surrounding area. This, of course, presents a fire hazard.

Figure 7-17 illustrates a rollover device that uses a float valve. This unit is actually built into a liquid-vapor separator. However, in this case, the float has a spring under it but still operates in the separator housing on top of the fuel tank. With the addition of the spring, the float valve seals off the orificed opening whenever the vehicle rolls over 90° or more. This prevents

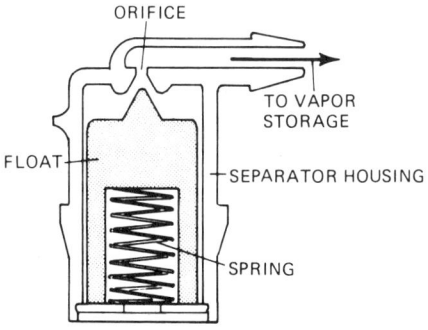

Figure 7-17. Orifice-type vapor separator with a spring-loaded float valve for rollover protection.

fuel from leaking into the vent line to the canister.

The modification made to the filler cap to prevent leakage after a rollover accident is to the pressure relief valve. Basically, the change just increased the opening pressure of this valve to stop liquid fuel from opening the unit if the tank should invert. All other functions of the cap remain the same.

The flow check valve in the fuel filter separator (Fig. 7-15) permits vapors to flow from the separator to the fuel tank. But it prevents liquid gasoline, in case of an accident, from flowing back through the separator and into the carburetor's float bowl. A supply of fuel from this source would pass through the open float needle valve in an inverted float bowl to create a serious leakage problem.

Liquid-Vapor Filter Separators

As previously stated, some EEC systems have a filter separator with a check valve. The separator itself (Fig. 7-18) has been in use on many vehicles for many years, even before the advent of the EEC system. The basic function of this component is to prevent engine flooding and/or vapor locking due to high underhood temperatures. These high temperatures can overheat the fuel in the line to the carburetor, causing it to expand to a point of vaporization. This results in excessive pressure within the fuel line that can unseat the carburetor needle valve and allow too much fuel to enter the unit. Consequently, the engine floods. If, on the other hand, the fuel in the line completely turns into a vapor, it can cause a vapor lock in the line that affects pump operation to a point where there is no fuel delivery to the carburetor.

The separator shown in Fig. 7-18 has

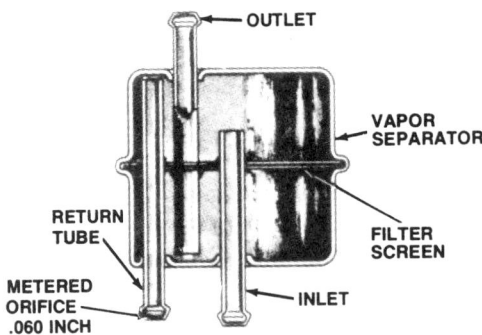

Figure 7-18. Common type of liquid-vapor filter separator without a rollover check valve (Courtesy of Chrysler Corp.).

three fittings: an inlet, outlet, and a vapor return. The inlet fitting connects via a hose and line to the fuel pump outlet. The separator's outlet fitting attaches directly into the hose or line to the carburetor. The last fitting has a metered orifice and connects by means of a vapor line back to the fuel tank.

When the engine is operating, fuel from the pump fills the vapor separator. Fuel from the bottom of the unit passes through the outlet fitting and into the fuel line to the carburetor. Any vapors that accumulate, whether the engine is running or shut off, rise to the top of the separator, where they pass through the orificed fitting and return to the fuel tank.

Carburetor Vents

All types of carburetors must have some form of external vent for the fuel bowl. This vent releases vapors from the bowl to prevent the buildup of pressure due to engine heat on the fuel in the float chamber. Otherwise, this pressure would cause percolation of the fuel within the bowl as well as engine flooding.

Vehicles built before the addition of the EEC system have a vent that opens directly into the atmosphere. However, on EEC-equipped vehicles using canister storage, the external carburetor vent connects into the canister itself by means of a hose (Fig. 7-19). While in many cases this vent is always open to the canister, some carburetors use mechanical linkage and a valve to open the vent, while still others use an electrically operated bowl vent valve.

Figure 7-20 shows a carburetor that uses a mechanically operated vent valve. This valve, activated through throttle valve linkage, remains closed during cruise and engine acceleration, but this same linkage opens the valve during normal hot-engine idle or when it is shut off.

Figure 7-21 is a schematic of a carburetor utilizing an electrically controlled, vacuum-operated float bowl vent valve. The operating mechanism for the valve itself consists of a diaphragm and electrical solenoid. The diaphragm connects directly to the vent valve and opens it

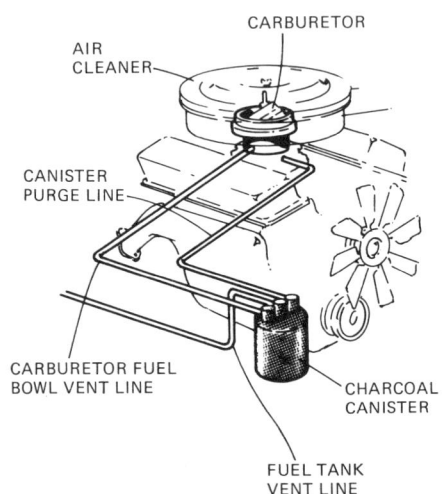

Figure 7-19. On a vehicle with a late-type EEC system, the carburetor's external vent connects into the charcoal canister.

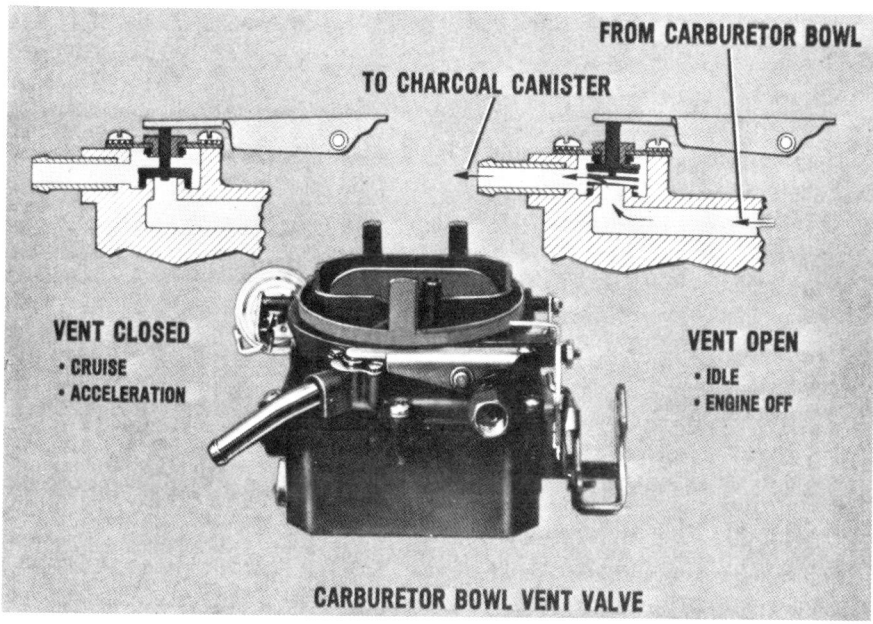

Figure 7-20. Carburetor with a mechanically operated bowl vent valve (Courtesy of Chrysler Corp.).

when intake manifold vacuum reaches a predetermined level.

The function of the solenoid is to maintain the vent valve in the open position, as long as the ignition key is on and the diaphragm has opened it. In other words, no matter what the intake manifold vacuum may be, during the various phases of

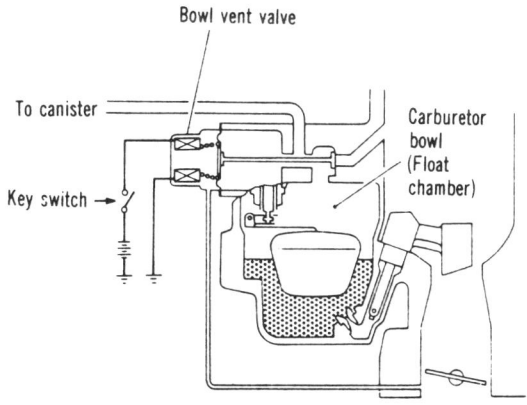

Figure 7-21. Electrically controlled, vacuum-operated float bowl vent valve (Courtesy of Chrysler Corp.).

engine operation, the solenoid holds the valve open as long as the ignition key is turned on.

The vent valve itself controls two port openings: one leading to the canister and the other to the carburetor's balance tube. The balance tube is nothing more than a passageway leading from the float bowl to an area in the air horn, usually above the choke valve. The other opening leads from the bowl itself to a carburetor fitting that accommodates a hose to the charcoal canister.

When the ignition key is off, the vent valve assumes the position shown in Fig. 7–21. Since the solenoid is deenergized and there is no engine vacuum, the valve is in the closed position, which seals off the port opening to the carburetor's air horn but opens the passage to the canister.

When the driver turns the ignition key and starts the engine, the solenoid activates, and the diaphragm pulls the valve open as intake manifold vacuum reaches a given amount. When the valve does open, it blocks the passageway to the canister but opens the port leading to the balance tube. The valve remains in this position as long as the ignition key is left on.

Charcoal Canisters

As mentioned earlier, the later-type EEC systems use one or more charcoal canisters instead of the engine's crankcase to store fuel tank and carburetor vapors. This metal or plastic canister (Fig. 7–22) fits underneath the vehicle's hood and holds a given quantity of activated charcoal granules. These granules store up to one-third of their weight in fuel vapors.

A typical canister will hold between 300 to 625 grams of charcoal. Each gram of activated charcoal has a surface area of

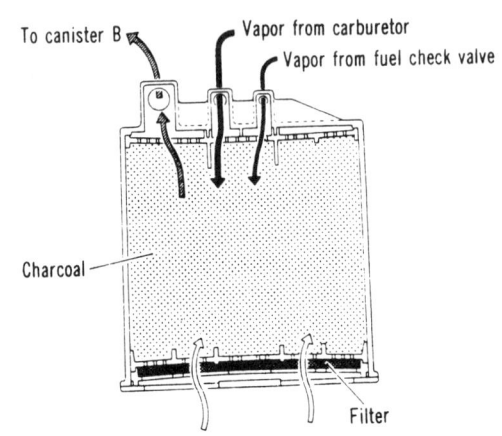

Figure 7–22. Typical charcoal canister (Courtesy of Chrysler Corp.).

1,100 square meters or more than a quarter of a mile. As a result, the exposed surface area of the charcoal in the canister is equivalent to 80 to 165 football fields, which is enough to store about a cup of liquid fuel when vaporized.

Because of its greater surface area and due to the fact that the fuel vapor molecules attach to the surface of the charcoal by absorption, the activated charcoal forms a good vapor trap. However, this attaching force is not very strong. Therefore, fresh air entering the canister's filter at its base and flowing through the charcoal can easily remove the vapor molecules from the granules.

As stated above, the canister does require a filtering device. This unit may be an oiled foam or fiberglass filtering element. But in either case, the device prevents dust or other contaminants from entering the canister in the purge air.

A typical canister will have three fittings located on its top (Fig. 7–23). A hose connected to one of these outlets carries the fuel vapors from the tank to the canister. A second hose attached to another canister fitting brings in carburetor vapors to

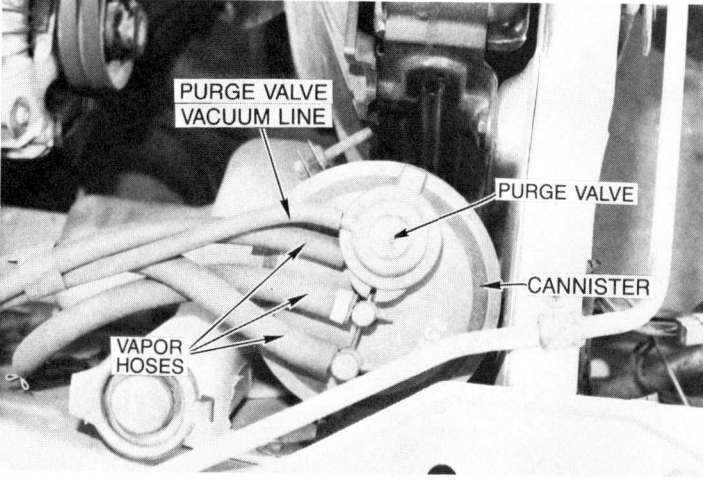

Figure 7-23. Typical charcoal canister usually has three fittings on its top.

the unit. The third fitting accommodates a purge hose that transports accumulated vapors within the charcoal into the intake manifold or air cleaner.

Some EEC systems have more than one canister (see Fig. 7-25). In this particular installation, canister A is the primary storage place for fuel vapors from the tank and carburetor during the heat-soak period. Canister B, on the other hand, functions as a temporary storage area for the fuel vapors removed from canister A during the purging process.

Carbon Elements

A carbon element is a recent addition to the basic EEC system. This device fits inside the standard air cleaner element between it and the throat of the carburetor (Fig. 7-24).

The function of the carbon element is to temporarily store fuel vapors remaining in the carburetor throat and intake manifold after the engine is shut down. Before the addition of the carbon element, these vapors could escape out of the carburetor air horn and enter the atmosphere through the air cleaner snorkel. Now on engine restart, the air passing through the air filter and carbon element purges these vapors and carries them back through the carburetor, intake manifold, and into the combustion chambers.

EEC System Operation

During the heat-soak period when the engine is not in operation, fuel vapors that collect in the thermo expansion space of the fuel tank pass through the two-way valve, fuel check valve, and enter canister A, where they are absorbed into the charcoal granules (Fig. 7-25). At the same time, since the bowl valve is open, any vapors above the level of the fuel in the bowl also pass into canister A and are absorbed into the granules.

The remaining fuel vapors in the throat of the carburetor and intake manifold cannot pass into canister A because there is no

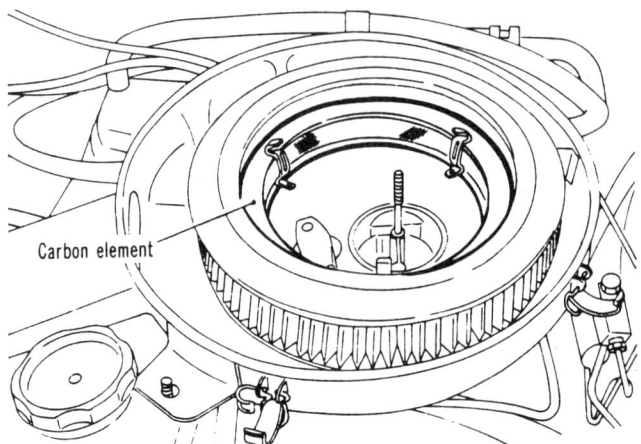

Figure 7–24. The carbon element fits inside the air cleaner element (Courtesy of Chrysler Corp.).

interconnecting passage or hose for this purpose. Instead, they are absorbed into the carbon element within the air cleaner housing.

When the driver starts the engine, purge air enters canister A through its filtering element. This air removes the fuel vapor molecules from the charcoal granules, carries them through canister B, and into the intake manifold. In this situation, canister B functions as a temporary storage for the fuel vapors moving into the intake manifold. On some engines, the canister vapors pass into the air cleaner housing instead of the intake manifold.

Types of Purging Methods

Constant Purge Method

Practically speaking, there are three types of purging methods used on domestic EEC systems to bring fresh air into the canister and carry the fuel vapors into the engine: constant purging, variable purging, and two-stage purging.

Two factors determine which method a given manufacturer will use. The first factor is the actual amount of fresh air that must pass through the canister in order to remove all the vapors and reactivate the charcoal granules. The second condition, which relates closely to the first, is that this air flow can have little effect either on the engine's air/fuel ratio or the driveability of the vehicle.

When a manufacturer uses the constant purge method, the rate of purging air through the canister remains fixed, regardless of the consumption of air by the engine. Manufacturers accomplish this by teeing into, for example, the PCV line to the carburetor, thereby using intake manifold vacuum to draw the air through the granules within the canister (Fig. 7–26). Even though intake manifold vacuum does vary with changes in engine load, an orifice in the purge line provides the system with a relatively constant air flow rate through the canister any time the engine is operating.

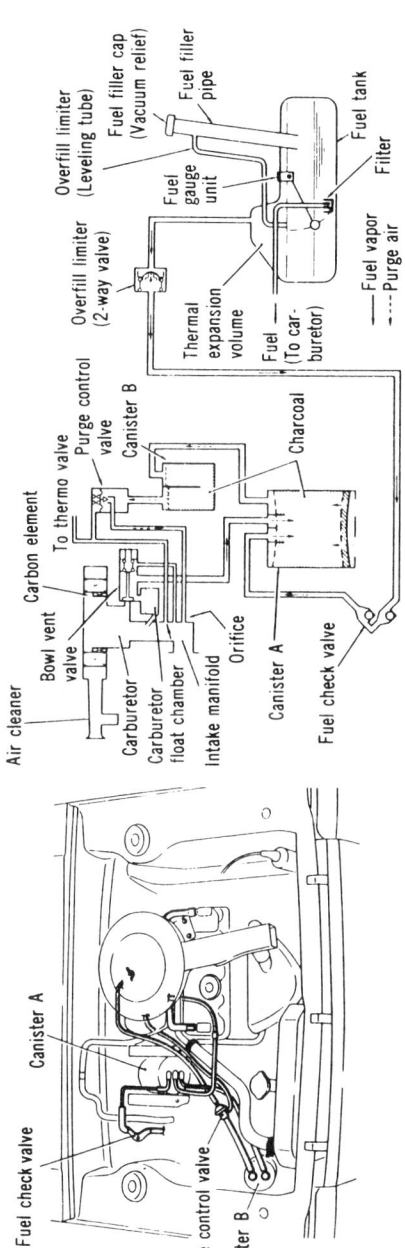

For Canada

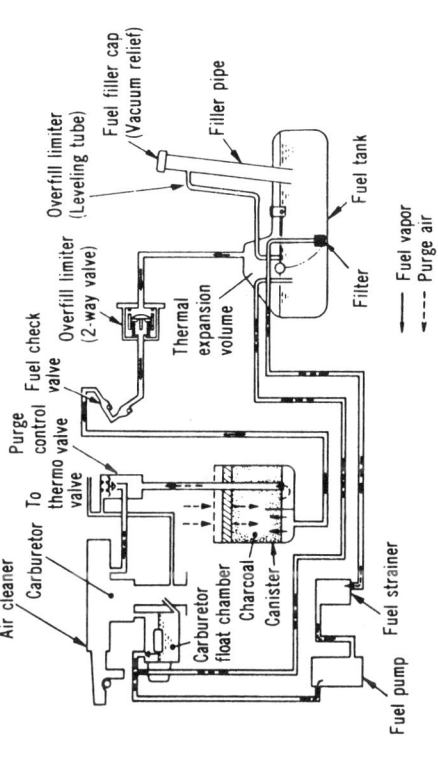

Figure 7-25. Operation of a typical EEC system (Courtesy of Chrysler Corp.).

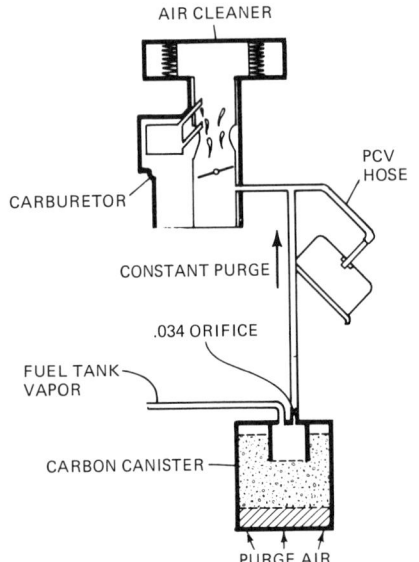

Figure 7-26. Schematic of a constant purge EEC system.

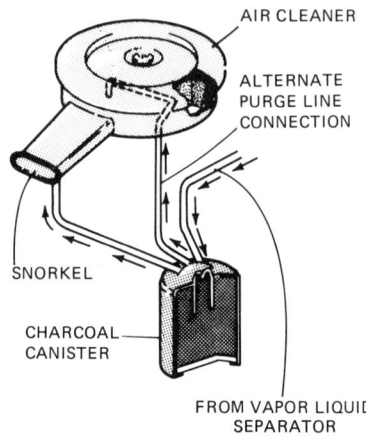

Figure 7-27. Schematic of a variable purge EEC system.

Variable Purge Method

In the variable purge system, the manufacturer connects the purge line to the air cleaner (Fig. 7-27). With this arrangement, the movement of air through the canister results from the intake air for the carburetor passing over a tube that projects into the air cleaner snorkel. The air flow itself creates a vacuum that moves the vapors out of the canister and into the airstream entering the snorkel.

This system is known as variable purge because the air flow entering the air cleaner regulates its action. In other words, the amount of purge air entering the canister from the atmosphere is in proportion to the amount of air moving into the engine through the air cleaner. Consequently, the more air that enters the engine, the greater will be the vacuum on the purge line and the amount of purge air entering the canister.

Figure 7-27 shows two alternate locations for the purge line connection at the air cleaner. A purge line that enters the air cleaner near the snorkel is acted upon by air velocity passing through the snorkel. As previously stated, this air velocity creates the low-pressure area that is necessary to cause atmospheric pressure to force air into the canister and system.

The other location for a purge line is on the clean side of the air cleaner element. In this situation, the difference in pressure or pressure drop across the air filter itself is sufficient to permit atmospheric pressure to force air through the canister. As in the snorkel-connected type, the amount of purge air depends upon the pressure drop or vacuum at the end of the purge line within the air cleaner. In other words, more air will flow when there is a greater vacuum at the end of the purge line than when only a slight vacuum exists.

Two-Stage Purging Method

In some instances, neither the constant or variable purge system provides the necessary air flow to completely purge the canister. To overcome this particular problem, some manufacturers use a two-stage purging process. This method involves the use of a purge valve located, in many cases, on top of the canister (Figs. 7–23 and 7–28). This valve operates by means of a ported vacuum signal that opens a second passage from the canister to the intake manifold. A ported vacuum signal, remember, is one taken from a passage above the throttle valve. Consequently, there is no vacuum in this passage when the throttle closes at idle, but the signal increases in proportion to the amount of throttle valve opening beyond this point.

When the ported vacuum signal reaches a predetermined level, it activates the purge valve. The valve, in turn, opens up a second passage from the canister to the intake manifold, thus permitting additional purge air to enter the canister. This air assists in reactivating the granules and carrying the vapors into the engine.

In an alternate design of EEC system utilizing a ported vacuum signal, the canister does not have a purge valve. Instead, the system has an additional ported vacuum connection on the carburetor responsible for purging the canister. This port is above the upper portion of the throttle valve, so there is no purge air through the purge line at idle. However, flow begins as soon as the throttle opens above the idle position. This system improves hot-idle quality by completely eliminating canister purging at idle.

Thermo Valve-Controlled Purging System

The EEC system illustrated in Fig. 7–25 uses a purge valve that is controlled by a thermo device. This purge valve has three port openings. The lower opening connects via a hose to canister B. The center fitting accommodates a hose connected into the intake manifold on some models and the air cleaner on others.

The upper port opening connects via a hose into a T fitting. This fitting has two hoses—one to the thermo valve and the other to a ported vacuum signal port on the carburetor.

A diaphragm fits inside the purge valve housing between the canister and air cleaner ports (Fig. 7–29). This spring-loaded unit and its attached valve react to the carburetor's ported vacuum signal. As a result, the valve remains closed at idle; this prevents fuel vapors from the canister from entering the intake manifold or air cleaner. In this engine configuration, this action is necessary for positive control of high CO emissions that are a particular problem under high ambient temperature conditions.

However, during off-idle operation of the engine, the ported vacuum reaches a

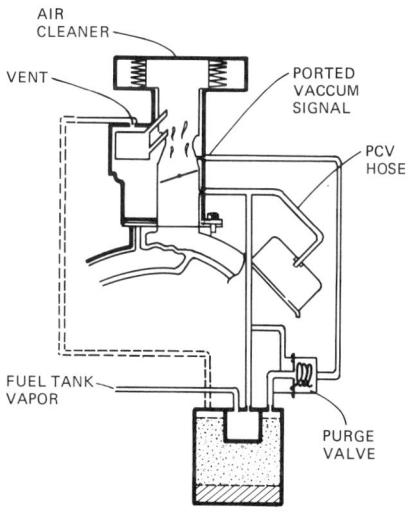

Figure 7-28. Diagram of an EEC system using two-stage canister purging.

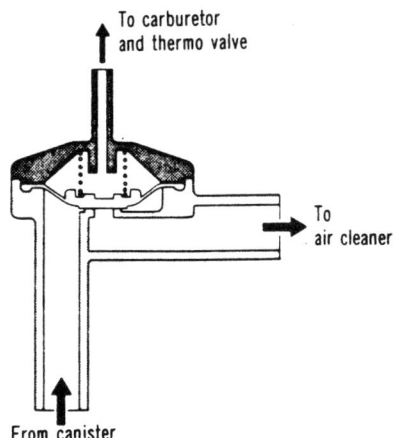

Figure 7-29. Details of a purge control valve (Courtesy of Chrysler Corp.).

sufficient level to open the purge control valve. This permits purge air to enter the lower canister and carry the fuel vapors out of the granules and into the engine.

The thermo valve is in this system for one reason. That is, this valve senses the engine's coolant temperature and closes the purge control valve when the temperature is lower than a preset value. This action prevents any canister vapors from entering the intake manifold or air cleaner, thus reducing CO and HC emissions during engine warm-up conditions. But when the engine temperature reaches a given degree, the thermo valve allows the purge control valve to open, and normal operation of the system begins.

Summary

1. All automobiles had to have an EEC-system starting in 1971.
2. An EEC system effectively reduces the escape of gasoline vapors from the fuel tank and carburetor into the atmosphere.
3. Automobile manufacturers have produced two types of EEC systems.
4. The canister storage type is the most popular EEC system.
5. An EEC filler cap is a sealed type that prevents any liquid fuel spillage or the escape of fuel vapors.
6. The EEC filler cap may contain a combination valve or just a vacuum relief valve.
7. When a filler cap's pressure valve is open, its vacuum valve remains closed.
8. If a low pressure builds up in the fuel tank, the vacuum valve opens, but the pressure valve remains closed.
9. Some EEC filler caps have an extended skirt and a two-step latching device.
10. An EEC system has some form of overfill-limiting device that prevents the normal complete filling of the fuel tank.
11. An overfill-limiting device may be in the form of a two-way valve, an expansion tank, or a specially designed filler tube.
12. An overfill-limiting two-way valve serves three functions.
13. The size of the small holes machined into an expansion tank prevents it from being filled during the refueling process.
14. It is not unusual for an automobile manufacturer to utilize several methods to prevent overfilling of the fuel tank.
15. A liquid-vapor separator stops any liquid fuel from reaching the charcoal canister.
16. There are three common types of liquid-vapor separators in use: the open-cell foam, standpipe, and float.

17. Rollover-leakage protection devices began to appear in EEC systems beginning in 1976.
18. Most rollover devices operate on the check-valve principle.
19. A rollover valve may be positioned at the midpoint in the vapor line or incorporated into the liquid-vapor separator valve on top of the fuel tank.
20. The modification made to the filler cap to prevent leakage after a rollover accident is to the pressure relief valve.
21. The basic function of a liquid-vapor filter separator is to prevent engine flooding and/or vapor locking due to high underhood temperatures.
22. All types of carburetors must have some form of external vent for the fuel bowl.
23. On later-model EEC systems, the carburetor vent connects to the canister.
24. Carburetor vent valves can operate either by mechanical linkage or an electrical solenoid.
25. A late-type EEC system uses one or more charcoal canisters to store fuel tank and carburetor vapors.
26. A typical canister holds between 300 to 625 grams of activated charcoal.
27. A charcoal canister has a filtering device.
28. The function of the carbon element is to temporarily store fuel vapors remaining in the carburetor throat and intake manifold after the engine is shut down.
29. During the heat-soak period, fuel vapors from the gasoline tank and carburetor float bowl enter the canister.
30. When the engine is not operating, any remaining vapors in the throat of the carburetor and intake manifold enter the carbon element.
31. Three types of purging methods are used to bring fresh air into the canister and carry the vapors into the engine.
32. The constant purge system uses a fixed rate of air flow to purge the canister.
33. In a variable purge system, the manufacturer connects the purge line to the air cleaner in one of two ways.
34. The two-stage system provides more air flow through the canister than either the constant or variable purge system.
35. Some EEC systems use a purge valve controlled by a thermo device.
36. The thermo valve in an EEC system closes the purge valve when coolant temperature is lower than a preset value.

Review Questions

The questions listed below will assist you in determining how well you remember the material contained in this chapter. Read each question carefully before choosing the answer. If you cannot answer the question, review the section in the chapter that covers the material.

1. Federal requirements for the installation of EEC systems began in _____.
   a. 1971
   b. 1972
   c. 1973
   d. 1974

2. An EEC filler cap may have a combination valve or just a(an) _____ valve.
   a. overflow
   b. relief

c. pressure
d. vacuum

3. A vacuum can form in the fuel tank due to the action of the _____.
   a. intake manifold
   b. fuel pump
   c. filler cap
   d. EEC system

4. The normal expansion space within a fuel tank amounts to about _____ percent of its capacity.
   a. 6 to 8
   b. 8 to 10
   c. 10 to 12
   d. 12 to 14

5. The device that prevents fuel from normally reaching the canister is the _____.
   a. two-way valve
   b. liquid-vapor separator
   c. flow valve
   d. check valve

6. Rollover devices began to appear on vehicles beginning in _____.
   a. 1972
   b. 1974
   c. 1976
   d. 1978

7. The _____ prevents engine flooding or a vapor lock due to high underhood temperatures.
   a. liquid-vapor filter separator
   b. carbon canister
   c. purge valve
   d. antipercolator valve

8. On later EEC systems, the carburetor vent connects via a hose to the _____.
   a. PCV valve
   b. purge valve
   c. crankcase
   d. canister

9. A typical charcoal canister will store about _____ cup of vaporized gasoline.
   a. one-half
   b. 1
   c. 1 and one-half
   d. 2

10. During the heat-soak period, carburetor throat and intake manifold vapors enter the _____.
    a. liquid-vapor separator
    b. engine crankcase
    c. carbon element
    d. charcoal canister

11. The _____ purge system uses a fixed rate of air flow through the canister.
    a. constant
    b. fixed
    c. variable
    d. two-stage

12. The _____ valve prevents some EEC systems from operating below a given coolant temperture.
    a. canister
    b. purge
    c. thermo
    d. coolant

For the answers to these review questions, turn to the Appendix.

Chapter 8

Evaporation Emission Control System Service

In normal service, an EEC system does not require a great deal of maintenance. The reason for this is twofold. First, generally speaking, the system has few complex moving parts that can wear out after a period of time. Second, the primary agents that the system handles are clean fuel vapors. Consequently, unlike the PCV system, there are few problems arising within the EEC system due to clogging of components.

The only exception to this rule is the filter inside the charcoal canister. Since normal atmospheric air passes through this element and it does contain given amounts of dirt, dust, and other contaminants, the filter requires periodic replacement or it becomes clogged. This, of course, prevents purge air from entering the canister and reactivating the charcoal granules.

Other than the replacement of the filter, the only other routine service work is to the vapor hoses. The frequent replacement of these hoses becomes necessary due to their exposure to high underhood temperatures, oil and grease, along with differing climatic conditions.

Unless a failure occurs to a system component, the maintenance intervals are usually provided by the vehicle manufacturer. The service interval may range from 12,000 to as much as 30,000 miles or 1 to 2 years on the various system components.

Also, since there are variations in basic system design for EEC devices among the many different manufacturers, the service intervals and procedures are not all the same. Therefore, when servicing a particular system, always refer to the manufacturer's specifications and instructions in the vehicle's service manual.

However, to aquaint you with some of these service techniques, this chapter will present some of the more common procedures used by the industry. These include a general visual inspection of the entire sys-

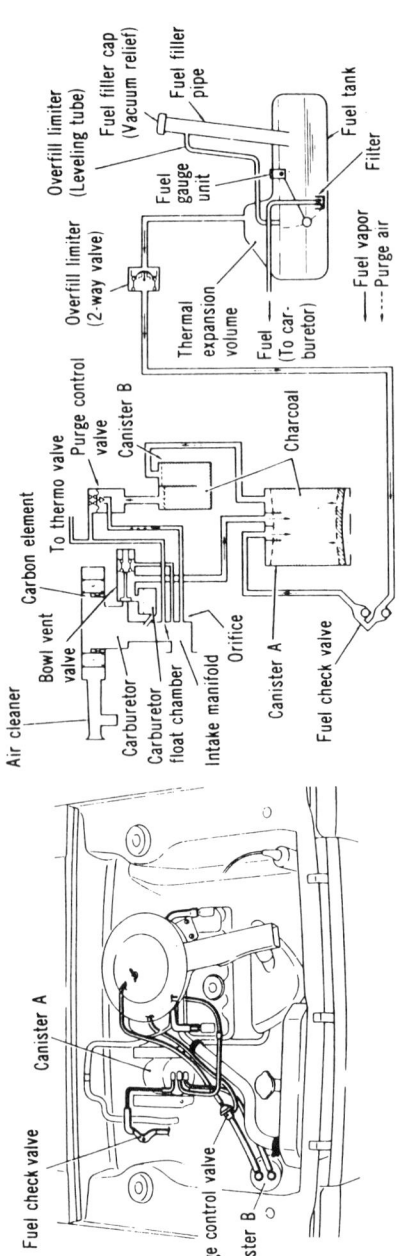
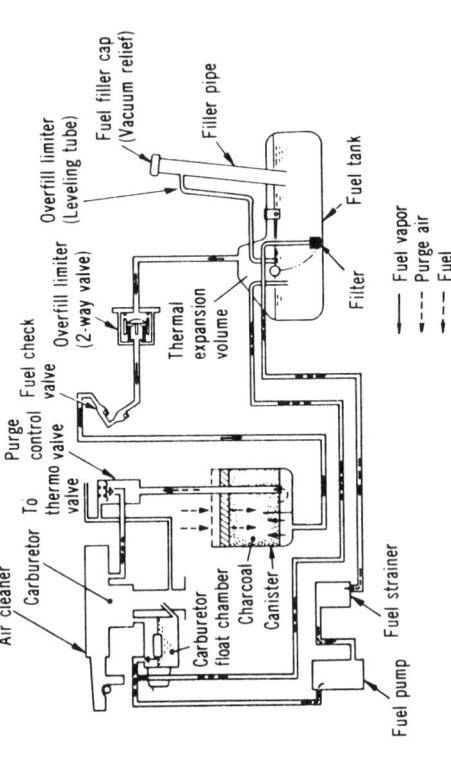

For Canada

Figure 8–1. Points to inspect on a typical EEC system (Courtesy of Chrysler Corp.).

tem, test of overall system operation with an infrared analyzer, and typical component replacement.

Inspection of the EEC System

To perform a visual inspection of a common type of EEC system (Fig. 8–1):

1. Raise the vehicle on an overhead hoist.

2. With a droplight, check the fuel tank and all its connections for signs of leakage.

3. If they are visible, inspect the liquid-vapor separator and/or liquid check valve for signs of leaks.

4. Inspect all vapor hoses for proper connection to components and signs of deterioration.

5. Lower the vehicle to the shop floor. Then, open the hood and install covers over both fenders.

6. If the vehicle has a charcoal canister, check the routing of each hose to make certain they are properly connected to the correct fittings. Also, check each hose for deterioration.

7. Inspect the canister's filter for obstructions. If the filter is dirty, replace it following the procedure outlined below.

8. Check the condition and routing of the vacuum hoses to the purge and thermo valve, if so equipped.

9. Remove the air cleaner cover, and inspect the carbon element for obstructions. If the element is dirty, replace it using the instructions listed below.

Checking the Condition of Other System Components

Fuel Cap

To check a typical EEC filler cap:

1. Loosen and then carefully remove the cap from the filler tube. While doing so, check the operation of the cap's two-step locking mechanism, if so equipped (Fig. 8–2).

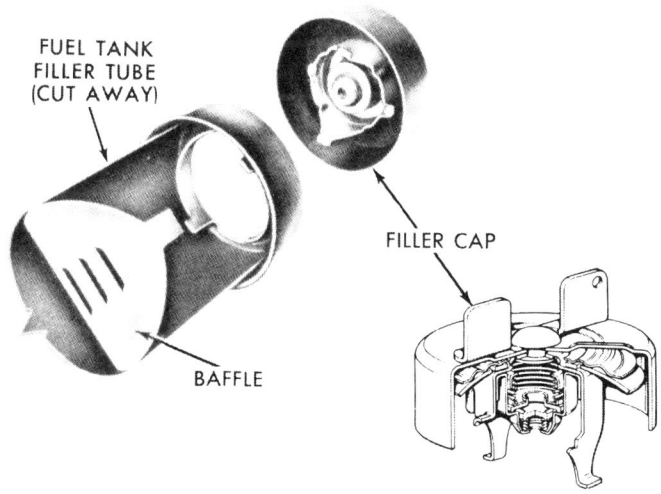

Figure 8–2. If the fuel cap has a two-step locking mechanism, make sure it is functioning properly (Courtesy of Chrysler Corp.).

2. Inspect the cap's seal or gasket for signs of deterioration.

3. Check the cap's combination or vacuum valve for signs of obstructions, wear, or damage.

4. If the fuel cap's seal or gasket, combination or vacuum valve, or its two-step locking mechanism is defective, replace the unit. Caution: Serious deformation or total collapse of the fuel tank will occur if you install a filler cap other than the one recommended by the vehicle manufacturer.

Thermo-Controlled Purge Valve

To inspect this particular type of purge valve (Fig. 8–3):

1. Start the engine and operate it sufficiently so that its coolant temperature is between 176 to 194 °F (80 to 90 °C).

2. Stop the engine and disconnect the purge hose from the air cleaner or intake manifold.

3. By mouth, attempt to blow air into the open end of the purge hose. If there is a definite obstruction to the entrance of air into the hose, the valve is closed and operating properly.

4. Connect a tachometer to the engine, following manufacturer's instructions.

5. Start the engine and slowly increase its speed to between 1,500 and 2,000 rpm. Block the throttle linkage so the engine constantly operates at this speed.

6. Again by mouth, blow into the end of the purge hose; it should now accept air because engine vacuum should have opened the purge valve.

7. If the purge valve is not open, check for a clogged or broken hose to the unit, intake manifold, or thermo valve. Also, check the thermo valve for serviceability. If all the vacuum hoses and the thermo valve are satisfactory, replace the purge control valve.

Overfill-Limiting Valve

To check the serviceability of this particular valve (Fig. 8–4):

1. Inspect the valve's body for cracks and leaks. Replace the assembly if it is defective.

2. Loosen and then slide back both the hose clamps on each end of the valve. Then, remove both hoses using a twisting and pulling action.

3. Loosen and remove the limiter bracket's retaining bolt and remove the assembly from the vehicle.

4. By mouth, blow air lightly into either the valve's inlet or outlet port. If air passes through the valve with a slight resistance, the overfill limiter is serviceable.

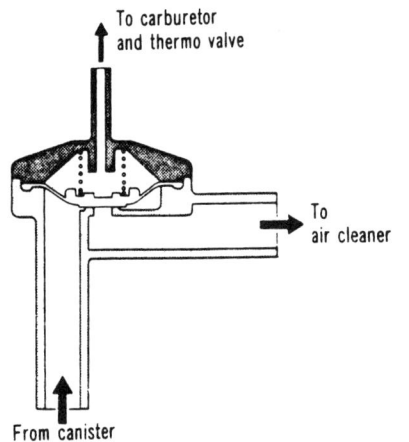

Figure 8–3. Checking the operation of a typical purge control valve (Courtesy of Chrysler Corp.).

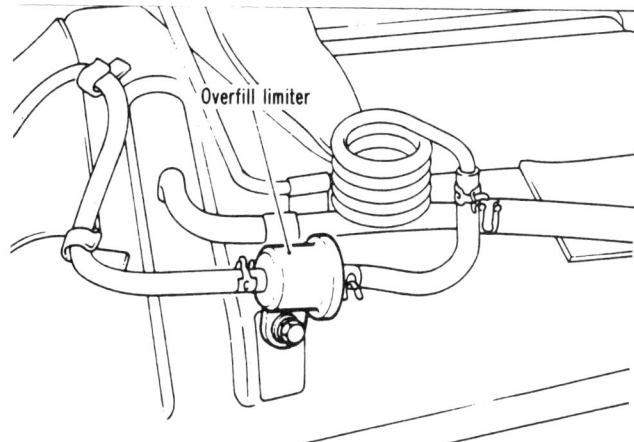

Figure 8–4. Checking an overfill-limiting valve (Courtesy of Chrysler Corp.).

5. Reinstall the limiter and tighten its attaching bolt to specifications.

6. Slide both hoses, if still serviceable, back over their respective fittings. Then, reinstall and tighten their hose clamps.

Testing the Operation of an EEC System with an Infrared Analyzer

To accurately test the operation of any EEC system with an infrared analyzer, follow these directions:

1. Following the steps outlined earlier, warm up and initially calibrate the infrared analyzer.

2. Install a tachometer onto the engine, following manufacturer's instructions.

3. Start the engine and permit it to reach normal operating temperature.

4. Insert the analyzer probe into the tail pipe. Then, note the HC and CO readings at engine idle.

5. Operate the engine at 2,500 rpm and note the HC and CO readings.

6. Disconnect the canister purge line at either the intake manifold or air cleaner and plug it.

7. Operate the engine again at idle and 2,500 rpm and note the readings on the HC and CO meters at both speeds.

8. Stop the engine and reconnect the canister purge line.

9. Disconnect the tachometer from the engine and remove the probe from the tail pipe.

Results and Indications

1. If after removing the purge line the HC and CO readings, both during idle and at 2,500 rpm, indicate a leaner mixture, the EEC system is functioning properly. Due to the differences in purging methods, on some systems the infrared analyzer may not indicate a mixture lean-out at idle with the purge hose removed. Therefore, before attempting to locate a malfunction where

none may exist, check to make sure the system is supposed to operate at idle.

2. If there is no change in the readings at both speeds, check for

a. a cracked or leaking purge hose;

b. a cracked, broken, or disconnected vacuum hose to the purge or thermo valve;

c. an inoperative purge or thermo valve;

d. a dirty canister filter;

e. a cracked, leaking, or disconnected canister vent hose;

f. an inoperative two-way or carburetor vent valve; or

g. a cracked, leaking, or disconnected vent hose at the fuel tank or liquid-vapor separator.

EEC System Component Replacement

Carbon Element

To replace a typical carbon element (Fig. 8-5):

1. Loosen and remove the air cleaner cover's attaching hardware. Then, lift the cover off the housing.

2. Unsnap the carbon element's retaining clips and remove the element from the air cleaner housing.

3. Position a new element within the air cleaner housing. Next, resnap all of its retainer clips.

4. Reinstall the air cleaner cover and its attaching hardware.

Canister or Filter Replacement

To replace either the charcoal canister or its filter, follow these steps (Fig. 8-6):

1. Mark all of the canister hoses as to their proper locations. This action prevents reinstalling a hose(s) on the wrong fitting after replacing the canister or filter.

2. Remove all the vapor hoses from the canister fittings.

3. Loosen all the hold-down clamps and remove the canister from its mount.

4. If so equipped, remove the canister base and pull the filter from the bottom of the canister. In some cases, the filter is just held in place by a retainer bar or ring.

5. Install the new filter element under the retainer bar or ring and install the canister base if so equipped.

6. Reinstall the canister into its mount and tighten its hold-down clamps securely.

7. Reinstall all vapor vent hoses to their proper canister fittings.

Vapor Vent Hose Replacement

As mentioned earlier, any EEC vent hose requires replacement if cracked or damaged. The hose used for this purpose is unique in that it is made especially to carry fuel vapor. Consequently, when replacing any vapor vent hose, use only the special rubber hose designated for this application. The hose itself is available in bulk form and usually has the label "EVAP" stamped on its outer surface. Never use regular vacuum hose for this purpose because it deteriorates when subjected to fuel vapors and will clog the system in time.

To replace a typical hose:

1. If so equipped, loosen and slide back the hose clamps on both ends of the hose.

2. Twist the hose slightly around its fittings and remove the hose using a straight, pulling action.

EEC SYSTEM COMPONENT REPLACEMENT

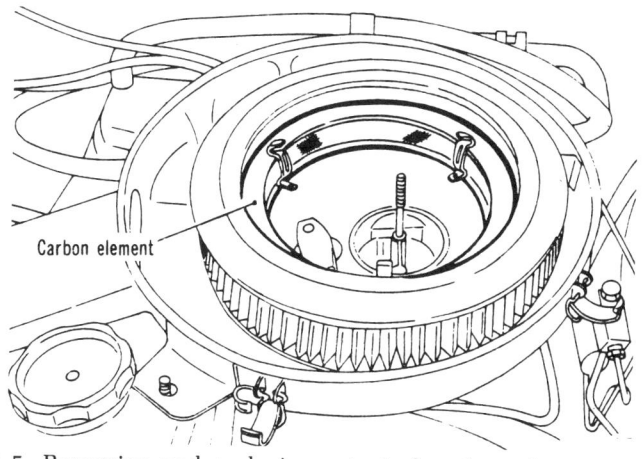

Figure 8-5. Removing and replacing a typical carbon element (Courtesy of Chrysler Corp.).

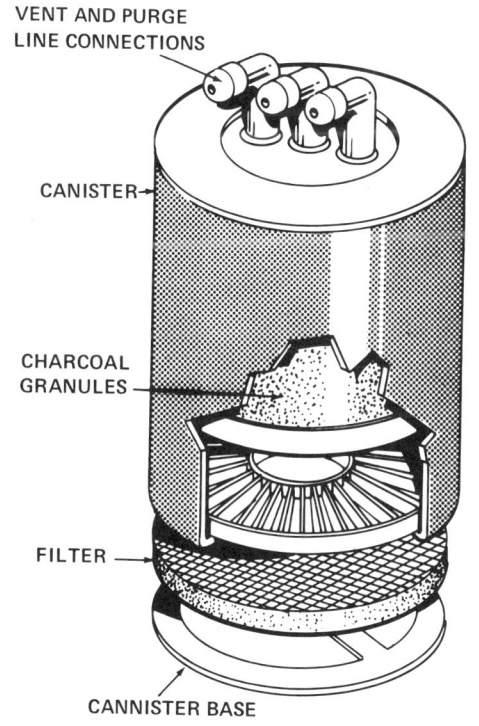

Figure 8-6. Replacing a charcoal canister or its filter.

3. Using the old hose as a guide, cut a new hose of the proper diameter to the correct length from a roll of hose stock.

4. Reinstall the clamps over the hose.

5. Install the ends of the hose over the fittings about 1 inch. Then, position and tighten each hose clamp securely about one-quarter inch from the ends of the hose.

Rollover Fuel Check-Valve Replacement

To remove and replace the fuel check valve shown in Fig. 8-7, follow this recommended procedure:

1. Remove the attaching bolt that holds the valve to the vehicle's toe board.

2. Loosen and slide back both hose clamps out of the way.

3. Disconnect each hose from the check valve using a twisting, pulling action.

4. If both hoses are in satisfactory condition, install them onto the new fuel check valve.

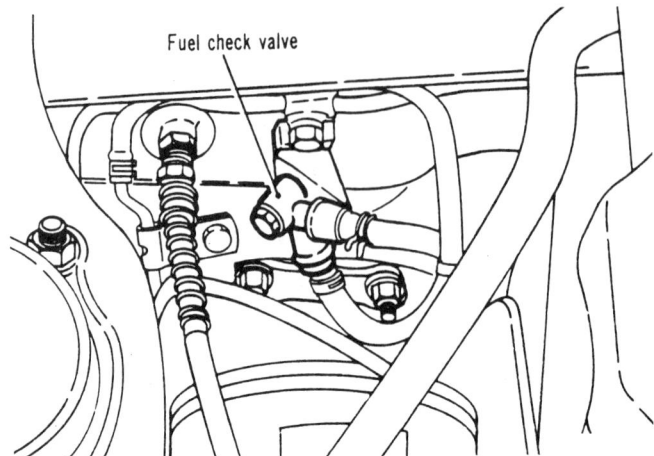

Figure 8-7. Removing and replacing a fuel check valve (Courtesy of Chrysler Corp.).

5. Reinstall the clamps over the ends of the hose and tighten securely.

6. Position the fuel check valve back onto the toe board as shown in Fig. 8-7. Next, reinstall and tighten the valve's attaching bolt to specifications.

Summary

1. In normal service, an EEC system does not require a great deal of maintenance for two reasons.
2. The canister filter needs periodic replacement, or it becomes clogged.
3. Service intervals on EEC systems range from 12,000 to as much as 30,000 miles or 1 to 2 years.
4. Always refer to manufacturer's specifications and instructions when performing maintenance on a particular EEC system.
5. To perform a visual inspection on most types of EEC systems, follow the directions outlined in this text.
6. Check the serviceability of the fuel cap, thermo-purge valve, and overfill-limiting valve using procedures outlined in this chapter.
7. You can test the operation of an EEC system with an infrared analyzer using the procedures listed in this chapter.
8. Compare the results of the infrared analyzer test with the data contained in the results and indications section of this chapter.
9. Replace a carbon element following manufacturer's instructions or those provided in this text.
10. Replace the charcoal canister or its filter using manufacturer's instructions or the steps outlined in this chapter.
11. When replacing an EEC vent hose, always use hose designated for this purpose.
12. Replace a typical vapor vent hose using techniques presented in this chapter.

13. Replace a rollover fuel check valve using the procedures specified in this text or manufacturer's instructions.

Review Questions

The questions listed below will assist you in determining how well you remember the material contained in this chapter. Read each question carefully before choosing the answer. If you cannot answer the question, review the section in the chapter that covers the material.

1. The EEC component that requires periodic replacement is the _____.
   a. canister filter
   b. canister valve
   c. check valve
   d. rollover valve

2. EEC system should have maintenance between 12,000 to 30,000 miles or every _____.
   a. 3 to 6 months
   b. 6 to 9 months
   c. 6 months to 1 year
   d. 1 to 2 years

3. When checking a thermo-controlled purge valve, coolant temperature should be between _____ °F.
   a. 136 to 154
   b. 156 to 174
   c. 176 to 194
   d. 196 to 214

4. If when testing an overfill-limiting valve air passes through the valve with some resistance, the valve is _____.
   a. partially clogged
   b. still serviceable
   c. inoperative
   d. stuck closed

5. You can test an EEC system with a (an) _____.
   a. infrared analyzer
   b. engine analyzer
   c. vacuum gauge
   d. enrichment tool

6. When replacing a carbon element, it is necessary to first remove the _____.
   a. charcoal canister
   b. air cleaner cover
   c. overfill limiter
   d. purge valve

7. In most cases replacement vapor vent hose has "_____" stamped on its outer surface.
   a. FUEL
   b. VAPOR
   c. EVAP
   d. VENT

For the answers to these review questions, turn to the Appendix.

Chapter 9

Engine Modification Systems

As a result of California's early emission control laws and federal legislation, manufacturers had to begin installing a variety of exhaust emission control equipment. These devices are installed on automobiles and light- to medium-size trucks so that their exhaust emissions will not exceed certain standards. Consequently, this equipment has the effect of reducing HC, CO, and NOx emissions.

To make it easier for the reader to understand the function, design and operation of the various controls, this text subdivides them into the following general groups: engine modification, spark-timing control, air injection, exhaust gas recirculation, and catalytic converter systems. But keep in mind that a given motor vehicle can have all or only a few of the devices installed by the factory, depending upon its model year, engine size, and engine type.

While this text covers all of these exhaust emission control systems, the next two chapters concentrate on engine modification. The remaining chapters will cover the rest of the devices.

Function of an Engine Modification System

With the advent of strict emission control standards, automotive engineers and scientists reasoned that the best way to control exhaust emissions was to burn the fuel as completely as possible inside the engine itself. This philosophy led to many design changes within the engine—built-in modifications to reduce the levels of HC, CO, and NOx emissions. However, as mentioned earlier, some of these design changes reduced engine performance and caused an increase in fuel consumption.

There have been a considerable number of alterations to modify a basic engine to

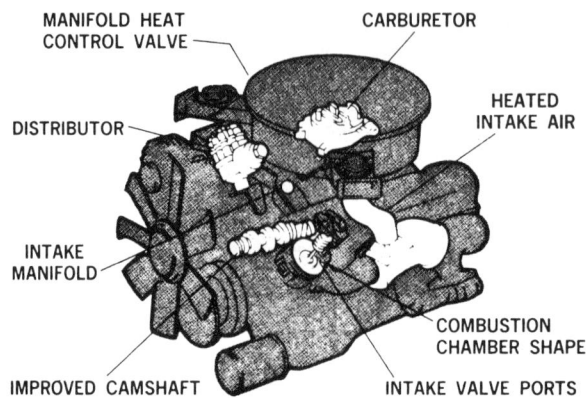

Figure 9-1. Typical modifications made to an engine in order to control exhaust emissions (Courtesy of Chrysler Corp.).

reduce its exhaust emissions. These changes include modifications to the combustion chambers, pistons, compression ratio, valves and valve seats, camshaft, intake manifold, and in many cases, the overall design of the combustion chamber shape itself. Along with these internal engine changes, there are also modifications to the carburetor, air cleaner, and the advance mechanisms of the ignition distributor (Fig. 9-1).

Engine Modifications

Combustion Chambers and Pistons

Certain portions of the combustion chamber, cylinder wall, and piston remain relatively cool during the burning of the air/fuel charge. Consequently, the burning mixture that is within close proximity to these areas does not reach ignition temperature. In other words, the normal flame front from combustion "snuffs" out as it approaches these cooler areas. As a result, these unburned hydrocarbons are ejected from the cylinder, along with some carbon monoxide and other gases on the exhaust stroke, to pollute the atmosphere. Figure 9-2 illustrates some of the common quench areas found in an engine's combustion chamber area.

Within the modified emission-controlled engine, the manufacturer eliminates the quench pockets and close clearance spaces by reworking the combustion chambers and pistons (Fig. 9-3). In this modified engine, the manufacturer has opened the closed or nearly closed areas near the ends of the combustion chambers. This action has fairly well eliminated the quenching of the burning air/fuel charge

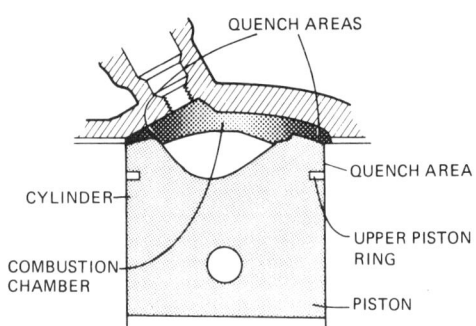

Figure 9-2. Common quench areas within an unmodified engine.

ENGINE MODIFICATIONS

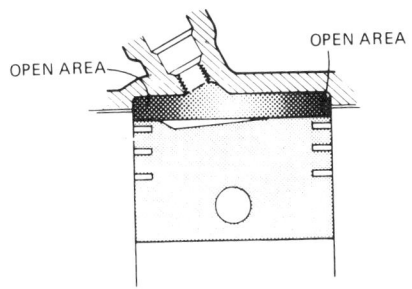

Figure 9-3. Combustion chamber and piston modified to reduce quench areas.

next to these metal surfaces, chilled below combustion chamber temperatures.

The reduction of the quench areas near the upper portion of the piston and cylinder wall, however, is a somewhat more difficult task. A typical approach to solving this problem is the repositioning of the top ring closer to the head of the piston. This action reduces the size of the small pocket formed between the top of the ring and the head of the piston but does not completely eliminate it. Therefore, a small quench area still exists between the upper ring and cylinder wall during the combustion process. However, the resulting HC emissions from this smaller quench area pocket are within tolerable limits.

In many engines, there is also another design change to the basic piston. This modification is to the head of the piston itself for the purpose of altering combustion chamber shape and reducing the engine's compression ratio. Remember, the head of the piston forms the lower portion of the combustion chamber; consequently, any changes in this area influence the combustion process within the chamber.

Reduced Compression Ratio

Another design feature of the emission-modified engine is a lower compression ratio. Engines with high compression ratios must burn high-octane leaded fuel and operate with high combustion chamber temperatures. As mentioned earlier, the main source of lead particulate emissions into the atmosphere is from the tetraethyllead added to fuel to raise its octane level. Also, the high combustion chamber temperatures create oxides of nitrogen, another primary air pollutant.

Since 1970, manufacturers have lowered engine compression ratios to an average of about 8:1 to permit the use of low-octane, low-lead, or unleaded fuel. This action resulted in reduced levels of lead particulates, NOx, and in some cases, even HC emissions.

Manufacturers have used several methods to reduce compression ratios within the emission-modified engine. These methods include altering the combustion chamber design, piston head, or engine stroke.

Modifications to Valve Ports and Seats

Manufacturers have redesigned the valve ports on some larger displacement engines to increase the turbulence of the air/fuel mixture (Fig. 9-4). This design alteration basically decreased the valve's operating angle, resulting in a more complete and balanced mixing of the air/fuel mixture. As a result, the engine has improved combustion and lower HC and CO emissions.

With the removal of tetraethyllead from fuel, a modification to the exhaust valve seats became necessary. In addition to controlling the combustion process, thus reducing detonation, the lead additive in gasoline also lubricates the valve seats. Without this form of lubrication, the rela-

ENGINE MODIFICATION SYSTEMS

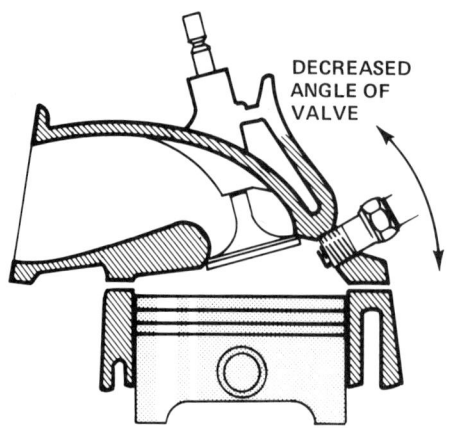

Figure 9-4. Decreased valve angle increases the turbulence of the air/fuel mixture.

tively soft cast-iron seat will wear out rather rapidly.

One method utilized by manufacturers to permit an engine to operate satisfactory on lead-free fuel is the induction hardening of the exhaust valve seats (Fig. 9-5). During engine production, the manufacturer heats, through the use of induction coils, all the exhaust valve seats within the cylinder heads to a temperature of 1,700 °F. Then, the seats are allowed to cool. This process hardens the seat to a depth of 0.05 to 0.08 inch, thus providing greater resistance to wear.

Modifications to Camshafts

Alterations were made to the camshaft of some engines in order to reduce NOx emissions (Fig. 9-6). A modified camshaft provides either a slight change in valve timing or extends valve overlap, which causes some dilution of the air/fuel mixture entering the engine on its intake stroke. This action lowers combustion efficiency and therefore peak operating temperatures. In other words, these alterations reduce the quality of the air/fuel mixture by not permitting complete purging of the exhaust gases from the cylinder. This, in turn, lowers the peak combustion chamber temperatures and, consequently, the formation of NOx.

Modifications to the Intake Manifold

To reduce CO levels, manufacturers have altered the design of the basic intake manifold in several ways. For example, to assure a more rapid vaporization of the air/fuel mixture during engine warm-up, the thickness of the manifold floor between the intake and exhaust runners

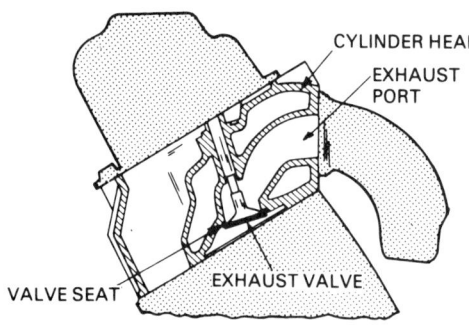

Figure 9-5. Induction-hardened valve seat within a typical six-cylinder engine (Courtesy of Chrysler Corp.).

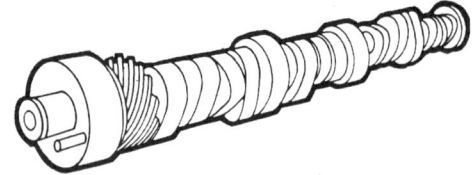

Figure 9-6. In order to reduce NOx emissions, some manufacturers modify the camshaft (Courtesy of Chrysler Corp.).

within the manifold was reduced. This lowered the time necessary for the exhaust gases to preheat the air/fuel mixture in the inlet runners. By increasing the heat transfer, the air/fuel mixture vaporizes faster. As a result, the engine can operate satisfactorily with leaner mixtures during its warm-up period.

Figure 9–7 shows another modification to an intake manifold. In this manifold, the manufacturer has separated the choke coil from the exhaust crossover passage by a stainless steel well. This well conducts exhaust passage heat faster than the one made from cast iron—the same metal as the manifold itself. This design causes faster choke opening and therefore a leaner warm-up mixture.

Figure 9–8 illustrates an alteration

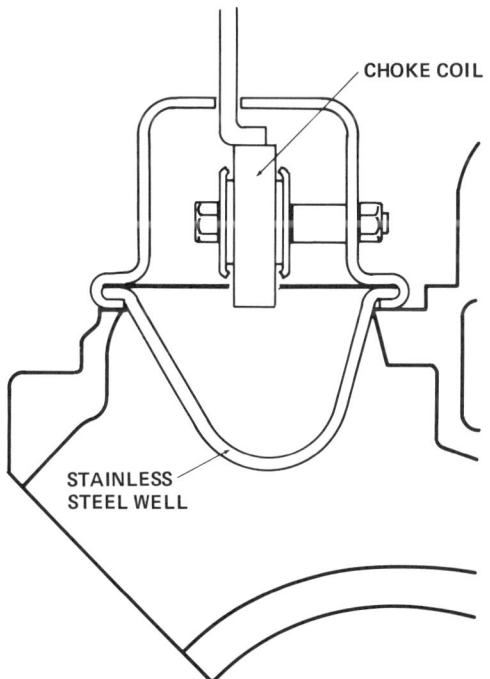

Figure 9–7. A stainless steel choke well conducts heat faster than one made of cast iron.

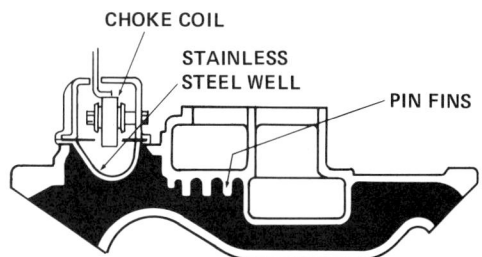

Figure 9–8. The addition of fins to the exhaust crossover passage, under the heat stove, increases heat transfer to the carburetor.

made to the exhaust crossover passage itself. In this situation, the manufacturer has added pin fins underneath the manifold's heat stove. These fins increase the heat transfer from the crossover passage to the carburetor's heat stove. This action improved fuel atomization, which permitted leaner air/fuel mixtures, which in turn improved combustion and lowered exhaust emissions.

Manufacturers also improved fuel delivery by changing the routing and length of the various intake manifold runners. Figure 9–9, for example, shows a single-plane intake manifold that replaces a two-level unit on some engine configurations. The single-plane manifold has all eight branches (runners) on the same level. With

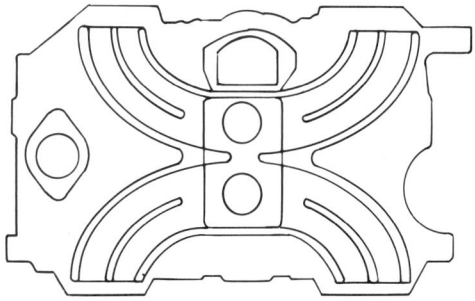

Figure 9–9. Single-plane manifold provides a more consistent air/fuel charge to the combustion chambers.

all the runners on the same level, cycle-to-cycle air/fuel delivery is more consistent, thus improving engine combustion and performance while reducing harmful emissions.

Runner length also has a lot to do with the efficiency of fuel flow in the intake manifold. Most emission-modified intake manifolds have runners of nearly equal length that actually carry the air/fuel mixture from the carburetor to the combustion chambers. As a result, a more equal air/fuel charge reaches each of the combustion chambers.

Combustion Chamber Shape Effects on Emissions

The shape of the combustion chamber has a direct effect on the efficiency of combustion and, therefore, the amount of emissions escaping from the engine during the exhaust stroke. Over the years, manufacturers have utilized many combustion chamber designs to improve fuel combustion and engine performance.

However, not all of these designs promoted a reduction in harmful emissions. Figure 9–10, for example, shows a wedge-type combustion chamber with its inherent quench areas. These quench areas are necessary, in many wedge-shaped combustion chambers, to reduce the possibility of detonation or spark knock.

In a wedge-shaped combustion chamber, the flame from the burning air/fuel mixture begins at the spark plug and then travels out in all directions. Finally, the flame front arrives at the chamber's quench area.

The compressed air/fuel mixture within the quench area is known as *end gas*. During the combustion process, this end gas is

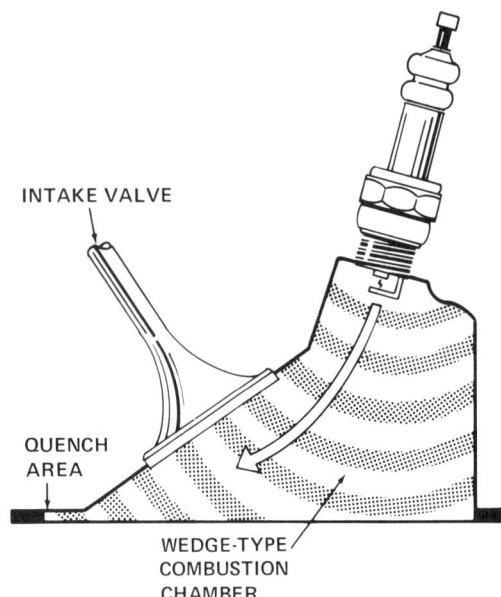

Figure 9–10. Flame front moving toward the quench area within a wedge-type combustion chamber.

subjected to increasing pressure and temperature, and it is this phenomenon that causes the end gas to detonate.

If the end gas explodes before the flame front reaches it, spark knock or detonation occurs. However, with the quench area in the chamber, the end gas loses heat as fast as the approaching flame front increases it. Therefore, the gas never gets hot enough to burn, which prevents detonation.

Next, the exhaust stroke of the piston sweeps the unburned air/fuel mixture out of the combustion chamber. Consequently, the exhaust gases from the engine contain hydrocarbons and carbon monoxide. Hydrocarbons come from the portion of the air/fuel charge within the quench area that did not burn at all. Whereas, CO emissions result from fuel within the mixture that started to burn but was not completely consumed in the process.

There are three general methods used to reduce the effects of quenching. The first, as already mentioned, is to open up these quench areas within the combustion chambers. The second is to design a combustion chamber that utilizes the principle of stratified charging, and the third is to reduce the surface area around the combustion chamber.

Stratified-Charge Combustion Chambers

In an engine designed to use the principle of stratified charging, the air/fuel mixture entering the combustion chambers is not uniform, before and while the pistons compress it. Instead, there are strata or layers that are relatively lean, and other strata or layers that are rich (Fig. 9-11).

The purpose of stratified charging of an engine is to concentrate the richest part of the air/fuel mixture around the spark plug, while the leaner portions of the charge are next to the cooler combustion chamber surfaces. The relatively rich mixture ignites first, and the resulting flame sets the leaner charge burning. As the flame front approaches the metal chamber surfaces, they quench it as already explained. However, the layers that do not burn have less hydrocarbons present; thus, the exhaust gases are relatively free of unburned fuel.

Therefore, on the average, an engine using stratified charging operates on a leaner air/fuel mixture. In addition, this same engine produces less HC emissions because there is less actual fuel concentrated within the mixture adjacent to the quench areas.

One way to achieve stratified charging is to provide the air/fuel mixture with a

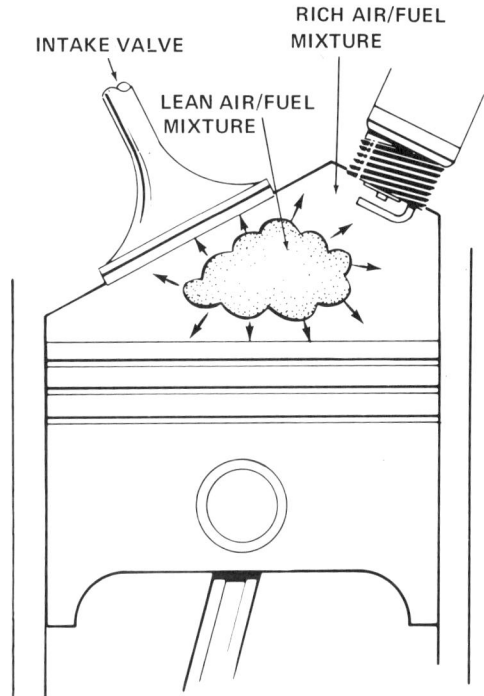

Figure 9-11. Principle of stratified charging.

swirling motion as it enters the cylinder. Manufacturers accomplish this by the careful placement of the intake valve and port within the combustion chamber (see Fig. 9-4). Supercharging an engine has about the same effect.

Another very efficient type of stratified-charge engine is the Honda Compound Vortex Controlled Combustion (CVCC) design. This particular engine incorporates a separate small combustion chamber located above the main chamber. This second mini chamber also contains a small intake valve and the spark plug (Fig. 9-12). Except for this feature, the CVCC engine is quite similar to any other four-stroke piston type.

However, the CVCC engine uses a two-stage combustion process during its normal four-stroke cycle. During its intake stroke (Fig. 9–12a), the carburetor delivers a very lean mixture to the main combustion chamber but supplies a very rich charge to the mini combustion chamber.

At or close to the end of the piston's travel on the intake stroke, both intake valves close (Fig. 9–12b). The piston then begins to travel upward within the cylinder and compresses both air/fuel charges in their respective chambers.

During the first stage of combustion, the spark plug ignites the rich air/fuel mixture in the small combustion chamber (Fig. 9–12c). The resulting flame front then moves down into the main chamber, where it ignites the leaner air/fuel charge (Fig. 9–12d).

Because the richest part of the total air/fuel charge begins burning first, there are less hydrocarbons in the mixture next to the cooler main combustion chamber walls. Consequently, there are less actual hydrocarbons not burned due to the quenching action, which of course lowers the engine's HC emission levels.

The expanding gases from the power stroke push the piston downward in the cylinder until it reaches bottom dead center (BDC) (Fig. 9–12e). At or near BDC, the exhaust valve opens.

Since there is still some pressure remaining in the spent gases within the combustion chamber above that of the atmosphere, the gases pour out of the open exhaust valve. However, as the piston begins its upward travel during the exhaust

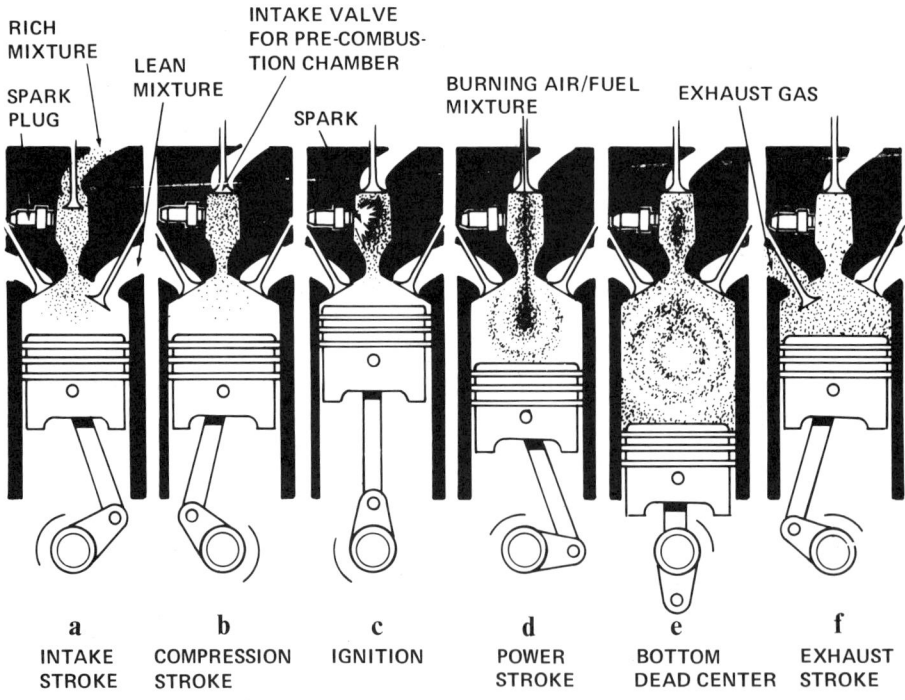

Figure 9–12. Design and operation of a Honda stratified-charge engine.

stroke (Fig. 9-12f) it also assists in pushing any remaining gases from the cylinder.

Reducing Emissions by Altering Combustion Chamber Area

The third method utilized to control the effects of quenching is the reduction in combustion chamber area. As previously mentioned, the surface of a combustion chamber is kept relatively cool by the engine's cooling system. These cooler areas quench the flame as it progresses through the air/fuel mixture; thus, the hydrocarbons in the fuel charge next to the metal never burn.

However, if the combustion chamber has a reduced surface area, there are less cool surfaces and therefore decreased quenching of the air/fuel mixture. Consequently, the engine emits less carbon monoxide and hydrocarbons within its exhaust gases.

In an engine with a hemispheric combustion chamber, the surface-to-volume (S/V) ratio is very low. This ratio is between the chamber's actual surface area S and its total volume V.

A sphere has the lowest possible S/V ratio while a wedge-shaped combustion chamber (see Fig. 9-10) has a relatively high S/V ratio.

The hemispheric combustion chamber, on the other hand, has a reduced surface area and therefore a low S/V ratio. Since with this design there is less surface to chill the air/fuel mixture and quench the flame front, the engine produces less HC and CO emissions in its exhaust gases (Fig. 9-13).

Jet Air System

Some late-model engines with hemispheric combustion chambers use a jet air system. This system has two functions. First, it provides air that scavenges the residual gases from around the spark plug to

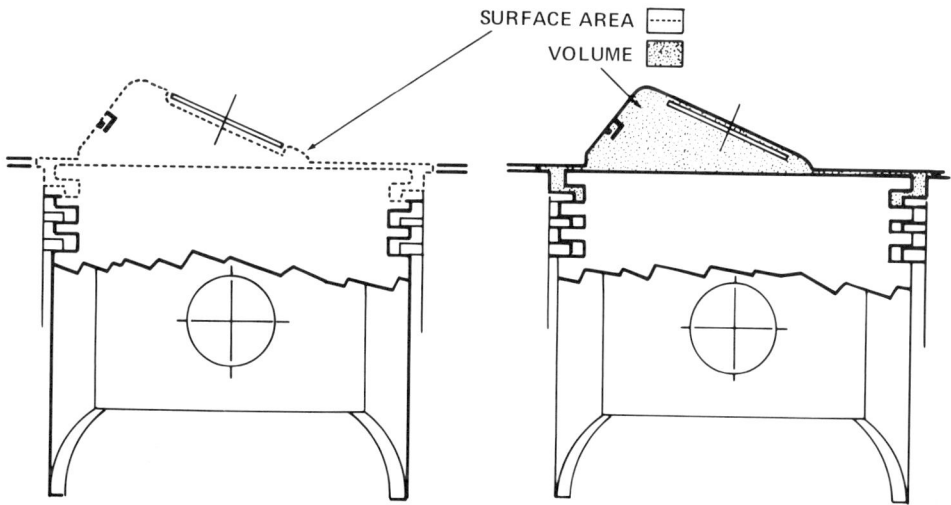

Figure 9-13. Hemispheric combustion chamber has a low S/V ratio.

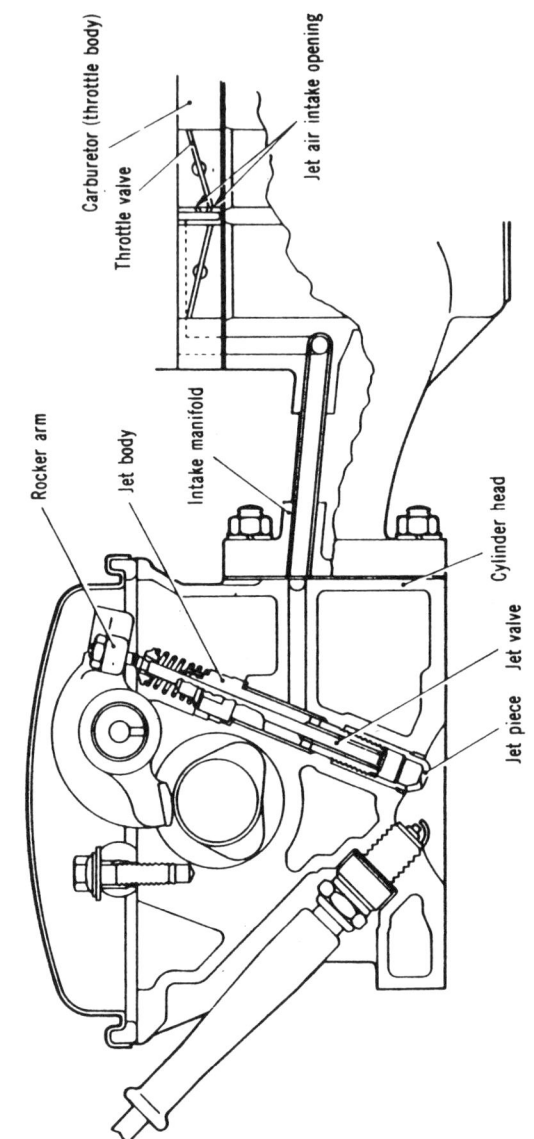

Figure 9-14. Components of a jet air system (Courtesy of Chrysler Corp.).

create a good ignition condition. Second, the system produces a swirl in the gases within the combustion chamber that continues throughout the compression stroke. This action improves the spreading out of the flame front after the initial ignition of the air/fuel mixture, thus assuring high combustion efficiency.

In addition to the regular intake and exhaust valves, an engine with a jet air system has a jet valve (Fig. 9–14). The jet valve controls the movement of air from a special carburetor passage into the combustion chamber.

To perform its function, the jet valve assembly consists of a jet valve and spring, jet body, along with a jet piece. The jet valve and its return spring fit inside a special bore within the jet body and are held in place by a retainer lock. The jet body has threads that screw into the jet piece, which is a press fit in the cylinder. The jet piece also has an opening that points toward the spark plug.

The same cam on the camshaft that operates the intake valve also actuates the jet valve through a common rocker arm. Therefore, the jet and intake valve open and close together (Fig. 9–14).

In addition to the other components, there is a special jet air passage provided in the carburetor, intake manifold, and cylinder head. When the engine is operating, air flows through the two intake openings provided near the primary throttle valves of the carburetor and moves through the passage in the intake manifold and head, past the jet valve and jet piece opening, and into the combustion chamber.

Jet Air System Operation

During the engine's intake stroke, the air/fuel mixture flows through the intake valve port and into the combustion chamber (Fig. 9–15). At the same time, additional air moves into the combustion chamber through the open jet valve. This air passes into the chamber because of the pressure difference produced between the two ends of the jet air passage (between the jet air opening in the carburetor throt-

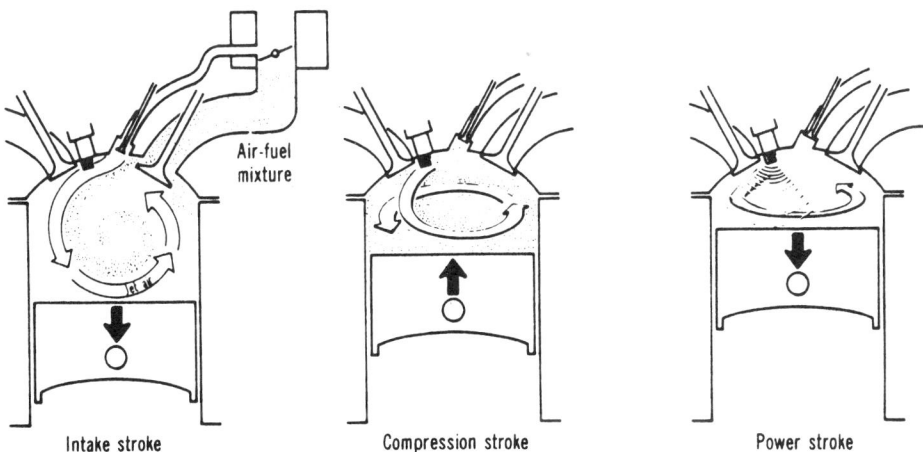

Figure 9–15. Operation of a jet air system (Courtesy of Chrysler Corp.).

tle bore and the opening in the jet piece) as the piston moves down.

When the throttle valve opening is small (during idle or light engine loads), a large difference in pressure exists as the piston moves down. This causes jet air to flow rapidly into the combustion chamber.

However, as the throttle valve opening increases and the engine accelerates, more air/fuel mixture passes through the intake valve port, so there is a reduction in the pressure difference between the two points in the jet air system. Consequently, less air moves through the jet air system and into the combustion chamber.

In other words, the action of the jet air system in creating a swirling condition within the combustion chamber dwindles with an increase in throttle valve opening. However, under these conditions, the intensified flow of the normal air/fuel mixture can now satisfactorily promote adequate combustion to reduce exhaust emissions at higher engine speeds.

Carburetor Modifications

The carburetor on an emission-controlled engine has modifications to help reduce the amounts of CO and HC emissions during the various phases of engine operation. The actual number of these carburetor design changes depends upon the design and size of the engine as well as the year model of the vehicle. In other words, these factors, for the most part, determine the allowable CO and HC emissions by a given vehicle. Therefore, some carburetors have more modifications than others.

Since all vehicle manufacturers have had to make some design changes in their carburetors, it would be impossible in the space provided to cover them all. Therefore, we present an overview of some of the main carburetor alterations within the following circuits: idle, high-speed light-load, and choke.

Idle Circuit Modifications

Mixture Adjustment Screws

To maintain a leaner air/fuel mixture during idle engine operation, a carburetor will have modifications such as fixed idle mixture adjusting screws and/or idle circuit restrictions, hot-idle compensator, and idle enrichment system. Some emission-equipped vehicles have idle mixture adjusting screws within the carburetor that the mechanic can only turn out a given amount. In other words, the manufacturer slots the threaded shanks of these screws and installs a lock pin into the carburetor casting. This prevents the turning out of the screws past a given point (Fig. 9–16).

However, there is a drawback to using this method of limiting the idle mixture. That is, the mechanic also cannot easily remove the screws to clean the idle circuit during a carburetor overhaul.

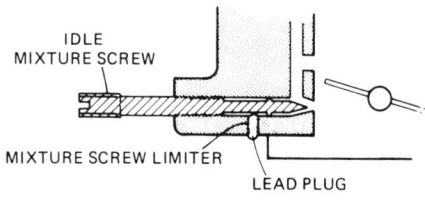

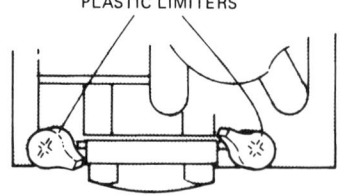

Figure 9–16. Limiter devices prevent the excessive movement of the mixture adjusting screws.

CARBURETOR MODIFICATIONS

Figure 9-16 also shows a carburetor with idle limiting caps installed over the mixture adjusting screws. These plastic caps perform about the same function as the stop pins on other carburetors. That is, the caps permit only about seven-eighths of a complete turn of the screws. With this arrangement, the vehicle's air/fuel mixture at idle is set to a precise amount at the factory. Then, these caps are installed over the adjusting screws to prevent the changing of the factory setting.

After their initial installation, a technician should only remove these caps for one of two reasons. First, it would be necessary to take the caps off in order to remove the adjusting screws during a carburetor overhaul. Second, the caps can also be removed if it becomes necessary to readjust the engine's idle mixture, beyond what is allowable by the limiter caps, to meet a given emission standard. However, in either case, the mechanic must put the caps back on after performing the carburetor adjustments.

Idle Circuit Restrictions

An alternate method of limiting the amount of idle enrichment is a restrictor placed into the idle circuit (Fig. 9-17). The restrictor may be used by itself or in conjunction with a fixed mixture adjusting screw or one with a limiter cap. In either case, the restriction helps to regulate the fuel flow more precisely into the engine during idle operation.

Hot-Idle Compensator Valve

Many carburetors now used on exhaust emission-controlled engines have a hot-idle compensator valve (Fig. 9-18). This particular valve allows additional air to enter the engine from below the throttle valve to prevent hot-idle stalling. This condition occurs frequently on exhaust emis-

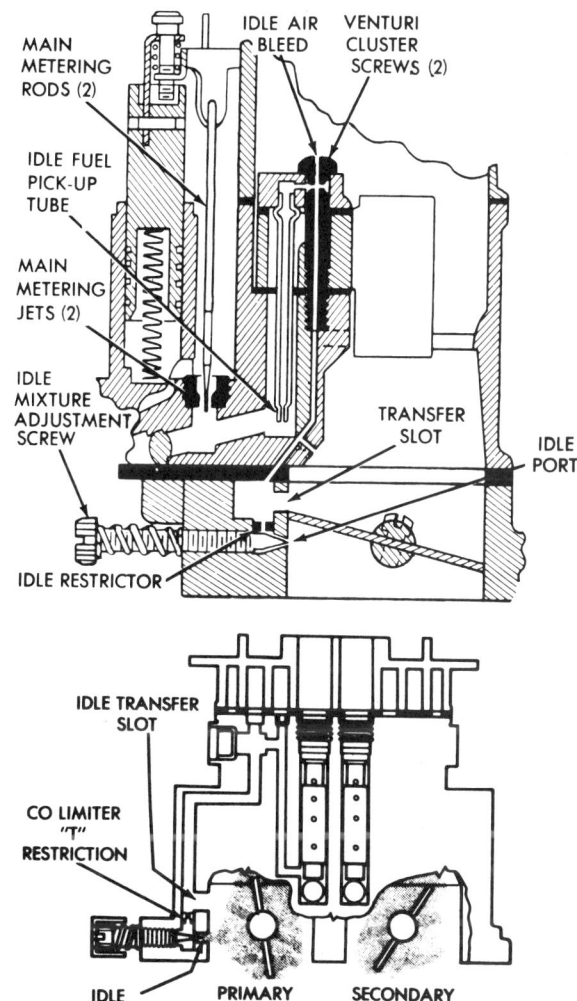

Figure 9-17. An idle circuit restrictor regulates the maximum amount of fuel flow into the engine during idle operation (Courtesy of Chrysler Corp.).

sion-controlled engines due to their higher underhood operating temperature.

Whenever the underhood temperature is high, the fuel in the carburetor's float bowl can boil; and the resulting fuel vapors can flow into the air horn via the internal vent or balance tube. This makes the

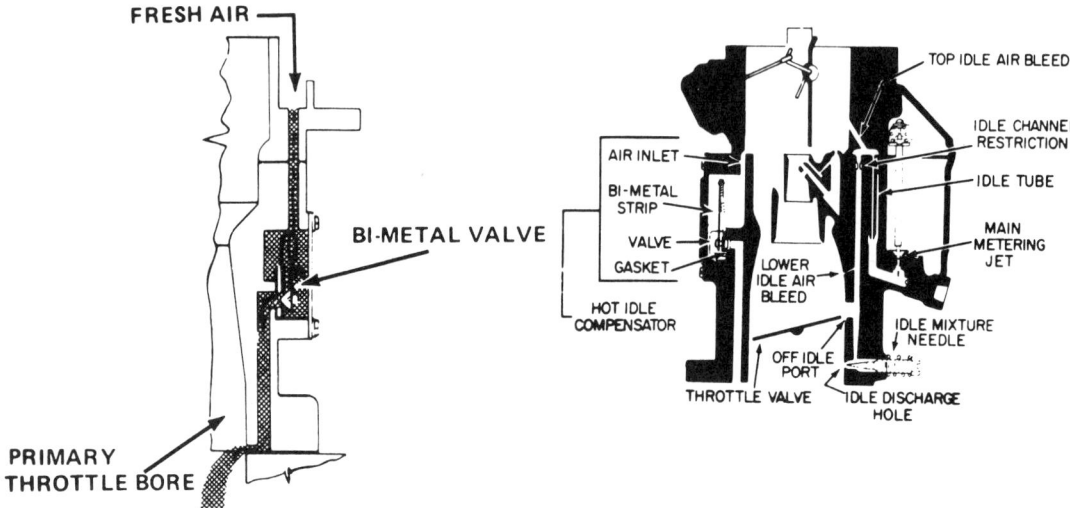

Figure 9–18. Typical hot-idle compensator valve (Courtesy of General Motors Corp.).

idle mixture too rich, causing a rough idle and raising both the CO and HC emission levels.

In operation, the bimetal strip closes the attached valve when underhood temperatures are below a predetermined amount. In this valve position, the normal air flow for the idle circuit passes around the edges of the throttle valves or through an air passage designed for this purpose.

However, when the underhood temperature reaches a preset amount, the bimetal strip lifts up and opens the attached valve. This action allows some additional air to pass through an extra air inlet and mix with the overrich air/fuel mixture.

This extra air flow serves two functions. First, it leans out the mixture to smooth out the engine idle and, at the same time, lowers both CO and HC emissions. Second, the leaner air/fuel mixture causes the engine speed to increase somewhat, stimulating the coolant flow, which assists in reducing the temperature. The valve once again closes as the underhood temperature decreases.

Idle Enrichment Systems

Some modern carburetors have an idle enrichment system (Fig. 9–19). This system increases the enrichment of the normal idle air/fuel ratio in order to reduce cold-engine stalling. The system itself consists of an idle-enrichment diaphragm and valve along with a thermo switch. The diaphragm receives its vacuum signal from the intake manifold through the thermostatic vacuum switch.

The thermo switch senses engine coolant temperature. For example, when the engine is cold, the switch is open; consequently, it permits passage of vacuum to the diaphragm, which moves a valve that blocks the idle enrichment air bleed. As a result, the idle circuit vacuum signal to the well increases. This causes additional fuel flow into the idle circuit.

During engine warm-up, the thermo

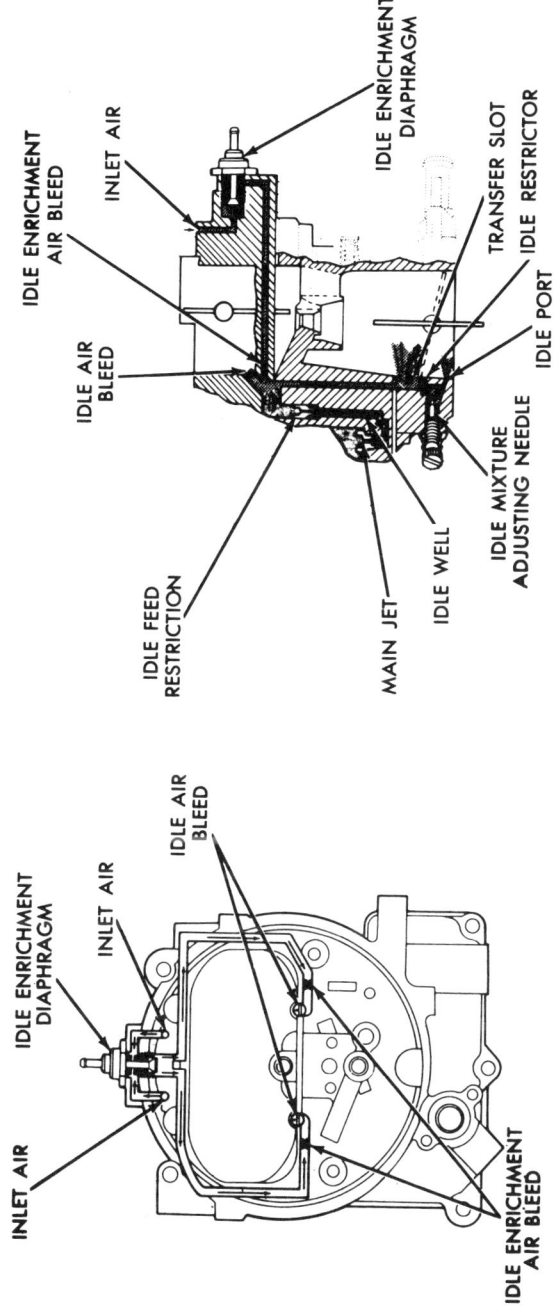

Figure 9-19. Carburetor with an idle enrichment system (Courtesy of Chrysler Corp.).

159

switch closes. Thus, it cuts off vacuum to the diaphragm and valve. This action opens the idle enrichment air bleed that, in turn, reduces the idle circuit vacuum signal to the idle well. Therefore, there is a decrease in fuel flow through the idle circuit.

Modifications to the High-Speed Light-Load Circuit

In some carburetors, manufacturers also provide a factory adjustment for the part-throttle (high-speed light-load) circuit. This adjustment is made at the factory to provide leaner mixtures during part-throttle engine operation. In the carburetor shown in Fig. 9–20, the part-throttle adjustment consists of a pin pressed into the side of the power piston and extends through a slot in the wall of the piston's bore.

When the power piston is down in its bore (economy position), the pin stops on top of the flat surface of an adjustment screw located in a cavity next to the power piston. This adjustment screw is stopped from turning by the tension of a spring beneath the head of the screw.

As previously stated, the adjustment screw is preset at the factory. During a production flow test, a technician moves the screw up or down, which in turn places the tapered metering rods operated by the power piston in the proper position in their respective metering orifices. This results in an air/fuel mixture sufficiently lean to meet emission control standards. This particular adjustment screw is preset at the factory and no attempt should be made to ever change the adjustment. If for some reason a float bowl replacement is necessary, the new assembly will have a preset adjusting screw.

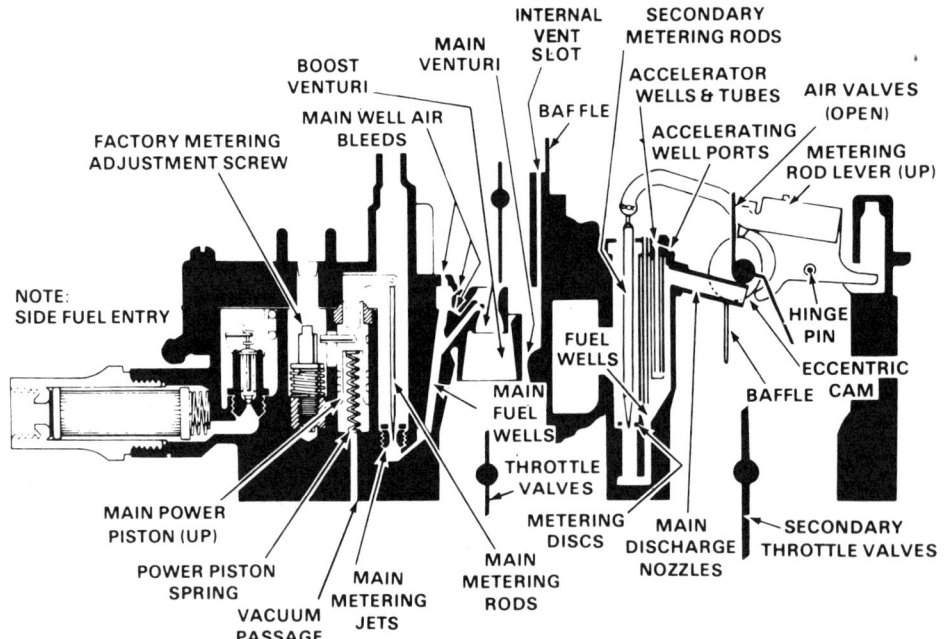

Figure 9–20. Factory-adjusted power piston (Courtesy of General Motors Corp.).

Altitude Compensation

Many carburetors built since 1975 have an altitude-compensating device in their main (high-speed light-load) circuit. The purpose of this device is to maintain, as closely as possible, the same air/fuel mixture when a vehicle is operating at high altitudes as it has at sea level.

Atmospheric pressure is greatest at sea level, but it reduces with increases in altitude. Consequently, there is less air entering the carburetor at higher altitude. As a result, the engine operates richer as the vehicle climbs to areas above sea level, resulting in poor driveability and high CO emissions.

While some altitude-compensating devices are mechanical, the most widely used is an automatic device with an aneroid bellows. An aneroid bellows is an accordion-shaped device that responds to changes in atmospheric pressure by expanding or contracting. For example, as a vehicle travels into a high-altitude area, the pressure decreases and the bellows expands. However, at sea level, the bellows contracts.

Figure 9-21 is a schematic of a Carter Thermo Quad carburetor that has a pressure-sensitive aneroid bellows. The bellows in this particular carburetor operates an air valve that opens an auxiliary air tube into the main metering system at high altitude. This action leans out the air/fuel mixture from the main metering system in proportion to the altitude the vehicle is operating at.

Choke Circuit Modifications

To control CO and HC emissions during engine warm-up, most manufacturers modify the automatic choke circuit on their respective engines. These modifications basically regulate just how long the choke valve remains on during engine warm-up. Although the modifications are somewhat different between automotive manufacturers, the design changes all do

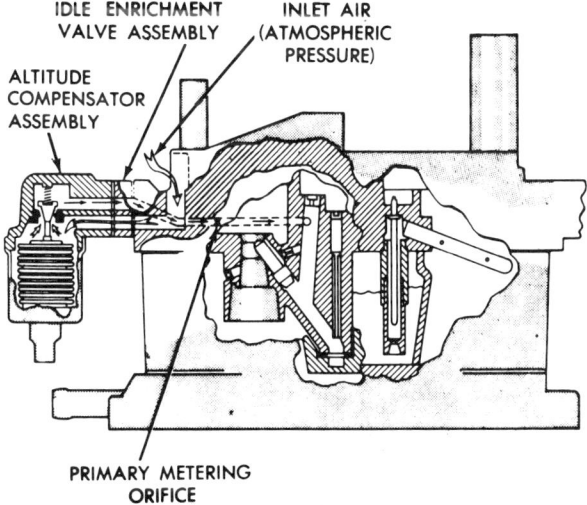

Figure 9-21. Carburetor with a pressure-sensitive aneroid bellows (Courtesy of Chrysler Corp.).

ENGINE MODIFICATION SYSTEMS

the same thing—supply heat faster or for longer periods to the choke valve's bimetallic spring. This opens the choke valve faster and permits it to stay open for longer periods even when the engine is shut down.

The most common modified-type choke spring heaters in use to control emissions are the electrical-assist, coolant-assist, and the choke hot-air modulator. Figure 9-22 shows a carburetor-mounted electrical-assist choke circuit. In this circuit, the bimetallic spring that closes the choke valve when the engine is cold receives its heat from two sources: exhaust manifold heat and a ceramic heater. For instance, when the engine is first started and cold, the bimetallic spring receives heat only from the exhaust manifold. However, as engine temperature reaches a predetermined amount, the electrical-assist choke mechanism turns on to supply additional heat, which opens the choke valve faster.

To accomplish this task, the heating element connects in series with a temperature-sensitive thermostatic switch. This switch has contacts that are open below about 60 °F (15 °C) to prevent the electrical-assist mechanism from operating. At this time, the choke's bimetallic spring receives its heat only from a pipe connected to the manifold.

When engine temperature rises to about 60 °F (15 °C) either during engine operation or starting, the temperature-sensitive contacts close. As a result, the choke heating element becomes operational and provides additional heat to quickly open the choke valve.

The electric power for this type of choke

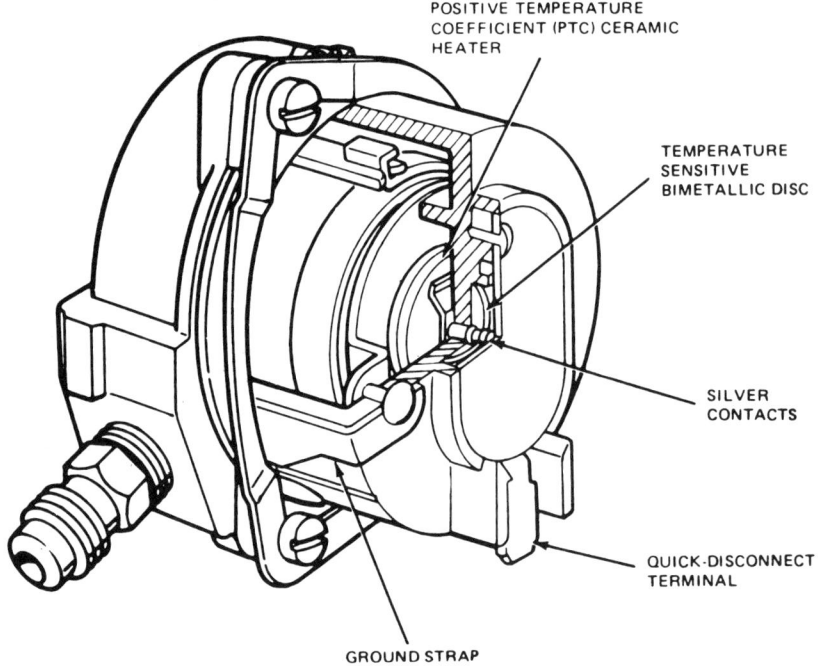

Figure 9-22. Typical carburetor-mounted electrical-assist choke system (Courtesy of Ford Motor Co. of Canada Ltd.).

circuit usually comes from the alternator. Therefore, the mechanism only receives power when the engine is operating, and the alternator is producing voltage and current flow. However, there are some carburetor-mounted electrical choke circuits that operate continuously. These units usually receive their power from the ignition switch whenever it is on.

Figure 9–23 shows an engine-mounted electrical choke heater. This assembly fits into a recess within the intake manifold and receives heat from the exhaust crossover passage at all times. But the device also has a heating element that activates at a given engine temperature to open the choke valve quickly.

In operation, with the engine coolant temperature below 80 °F (27 °C), the bimetallic choke coil reacts only to exhaust heat from the crossover passage within the manifold itself. However, as the engine coolant temperature reaches approximately 80 °F, a control switch mounted on the engine turns on the heating element to cause a fast and positive choke opening. This control unit then turns the heating element off after about 5 minutes of operation.

In some installations, a dual-acting oil-pressure switch controls the operation of the electrical-assist choke. In this instance, engine oil pressure closes the switch contacts, permitting current to flow through the control switch to the choke heater. However, when the engine is off, the pressure switch opens, preventing the electric choke heater from operating. This action prevents the choke heater from prematurely opening the choke valve if the ignition key has been left on for a period of time before the engine starts.

Coolant-Assisted Choke Mechanisms

For many years, some manufacturers have utilized engine coolant to heat the bimetallic spring of carburetor-mounted choke assemblies. In some earlier designs, these assemblies used coolant along with exhaust gas temperature to heat the bimetallic spring that opens the choke valve. However, in some installations, coolant alone was the heat medium. Now, many coolant-assisted units work in conjunction with an electrical heater.

In any case, coolant passes through a special compartment in the choke housing (Fig. 9–24). The housing in turn becomes hot as the engine reaches normal operating temperature. This causes the enclosed bimetallic spring to unwind and open the choke valve.

The main benefit derived from using a coolant-operated choke is that this unit reduces overchoking of the engine after it has been shut down for a short period. The reason for this is that the coolant holds heat longer than air; consequently, the choke coil remains warm for an increased

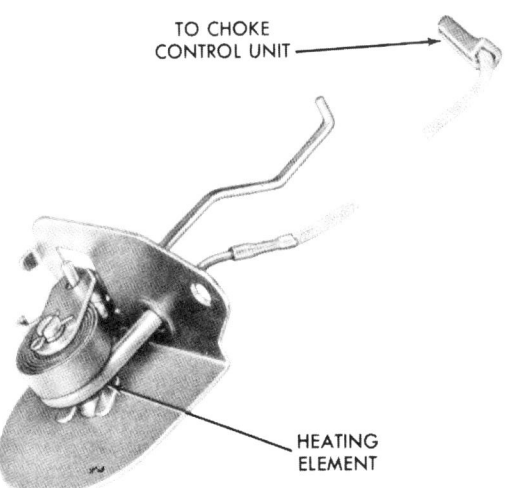

Figure 9–23. Electrical-assist choke heater mounted on the intake manifold (Courtesy of Chrysler Corp.).

period of time. Therefore, the engine receives a lesser amount of choking action on restart.

Choke Hot-Air Modifier

A few later-model General Motors engines have a modified version of an automatic choke system, heated alone by exhaust gas temperature (Fig. 9-25). In this particular system, the heat for the automatic choke assembly comes from a heat stove mounted into the intake manifold around which exhaust gases flow. The heat from the exhaust gases radiate into the tubing that forms the stove itself.

A small amount of vacuum, applied to the choke housing, draws the heated air from the stove and into the housing. This air in turn passes around the bimetallic spring that causes the choke valve to open.

The main difference between this later system and the earlier type is that the early assemblies permitted cool air from outside the air cleaner to pass through the heat stove tubing, where it received heat from the exhaust gases. However, in the modified system, the manufacturer connects the open stove tube to the air cleaner.

Connected to the end of this cool air hose fastened to the air cleaner is a hot-air

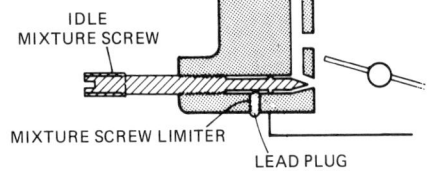

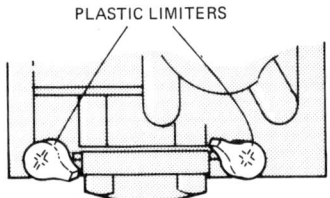

Figure 9-24. Typical coolant-assisted carburetor-mounted choke assembly.

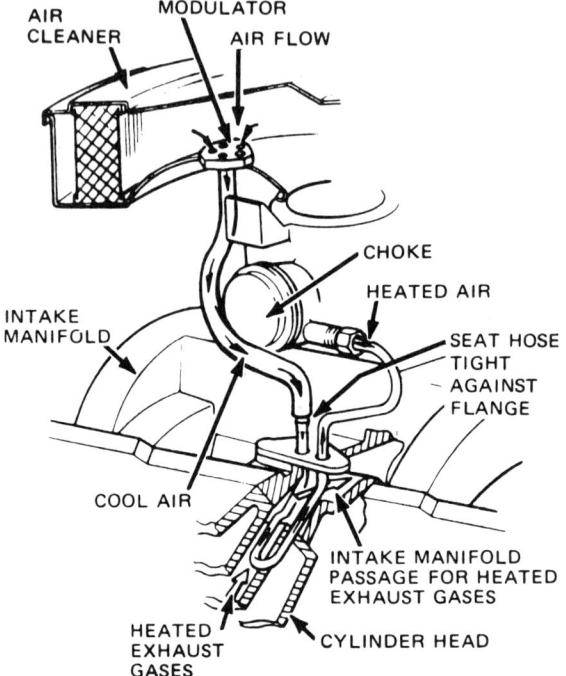

Figure 9-25. Hot-air modified choke system, used on a carburetor-mounted automatic choke (Courtesy of General Motors Corp.).

modifier check valve (CHAM-CV). This valve controls just how much air passes through the hose to the heat stove during periods of differing air cleaner temperatures. For example, at air cleaner temperatures below about 68 °F, the check valve closes. Consequently, the air that enters the heater tubing must pass through a tiny hole in the modifier. This action restricts the hot air flow over the bimetallic coil and results in a slower opening of the choke valve during initial engine warm-up.

As the temperature within the air cleaner rises above about 68 °F, the modifier check valve opens to permit more air flow. This permits the bimetallic coil to now begin unwinding faster, which in turn accelerates the opening of the choke valve.

Tamperproof Automatic Chokes

Many late-model carburetors now come from the factory with tamperproof automatic choke mechanisms. In other words, the factory adjusts all choke-related parts and then installs devices that prevent an easy adjustment in the field.

Figure 9–26 shows a typical tamperproof automatic choke mechanism. In this particular unit, the choke cover is riveted onto the choke body; therefore, a mechanic cannot alter choke bimetallic spring tension on the valve without removing the rivets.

This mechanism also has a stroke adjusting screw. This screw is necessary to adjust the initial position of the choke valve. However, the manufacturer coats the head of this screw with wax to prevent just anyone from disturbing the factory adjustment.

There are only two reasons for tampering with these devices. First, a mechanic will have to at least remove the rivets and choke cover for major carburetor service or overhaul. Second, in some cases, it may be necessary to move the stroke adjusting screw to change the choke valve calibration to meet local or state emissions standards.

Electronically Controlled Carburetors

The carburetor shown in Fig. 9–27 is an electronically controlled unit, utilized on a V-8 engine that has a specially designed

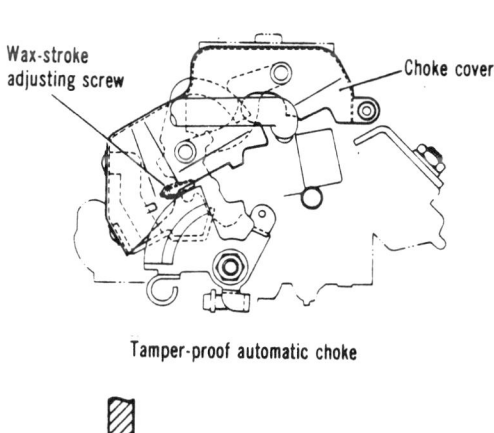

Tamper-proof automatic choke

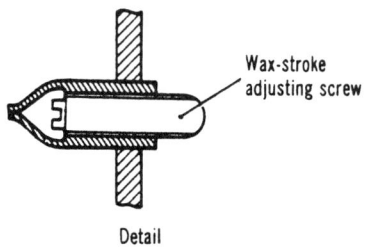

Detail

Figure 9–26. Tamperproof automatic choke mechanism (Courtesy of Chrysler Corp.).

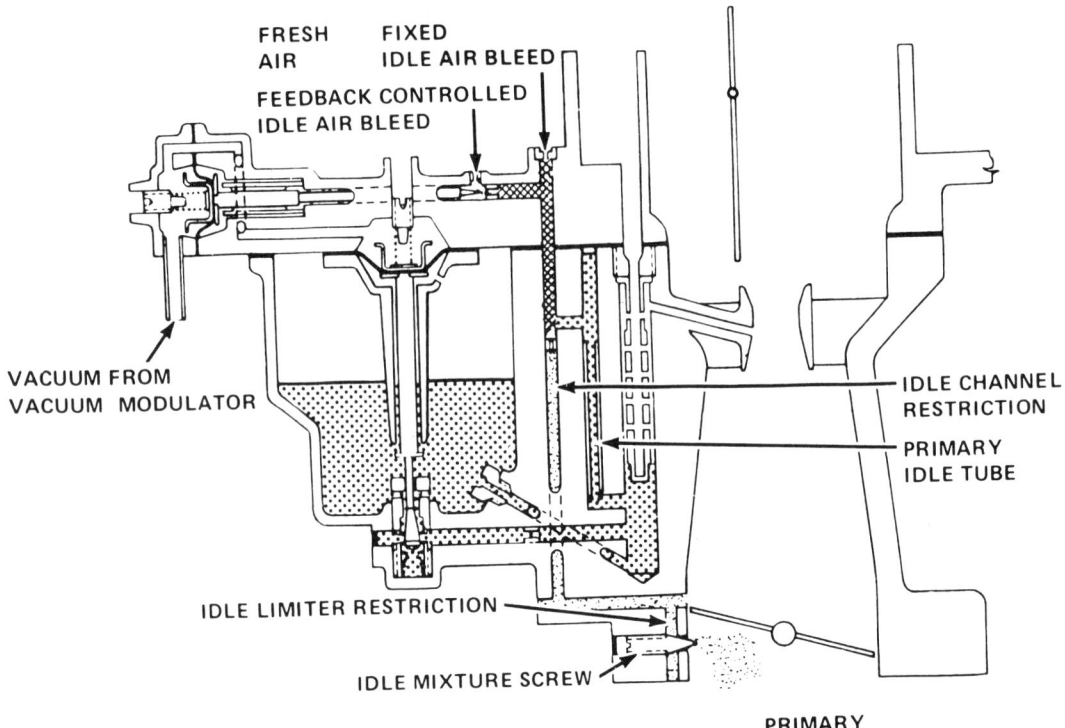

Figure 9-27. A carburetor that incorporates an electronic control of the air/fuel mixture from the idle and main metering systems (Courtesy of General Motors Corp.).

catalytic converter. This carburetor has vacuum diaphragms in the idle and main metering circuits, which provide precise control of the air/fuel mixture from these particular systems. This action reduces both CO and HC emissions from the vehicle's exhaust.

Figure 9-28 illustrates the remaining components of the electronic fuel control system. These units include an oxygen sensor in the exhaust pipe or manifold, a manifold vacuum switch, a coolant temperature switch, a vacuum modulator, and an electronic control unit (ECU).

In operation, signals from the oxygen sensor, manifold vacuum switch, and coolant temperature switch are fed to the ECU. The ECU in turn directs a signal to the vacuum modulator indicating just how much vacuum to feed to the vacuum diaphragms within the carburetor's idle and main metering circuits. The diaphragms react to the incoming vacuum signals and provide a precise control of the intake air/fuel ratio required by the catalytic converter for good emission control.

Carburetor-Assist Devices

Along with the various circuit changes within the carburetor itself, manufacturers have also added one or more assist devices onto the carburetor. These units primarily serve the purpose of further reducing exhaust emissions or improving the drivea-

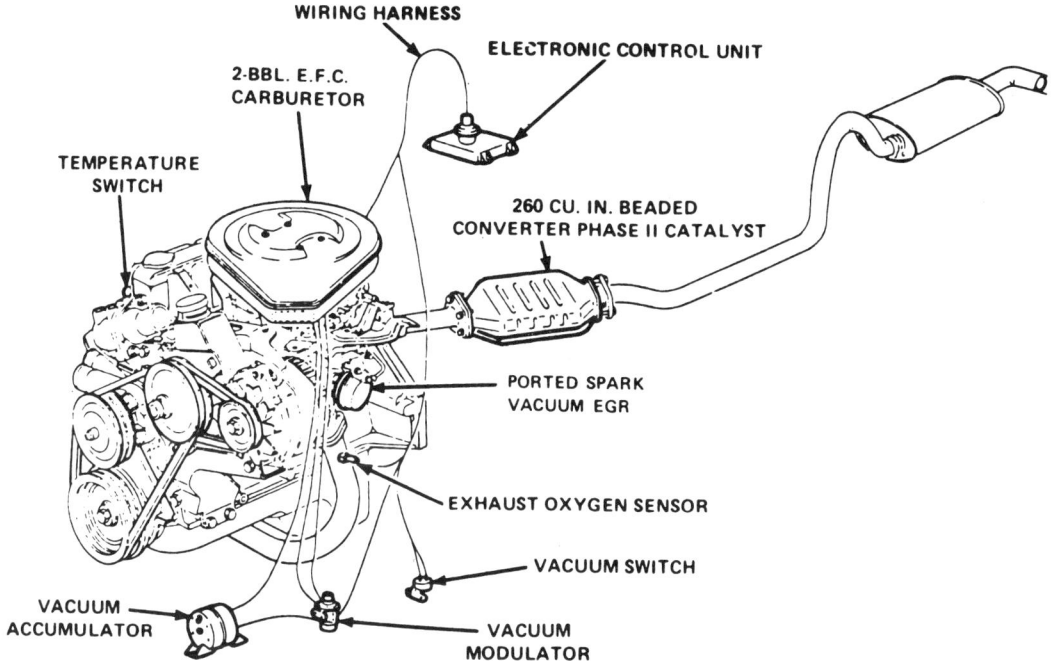

Figure 9-28. Components that support the electronic fuel control system (Courtesy of General Motors Corp.).

bility of the vehicle. These assist devices include such units as a choke vacuum brake, choke-delay valve, deceleration valve, dashpot, antidieseling or air-conditioning solenoid, and a jet valve.

Choke Vacuum Brake

Almost all automatic carburetor-mounted chokes have a metal vacuum piston that opens the choke slightly against bimetallic spring tension after the engine starts. This action provides the additional air flow through the carburetor, thus preventing the engine from stalling due to an overly rich mixture.

However, engines which have a manifold-mounted bimetallic spring (see Fig. 9-23) use a vacuum brake assembly for the same purpose (Fig. 9-29). This unit is nothing more than a small diaphragm and spring mounted inside a plastic or metal housing. The diaphragm has an attached rod that connects through linkage to a choke valve shaft. Consequently, any movement of the diaphragm will cause the rod and attached linkage either to open the choke valve slightly against bimetallic spring tension or permit the spring's tension to close the valve.

The diaphragm chamber on one side connects via a fitting and hose to the intake manifold; the other side is open to the atmosphere. Therefore, when the engine is operating under moderately low-load conditions, its vacuum pulls the diaphragm to the left. This action causes the diaphragm rod and its attached linkage to open the choke valve slightly against bimetallic spring tension. As a result, additional air

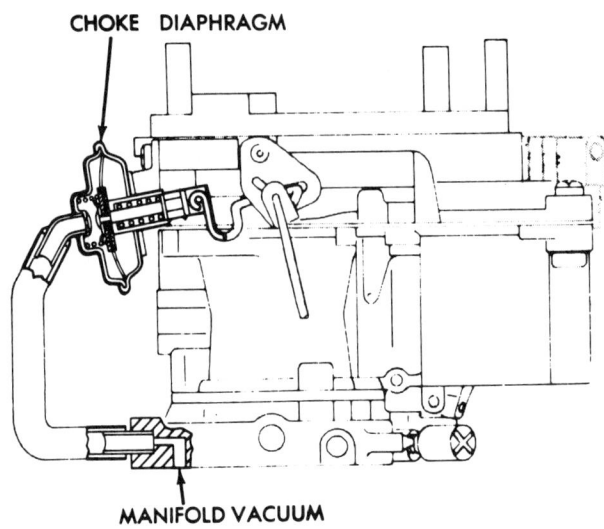

Figure 9–29. Typical vacuum brake assembly (Courtesy of Chrysler Corp.).

flow enters the carburetor through the air horn to prevent stalling while the engine and bimetallic spring are cold.

If engine vacuum drops to zero or a preset value as during acceleration, the spring on the vacuum side of the diaphragm pushes it more toward the right (Fig. 9–29). This movement through the diaphragm rod and linkage reduces the pulling action on the choke shaft. Consequently, if the bimetallic spring is still cold, it will once again slightly close the choke valve. When the engine is cold, this action enriches the air/fuel mixture necessary for cold-engine acceleration. However, if the choke's bimetallic spring is hot and the choke valve is already open, the decreased vacuum on the diaphragm has no affect on the valve's position.

On some late-model Rochester carburetors, there are two vacuum brake diaphragms to provide better mixture control during engine warm-up (Fig. 9–30). These two assemblies provide more choking of the engine when it is cold and less as it warms up. Thus, the devices reduce emissions during this period of engine operation.

Choke-Delay Valve

Many modern choke circuits with either a vacuum piston within the carburetor-mounted assembly or a vacuum brake, like the one just described, have a choke-delay valve. Figure 9–31 illustrates one version of this type of valve. The valve itself delays the opening of the choke valve for a period of time to improve driveability of the vehicle during cold-engine warm-up.

The delay valve fits into the vacuum hose between the intake manifold and either the line to the choke vacuum piston or vacuum brake. Inside the valve, there is a sintered-steel fluidic restrictor that filters intake manifold vacuum to either one of these units, making it take longer for them to act on the choke valve. As a result, vacuum must be in the hose for 15 to 30 seconds before it affects the operation of

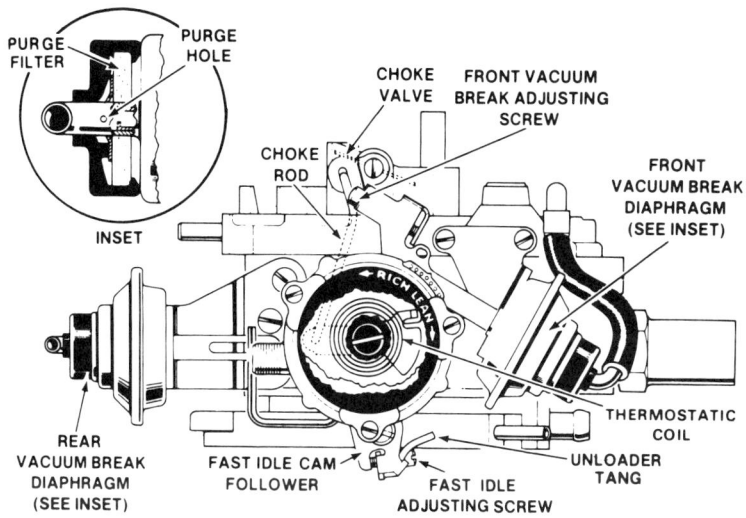

Figure 9–30. A carburetor that uses two vacuum brake assemblies to control choke operation (Courtesy of General Motors Corp.).

either the choke piston or brake diaphragm.

The check valve within this unit permits immediate release of the vacuum to let the choke close quickly, anytime the vacuum in the manifold drops to zero. This action is necessary to permit the choke valve to close and provide an enriched air/fuel mixture if the engine should stall when cold. If the choke does not close under these circumstances, the engine can be hard to restart.

Deceleration Valve

The carburetor used on some late-model engines has a deceleration (decel) valve (Fig. 9–32). This valve prevents the engine cylinders from misfiring during deceleration and sending an unburned charge of hydrocarbons through the exhaust and into the atmosphere. The valve accomplishes this task by providing additional air and fuel to the combustion chambers during this phase of engine operation.

The valve shown in Fig. 9–32 fits onto or near the intake manifold. One end of the valve connects into the manifold—either directly or by means of a hose. The other end connects to a deceleration section in the carburetor, which contains a fuel pickup tube and an air bleed.

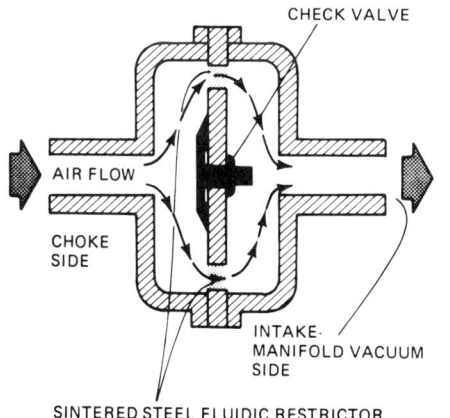

Figure 9–31. Choke-delay valve slows the opening of the choke valve.

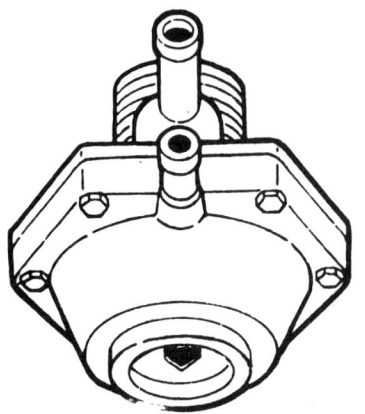

Figure 9-32. Common type of decel valve (Courtesy of Ford Motor Co. of Canada Ltd.).

A diaphragm within the decel valve reacts to intake manifold vacuum via a vacuum fitting. For example, when a given engine decelerates, its intake manifold vacuum is high; and the spring-loaded diaphragm opens the decel valve within the assembly. This action permits additional air and fuel to pass through the valve, as long as the vacuum remains above 20 inches Hg. The time period involved in the process is about 5 seconds during normal engine deceleration.

When engine vacuum drops back down to normal, the diaphragm spring seats the decel valve. This cuts off the additional air and fuel entering the engine, and the normal carburetor circuitry once again provides the necessary mixture for proper engine operation.

Dashpot

Manufacturers have installed dashpots on vehicles for many years. When used on vehicles with automatic transmissions, this device primarily keeps the engine from stalling due to an overly rich mixture on rapid engine deceleration. In other words, if the driver snaps the throttle valve open and then closed, the engine receives a very rich mixture before its air flow is cut off. This situation creates an unbalanced condition in which there is too much fuel and insufficient air. Therefore, the engine stalls.

On vehicles with a standard transmission, this problem does not happen because this transmission provides a direct connection between the drive axles and the engine; so the momentum of the vehicle keeps the engine running. However, under these conditions, the engine produces high HC exhaust emissions.

A dashpot takes care of both problems by controlling the rate at which the throttle valve closes. In other words, the dashpot prevents the throttle valves from closing immediately after the driver releases the accelerator pedal. As a result, the engine naturally receives enough air to consume the rich mixture without stalling or producing excessive HC emissions.

Although manufacturers have utilized hydraulic and magnetic dashpots, the most common one consists of a spring-loaded diaphragm inside a housing (Fig. 9-33). A rod attached to the front of the diaphragm extends from the metal housing and contacts the throttle valve arm. When the throttle valve opens during vehicle acceleration, the arm moves away from the rod;

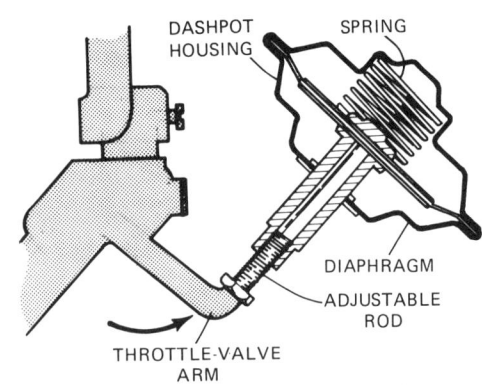

Figure 9-33. Typical throttle return dashpot.

the diaphragm spring forces the rod from the housing as far as possible.

As the spring moves the diaphragm and rod outward, air enters the housing behind the diaphragm through an air-bleed hole. Then, when the throttle valve closes during deceleration, its arm contacts the rod again. However, in order for the throttle valves to close completely, the air must force the rod back into the housing.

In order for this to occur, the throttle arm must overcome two obstacles: the tension of the spring and the air within the housing. The force of the spring is easy enough to overcome because it has only a small tension, just sufficient to extend the rod. However, a greater force is necessary to cause the rod and attached diaphragm to push the air out of the housing through the small bleed hole.

The hole itself acts as a restriction in this case. Therefore, the air escapes slowly. This action retards the movement of both the arm and throttle rod, which in turn slows the closing action of the throttle valves.

Antidieseling Solenoid

To reduce the effect of engine run-on or dieseling, many vehicles now utilize an antidiesel solenoid. *Dieseling* or run-on in a gasoline engine is a condition in which extreme conbustion chamber heat continues to ignite excess fuel after the driver turns off the ignition switch. As a result, the engine wants to keep running.

Dieseling of late-model emission-controlled engines is the direct result of higher overall operating temperatures, lean air/fuel mixtures, and in some cases, retarded ignition timing at idle. All of these factors bring about very high combustion chamber temperatures—high enough at times to ignite an air/fuel mixture.

In addition, due to the leaner air/fuel mixtures, late-model engines must idle faster or they will stall, so the throttle valves are open slightly for curb-idle operation. Consequently, engine vacuum can continue to pull a fresh charge of air/fuel mixture from the carburetor's idle port, even with the ignition switch off, as long as the engine is turning over. The high temperatures within the combustion chambers ignite this mixture, and the engine continues to run.

The antidiesel solenoid (Fig. 9-34) overcomes this condition by allowing the throttle valves to close beyond the normal curb-idle position. To do this, the solenoid provides the normal stop position for the throttle lever or adjusting screw during curb-idle engine operation. In other words, when the driver turns on the ignition switch and starts the engine, the solenoid energizes; and the solenoid plunger moves out to contact the throttle lever or idle speed adjusting screw on the throttle valve shaft. This holds the throttle valves open slightly for normal curb engine idle.

When the driver turns off the ignition switch, the solenoid receives no electrical current. Consequently, the plunger retracts and permits the throttle lever or adjusting screw to move toward the closed position until it strikes the closed throttle valve stop. This action completely closes the throttle valves that block the air flow into the intake manifold. As a result, the engine stops running due to a lack of air flow.

Air-Conditioning Idle Solenoid

Many automobiles with air conditioning have a special idle solenoid (Fig. 9-35). This device looks much like the antidiesel solenoid, but it operates at a different time and for a different reason.

The air-conditioning idle solenoid ener-

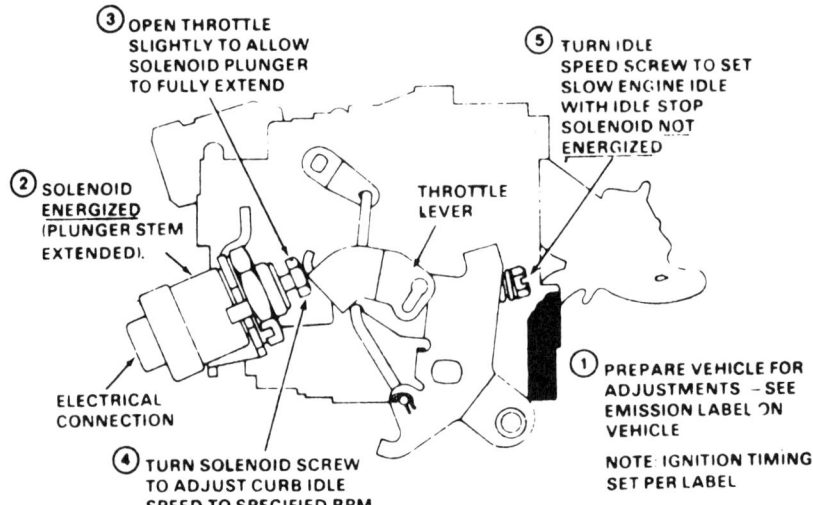

Figure 9-34. Antidieseling solenoid allows the throttle valve to close farther when the engine is shut off (Courtesy of General Motors Corp.).

gizes as the driver turns on the air conditioner. The plunger of the solenoid then moves forward to contact a special bracket of the throttle valve shaft. This action raises the normal curb-idle speed to prevent the engine from stalling due to the increased load of the compressor.

When the driver turns the air-conditioning system off, the solenoid deenergizes. The solenoid plunger then retracts, allowing the engine to return to its normal curb-idle rpm.

Actually, the engine rpm is nearly the same in both situations. However, the sole-

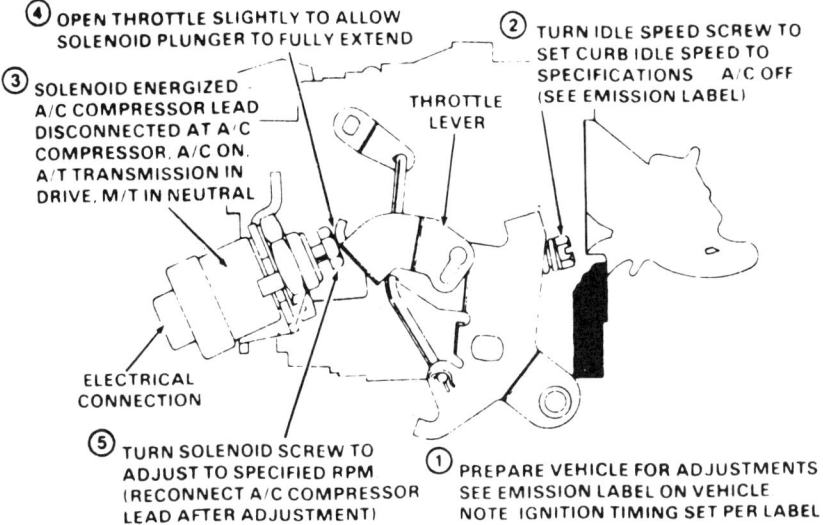

Figure 9-35. An air-conditioning solenoid increases normal engine idle when the driver turns on the air conditioner (Courtesy of General Motors Corp.).

CARBURETOR-ASSIST DEVICES

noid plunger must open the throttle valve a little more when the air conditioner is on in order to compensate for the increased compressor load on the engine, thereby maintaining a fairly stable curb idle at all times.

Jet Air Control Valve System

As stated earlier, some engines utilize a jet air valve. This valve provides additional air to the combustion chambers in order to promote more complete burning of the air/fuel charge, thereby reducing exhaust emissions.

Along with the jet valve, its operating mechanism, and interconnecting passages, many carburetors will also have a control valve system. This jet air volume control system supplies more air to the jet air valve passage during engine warm-up operation with the choke valve closed. This action decreases both CO and HC emissions during this phase of engine performance.

The system itself consists of a jet air control valve and a thermo valve (Fig. 9–36). The control valve mounts on the carburetor and opens by its ported vacuum to allow additional air to flow into the jet air passage. Thus, the valve, by supplying additional air, controls the air/fuel ratio during engine warm-up so that it is not too rich.

The thermo valve controls the operation of the jet air valve on the carburetor. This is necessary to enhance vehicle driveability under certain operating conditions. In other words during engine warm-up, the thermo valve prevents the control valve from supplying additional air into the jet air passage just after a cold engine starts or after the choke valve opens fully. The extra air at these times along with the operation of the EGR system would make vehicle driveability unsatisfactory.

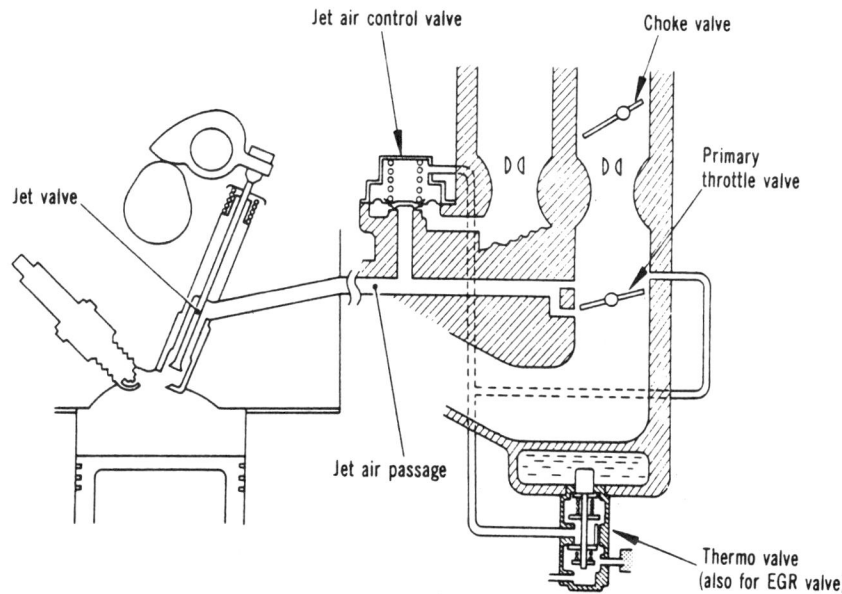

Figure 9–36. Common type of jet volume control system (Courtesy of Chrysler Corp.).

The thermo valve fits into the engine's cooling system; thus it reacts to engine temperature. For instance, the thermo valve detects cooling system temperature either below or above a preset value and suspends control valve operation during these periods. Consequently, the jet air control valve supplies extra air to the jet air passage only during given phases of an engine's warm-up period.

Thermostatically Controlled Air Cleaners

Function

Another integral part of the exhaust emission control system is the thermostatically controlled air cleaner (Fig. 9-37). This device continuously preheats the air coming into the carburetor on severe winter days and during the engine warm-up period in warmer climates, causing improved fuel vaporization and distribution. This action results in a leaner fuel requirement, which permits the use of a weaker fast-acting choke bimetallic spring and overall leaner carburetor calibration. Consequently, the thermostatic air cleaner, modified choke spring, and leaner carburetor mixtures bring about a reduction in CO and HC emissions during starting and the warm-up period. Finally, this type of air cleaner also reduces carburetor icing during cold-weather operation that can cause engine stalling.

Design of a Thermostatically Controlled Air Inlet System

Figure 9-38 shows a typical heated air inlet system. This system consists of a manifold shroud, heated air tube, vacuum motor, and a temperature-sensing vacuum valve. The manifold shroud fits over one of the exhaust manifolds. On one end of this shroud is an opening that accommodates the heated air tube. The shroud itself acts as a heat stove; that is, the shroud absorbs

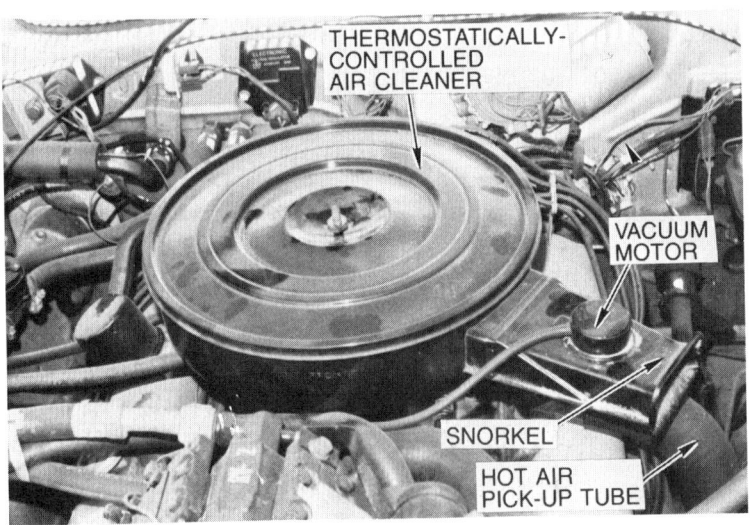

Figure 9-37. Typical thermostatically controlled air cleaner assembly.

THERMOSTATICALLY CONTROLLED AIR CLEANERS

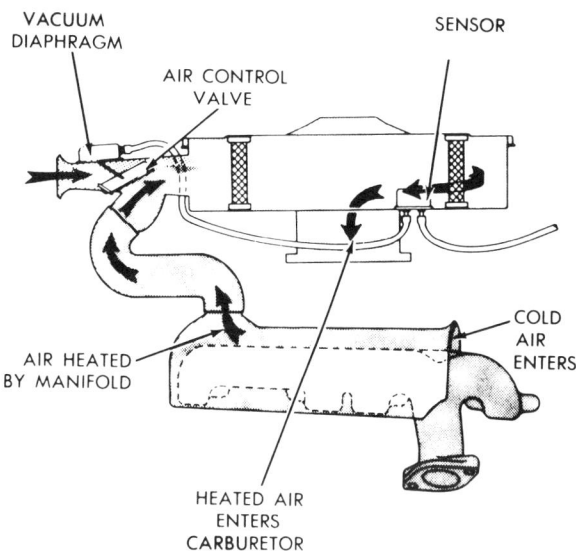

Figure 9-38. Schematic of a thermostatically controlled air inlet system (Courtesy of Chrysler Corp.).

heat from the exhaust manifold and, in turn, preheats the air coming into the shroud chamber. This heated air passes through the heat tube to the air cleaner snorkel.

The vacuum motor fits on top of the snorkel. This motor is nothing more than a spring-loaded diaphragm, acted upon by intake manifold vacuum. The diaphragm itself connects through linkage to the heat control door inside the snorkel and will activate this door when the air coming into the air cleaner is below a predetermined temperature.

The door itself controls the air flow into the air cleaner from two different sources. For example, when the door closes, heated air from the shroud is cut off, and the snorkel is wide open to outside air flow. However, when the door opens, only air passing through the shroud and heat tube enter the air cleaner.

The temperature-sensing vacuum valve mounts inside the air cleaner. A vacuum hose to the intake manifold carries engine vacuum through this valve to the vacuum motor. The temperature-sensitive valve itself controls the opening and closing of the door in the snorkel by regulating the strength of the vacuum signal to the vacuum motor, according to air temperature.

Operation of the Thermostatically Controlled Air Cleaner

Figure 9-39 illustrates the two positions of the heat control door. When the temperature in the air cleaner is below about 85 °F (29 °C), the temperature-sensitive vacuum valve is in the closed position. Thus, it permits full manifold vacuum to the vacuum motor diaphragm. Through linkage the diaphragm in turn opens the heat control door (Fig. 9-39). This door position cuts off cool air flow through the open end of the snorkel and permits only heated air

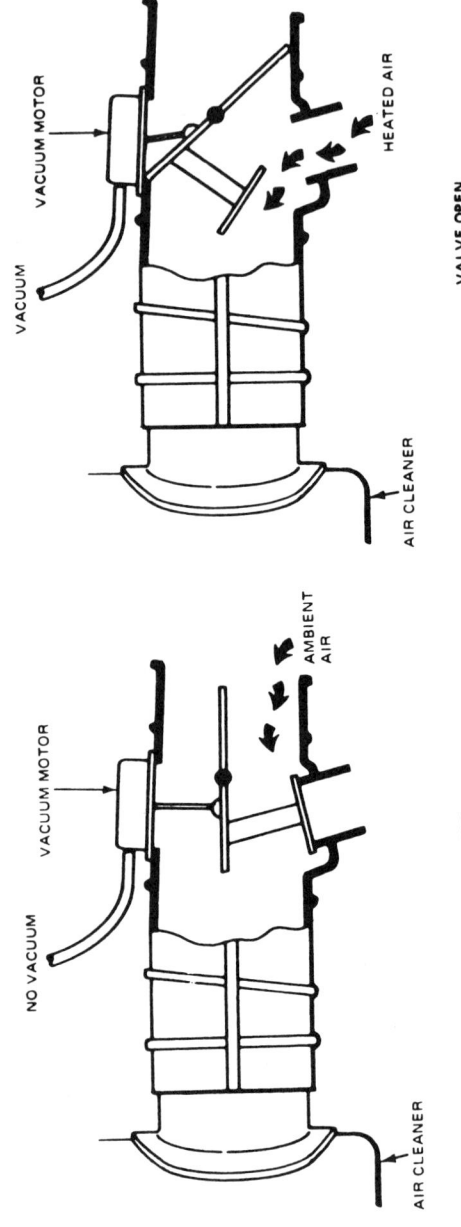

Figure 9–39. Operation of a thermostatically controlled air cleaner (Courtesy of Ford Motor Co. of Canada Ltd.).

from the manifold shroud to pass into the air cleaner.

As the air temperature in the cleaner reaches about 128 °F (53 °C), the temperature-sensitive vacuum valve opens. This action permits the valve to bleed off the vacuum signal to the vacuum motor. Then, the diaphragm spring within the motor causes the heat control door to close (Fig. 9–39). This door position permits outside cooler air to pass through the snorkel and enter the air cleaner and cuts off heated air from the shroud.

Between 85 °F (29 °C) and 128 °F (53 °C), the heat control door actually assumes a position based on the actual temperature within the air cleaner. In other words, at temperatures between these two figures, the door will be moving slowly from the open to the closed position, with the amount of vacuum bleeding by the vacuum valve and the diaphragm spring controlling the actual door position.

Under a heavy acceleration condition in which zero or very low vacuum exists in the intake manifold, the heat control door closes. This door position permits maximum outside air flow through the snorkel to the engine. The door moves to this position due to the loss of vacuum to the diaphragm. Consequently, its spring moves the diaphragm, linkage, and heat control door to the closed position.

Spark-Advance Control Devices

In addition to the thermostatic air cleaner, vehicles with engines modified to control exhaust emissions also usually have some form of spark-timing control device. The design and operation of these particular units varies among the various manufacturers. (Spark-timing devices will be discussed in greater detail in Chapters 11 and 12.)

Increased Operating Temperatures

Another alteration to the basic engine, made by vehicle manufacturers to reduce both HC and CO emissions by more complete combustion of the air/fuel charge, is an increase in its operating temperature. As previously mentioned, the temperature of the combustion chamber has a direct effect on just how well the burning process consumes the hydrocarbons within the air/fuel mixture. In other words, the combustion process burns more of the hydrocarbons from the mixture at higher temperatures. However, if the temperature is too high, NOx emissions increase; therefore, manufacturers use a slightly increased minimum operating temperature to help reduce HC emissions without adversely affecting NOx levels.

Also, at higher engine operating temperatures, the intake manifold more easily vaporizes the air/fuel mixture. As the fuel in the charge becomes more vaporized, distribution of the mixture to all the various cylinders is more uniform, and the combustion process consumes the air/fuel mixture more completely. This reduces CO emissions.

For these reasons, the minimum operating temperatures are now higher than they were before the advent of exhaust emission control systems. Manufacturers raise the minimum operating engine temperature through the use of thermostats with higher initial opening temperatures. Therefore, the use of a thermostat with a lower opening temperature in an exhaust emission-controlled engine results in an increase in both HC and CO emissions.

Vacuum-Operated Heat-Riser Valves

Many vehicles now use a vacuum-operated heat-riser valve (Fig. 9-40) instead of the previous type, which functioned through the use of a bimetallic spring. Manufacturers use this type of heat-riser valve because it provides more precise control of intake manifold heating, which improves the vaporization of the engine's air/fuel mixture. This action reduces HC and CO emissions while improving the driveability of the vehicle.

Automobile manufacturers use several different names for this device. For example, General Motors calls it an early fuel evaporation (EFE) valve, Ford a vacuum-operated heat control valve (HCV), and Chrysler a power heat control valve.

All of these units operate in about the same manner and have much the same style of construction. The device consists basically of a valve operated by a vacuum motor. The rotating heat-riser valve operates inside a cast-iron body, which fits between the exhaust manifold outlet and the exhaust pipe.

Linked to the shaft that extends through the cast-iron body and attaches to the valve is a diaphragm within a vacuum motor. This diaphragm activates the shaft and valve by reacting to a vacuum signal supplied by the intake manifold. When the diaphragm receives a full or high manifold vacuum signal, it closes the valve. However, when the signal is low or is cut off

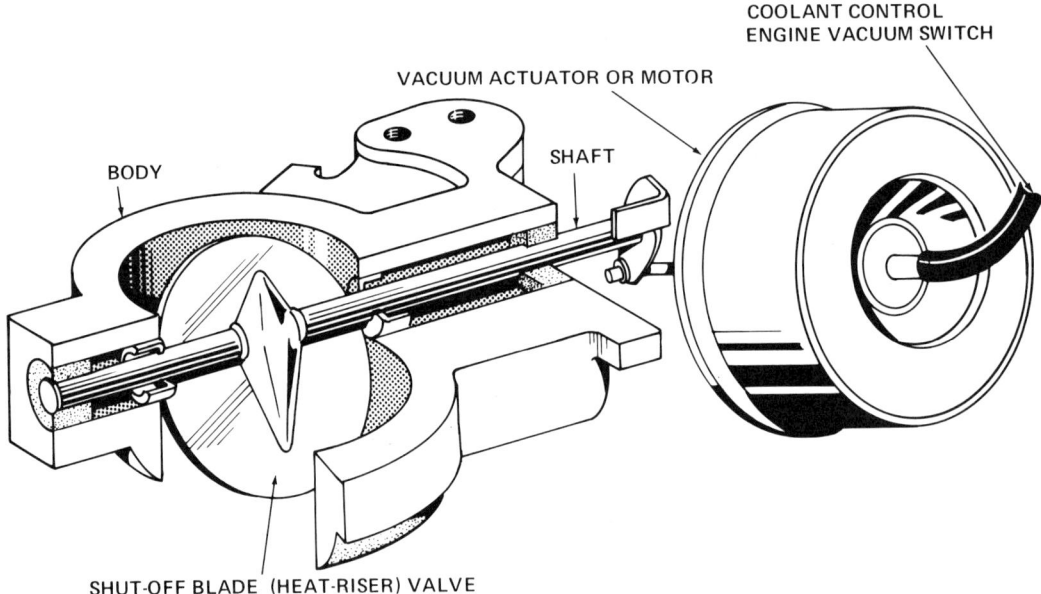

Figure 9-40. Vacuum-operated heat-riser valve improves the vaporization of the air/fuel mixture, which reduces HC and CO emissions.

completely to the motor, a spring behind the diaphragm pushes the heat-riser valve open.

A switch or valve controls the vacuum signal to the diaphragm within the motor. This switch or valve reacts to either oil or coolant temperature (Fig. 9–41). When, for instance, the engine is cold, the oil temperature switch or coolant-operated valve permits the application of a high intake manifold vacuum signal to the motor's diaphragm. This signal causes the diaphragm to move, closing the riser valve. The valve then directs exhaust gas to the heat stove within the intake manifold, which heats the manifold runner floor to improve vaporization of the air/fuel mixture.

As the engine warms up, the oil temperature switch or the coolant-operated valve cuts off vacuum to the motor. With no vacuum signal to the motor, the diaphragm spring pushes the heat-riser valve open. All the exhaust gases now flow out of the exhaust manifold and into the exhaust pipe. However, by this time, the engine has developed enough heat that the intake manifold remains warm due to radiation.

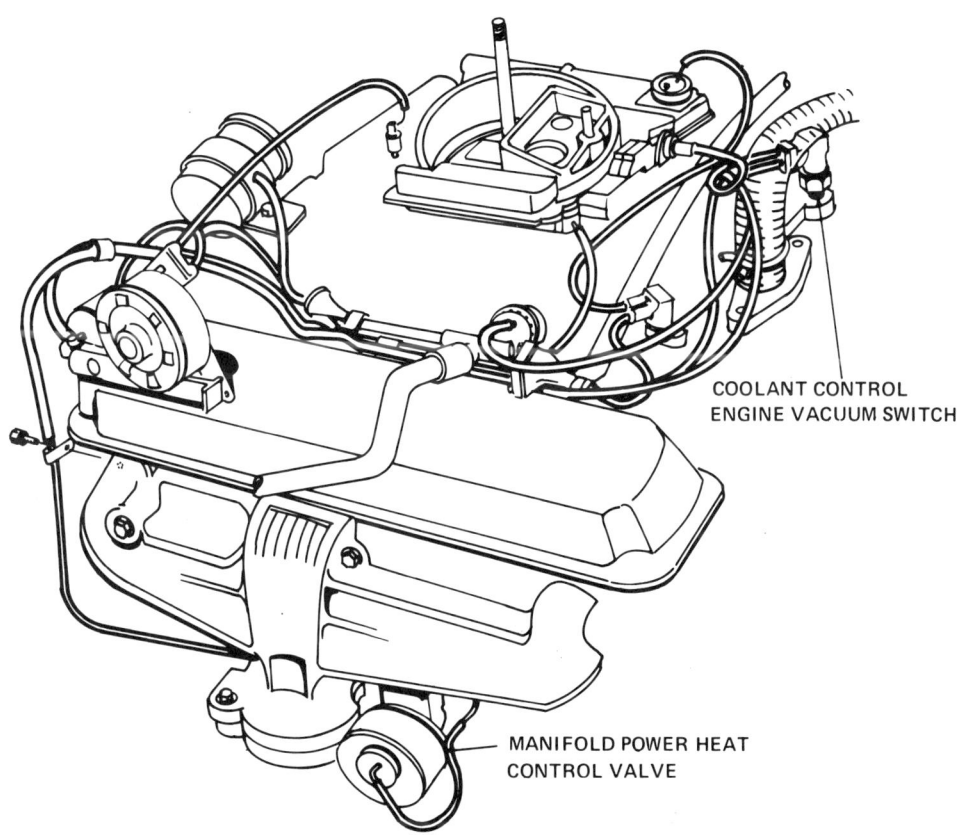

Figure 9–41. Oil temperature switch or a coolant-operated valve controls the vacuum signal to the motor's diaphragm.

Summary

1. Exhaust emission control equipment has the effect of reducing HC, CO, and NOx emissions.
2. The best way to control exhaust emissions is to burn the fuel as completely as possible inside the engine itself.
3. There have been a considerable number of alterations to the basic engine to reduce its exhaust emissions.
4. Quench areas within the combustion chamber, cylinder wall, or piston cause unburned hydrocarbons to eject from the engine along with some carbon monoxide.
5. Manufacturers eliminate quench pockets and close clearance areas by reworking the combustion chambers and pistons.
6. Manufacturers also modify the head of the piston to alter the engine's compression ratio.
7. Exhaust emission-modified engines have a lower compression ratio.
8. Manufacturers have redesigned the intake valve ports on some larger displacement engines to increase the turbulence of the air/fuel mixture.
9. With the removal of the tetraethyllead from gasoline, a modification to the exhaust valve seat became necessary.
10. A modified camshaft provides either a slight change in valve timing or extends valve overlap.
11. To reduce CO levels, manufacturers have altered the design of the intake manifold in several ways.
12. The shape of the combustion chamber has a direct effect on the efficiency of combustion and, therefore, the amount of emissions escaping from the engine during the exhaust strokes.
13. Wedge-type combustion chambers have quench areas, necessary in some cases to reduce the possibility of detonation.
14. There are three general ways used to reduce the effects of quenching.
15. A stratified-charge engine uses an air/fuel mixture that is not uniform.
16. The purpose of stratified charging of an engine is to concentrate the richest part of the mixture around the spark plug.
17. A very efficient type of stratified-charge engine is the Honda Compound Vortex Controlled Combustion powerplant.
18. The CVCC engine uses a two-stage combustion process.
19. The combustion chamber that has a reduced surface area emits less HC and CO emissions within its exhaust gases.
20. The hemispheric-type combustion chamber has a low S/V ratio.
21. The jet air system provides air that scavenges the residual gases from around the spark plug to create a good ignition condition while providing a swirling action to the gases within the combustion chambers.
22. In addition to a regular intake and exhaust valve, an engine with a jet air system has a special jet valve.
23. The carburetor on emission-controlled engines has modifications to help reduce the amounts of CO and HC emissions during the various phases of engine operation.
24. Some engines have idle mixture adjusting screws within their carburetors that the mechanic can only turn out a given amount.

25. Some manufacturers limit the travel of the idle mixture screws through the use of plastic limiter caps.
26. Another method of controlling the amount of enrichment is a restrictor placed into the idle circuit.
27. The hot-idle compensator valve allows additional air to enter the engine from below the throttle valves to prevent hot-idle stalling.
28. An idle enrichment system increases the enrichment of the normal air/fuel ratio at idle in order to reduce cold-engine stalling.
29. Some manufacturers provide a factory adjustment for the part-throttle carburetor circuit.
30. Many carburetors built since 1975 have an altitude-compensating device in their main metering circuit.
31. To control CO and HC emissions during engine warm-up, most manufacturers modify the automatic choke circuit.
32. The most common types of modified choke spring heaters in use are the electric-assist, coolant-assist, and the hot-air modulator.
33. Many late-model carburetors now come from the factory with tamper-proof automatic choke mechanisms.
34. Some engines utilize an electronically controlled carburetor with a specially designed catalytic converter to control exhaust emissions.
35. Carburetor-assist devices primarily serve the purpose of further reducing exhaust emissions or improving the driveability of the vehicle.
36. Typical carburetor-assist devices include such units as the choke vacuum brake, choke-delay valve, deceleration valve, dashpot, antidieseling or air-conditioning solenoid, and a jet valve.
37. The thermostatically controlled air cleaner preheats the air coming into the carburetor.
38. A heated air inlet system consists of a manifold shroud, heated air tube, vacuum motor, and a temperature-sensitive vacuum valve.
39. In addition to the thermostatically controlled air cleaner, vehicles with engines modified to control exhaust emissions usually have some form of spark-timing control system.
40. Manufacturers raise the operating temperature of an engine to reduce both HC and CO emissions.

Review Questions

The questions listed below will assist you in determining how well you remember the material contained in this chapter. Read each question carefully before choosing the answer. If you cannot answer the question, review the section in the chapter that covers the material.

1. The cooler metal surfaces near the edges of a combustion chamber are known as _____ areas.
   a. quench
   b. cool
   c. hot
   d. swirl
2. The compression ratio of an exhaust emission-controlled engine is usually about _____.
   a. 11:1
   b. 10:1
   c. 9:1
   d. 8:1

3. The type of combustion chamber design that has inherent quench areas is the _____ type.
   a. hemispheric
   b. stratified
   c. wedge
   d. dome

4. A very efficient stratified-charge engine is built by _____.
   a. General Motors
   b. Honda
   c. Chrysler
   d. Ford

5. The combustion chamber type that has a low S/V ratio is the _____.
   a. flat
   b. hemispheric
   c. wedge
   d. round

6. A jet air valve opens due to the action of the _____.
   a. camshaft
   b. crankshaft
   c. piston
   d. rod

7. The devices installed on some carburetors to prevent excessively rich mixtures are the _____.
   a. wax covers
   b. plastic screws
   c. limiter caps
   d. limiter screws

8. Some carburetors built since _____ have an altitude-compensating device.
   a. 1974
   b. 1975
   c. 1976
   d. 1977

9. Most altitude-compensated carburetors use a (an) _____ to alter the engine's air/fuel ratio.
   a. electric solenoid
   b. bimetallic spring
   c. mechanical linkage
   d. aneroid bellows

10. The choke system that keeps the bimetallic spring warm longer uses _____ as its heat medium.
    a. engine oil
    b. electric power
    c. engine coolant
    d. exhaust gas

11. A (an) _____ carburetor is used with a specially designed catalytic control system to reduce exhaust emission on some engines.
    a. electronically controlled
    b. vacuum-controlled
    c. mechanically controlled
    d. both b and c

12. The assist device that opens the choke valve slightly after the engine starts is the _____.
    a. choke unloader
    b. vacuum brake
    c. delay valve
    d. throttle dashpot

13. The unit that retards the complete closing of the throttle valve is the _____.
    a. brake
    b. linkage
    c. solenoid
    d. dashpot

14. The device that prevents engine run-on in some vehicles is the _____.

a. antidiesel solenoid
b. air-conditioning solenoid
c. carburetor dashpot
d. vacuum brake

15. The _____ preheats the air coming into the carburetor.
a. engine shroud
b. thermostatic air cleaner
c. carburetor heater
d. air flow heater

16. The position of the air cleaner's preheat door during acceleration is _____.
a. closed
b. open

For the answers to these review questions, turn to the Appendix.

Chapter 10

Testing and Servicing Engine Modification Systems

As mentioned in the last chapter, an engine modified to reduce exhaust emissions has a number of alterations that permit it to burn the fuel more completely during its combustion process. The modified engine continues to reduce emissions internally as long as it is in good mechanical condition. In other words, other than testing the mechanical condition of the engine, normal lubrication, and cooling system service, it requires no other preventive maintenance to preserve its inherent lower emission levels.

There are several established methods used by the industry to test the mechanical condition of the engine. For example, some technicians still utilize either a compression or leakage tester to measure the efficiency of each cylinder to build and hold compression. However, it is now a very common practice to use the cylinder balance tester section of an engine analyzer (Fig. 10–1) to electronically measure the power output of each engine cylinder. Chapters 3 and 4 explained how to use both the vacuum gauge and the infrared analyzer for troubleshooting an engine for mechanical malfunctions.

But there are a large number of devices added to the basic modified engine, either to further reduce exhaust emissions or improve the driveability of the vehicle, that do require periodic adjustment or servicing. The number, type, and design of these devices depend largely upon the particular manufacturer, the vehicle's engine size and type, or the year model of the vehicle. Consequently, when attempting to test or service one of these particular units, it will be necessary in most cases to use factory instructions and specifications.

For this reason, this chapter only presents samples of typical procedures. Those included are: adjusting a jet valve, the air/fuel mixture with an infrared analyzer and propane enrichment equipment, testing a

Figure 10-1. Engine analyzer with a cylinder balance tester incorporated into its console.

choke-delay valve, adjusting a carburetor dashpot or antidiesel and air conditioning solenoids, and testing thermostatic air cleaner assemblies.

Adjusting a Jet Air Valve

The jet air valves used on some engines require periodic adjustment to maintain their proper operating clearances. If this preventive maintenance is not done when specified and these valves develop incorrect clearances, the engine will begin to mechanically malfunction and produce excessive emission levels.

To adjust the jet valves on a typical Chrysler-built engine, follow this recommended procedure (Fig. 10-2):

1. Start and warm up the engine until its coolant temperature reaches 176 to 194 °F (80 to 90 °C).

2. Position number-one piston at top dead center (TDC) on its compression stroke.

3. Remove the valve cover and fully loosen the intake valve's side adjusting screw of number-one cylinder, two or more turns.

4. Loosen the locknut on the number-one jet valve's side adjusting screw.

5. Back out the jet valve adjusting screw (counterclockwise). Then, place a 0.006-inch (0.15-mm) feeler gauge between the top end of the jet valve stem and the bottom of the adjusting screw.

6. Turn down the adjusting screw clockwise until its bottom end touches the feeler gauge. Note: Since the jet valve spring is rather weak, use special care not to force the valve in, especially when the adjusting screw is hard to turn.

7. While holding the adjusting screw in position with a screwdriver, tighten its locknut securely.

8. With the 0.006-inch (0.15-mm) feeler gauge recheck the jet valve clearance. The gauge should just slide through the clearance with a slight drag.

9. Readjust the intake valve clearance.

10. Repeat these steps on the remaining jet valves, following the engines's firing order.

Adjusting the Air/Fuel Ratio With an Infrared Analyzer

When adjusting a carburetor using an infrared analyzer, you must remember two things. First, each turn of the adjusting screw affects the air/fuel mixture. For example, when you turn an adjustment screw counterclockwise (Fig. 10-3), the air/fuel mixture will enrich and the CO level will increase. On the other hand, a clockwise movement of the screw will lean out the mixture (Fig. 10-4), and the CO level will decrease. However, should you make the mixture so lean as to cause one or more

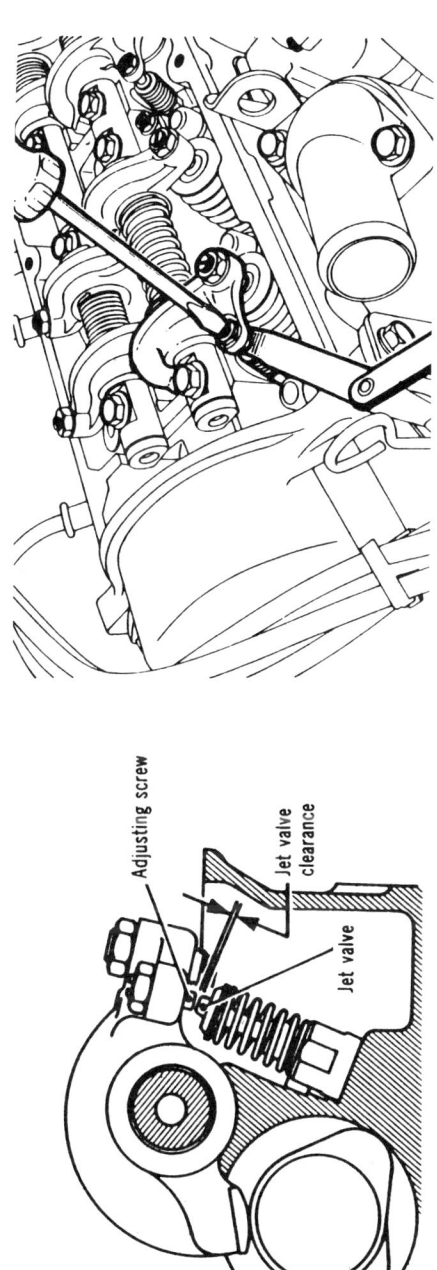

Figure 10–2. Adjustment of the jet valve clearance (Courtesy of Chrysler Corp.).

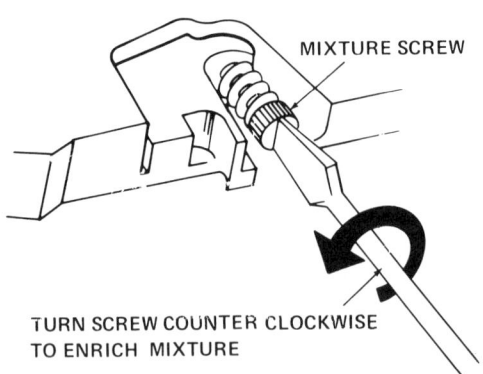

Figure 10-3. Adjusting a carburetor mixture screw to enrich the air/fuel mixture.

cylinders to misfire, the HC level will increase.

Second, always perform all idle and mixture adjustments as per manufacturer's recommended procedures. For instance, when the manufacturer provides a CO setting, use it and the analyzer to make the adjustment. On early model vehicles or those where CO specifications are not available, use a setting that is below state or federal standards and normal for the year and make of the vehicle being tested. If you are not able to obtain the desired HC and CO levels by following the manufacturer's instructions, there may be other problems such as faulty ignition, engine compression, air injection, or catalytic converter action. In this situation, you should test out all of these components to locate the cause of the problem before condemning the carburetor.

Although the adjustment procedures do vary somewhat among manufacturers, the one presented below is very commonly employed by the industry and is especially useful when adjusting an overhauled or repaired carburetor. (Chapter 4 explained how to adjust a carburetor's mixture for lean-best idle and the lean-drop procedure, through the use of a tachometer.)

To adjust a typical carburetor's air/fuel ratio at idle using an infrared analyzer, follow these directions:

1. As necessary, remove the limiter caps from the mixture adjusting screws (Fig. 10-5). Be careful not to damage the screws during the process.

2. As designed by the manufacturer, disconnect the air injection hose to the manifold; remove and plug the vacuum hoses to the EGR valve and vapor storage canister; or turn the headlights and air conditioner on.

3. Lightly seat each mixture adjusting screw. Then, back out each screw the number of turns specified by the manufac-

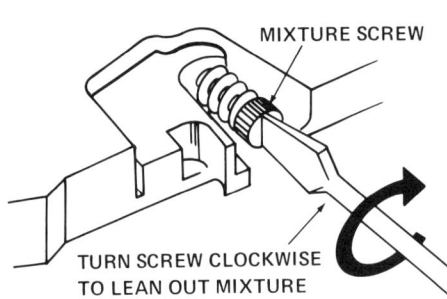

Figure 10-4. Adjusting a carburetor mixture screw to lean out the air/fuel ratio.

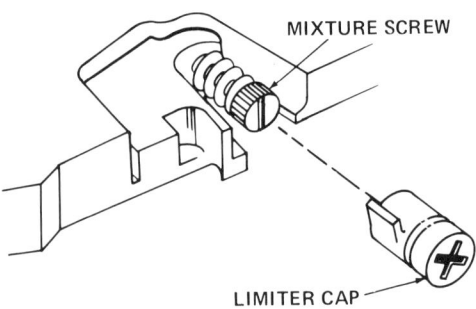

Figure 10-5. Removing the plastic limiter caps prior to adjusting the mixture screws.

turer. If no specifications are available, turn each screw out about two and a half turns or until it reaches its internal stop, if so equipped.

4. Start the engine and permit it to reach normal operating temperature.

5. Warm up and calibrate the infrared analyzer and insert its probe into the tail pipe or access hole within the catalytic converter. Then, depress the analyzer's test button.

6. Using a tachometer, adjust engine idle speed to manufacturer's specifications (Fig. 10-6).

7. Turn the mixture adjusting screw in until there is a definite drop in rpm on the tachometer (Fig. 10-7). Then, slowly rotate the screw counterclockwise until the tachometer needle stabilizes on the highest rpm reading (Fig. 10-8). Next, repeat the process on the other adjusting screw for two- and four-barrel carburetors.

8. If the stabilized rpm is not now to specifications, adjust the curb-idle speed with the idle speed screw or solenoid.

9. Repeat the adjustment of each adjustment screw on two- and four-barrel carburetors to make sure that both primary idle circuits are in balance. At this point, the engine is operating at lean-best idle.

10. Install the air cleaner.

11. Check the HC and CO readings on the analyzer. If the readings are too high, gradually lean out the mixture by turning the adjusting screws in, one at a time, one-sixteenth of a turn at a time. Continue to adjust the mixture screws in this manner equally until you obtain the lowest HC reading and the smoothest possible idle. Note: It will be necessary for you to wait for the HC and CO readings to stabilize after each adjustment. This can require as much as 7 to 10 seconds from the time you

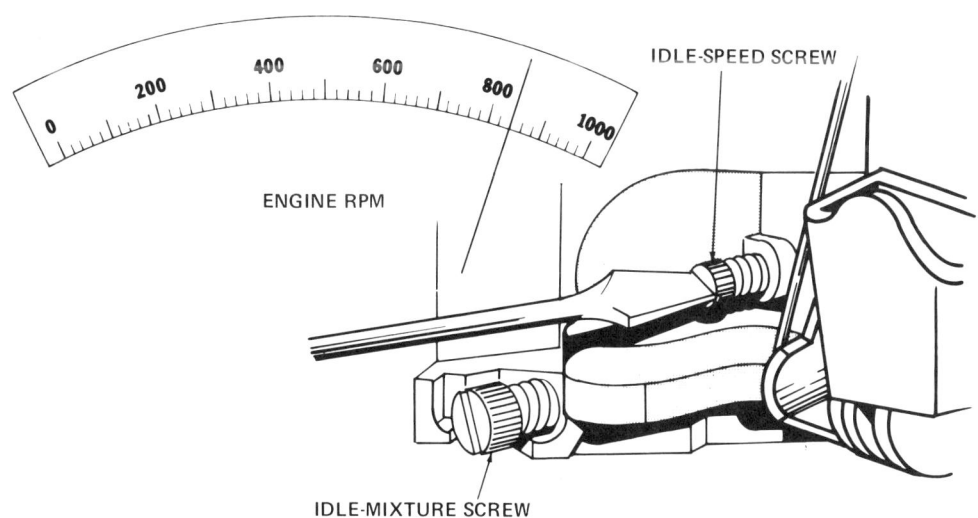

Figure 10-6. Adjusting the idle speed screw until the engine rpm is to specifications on a tachometer.

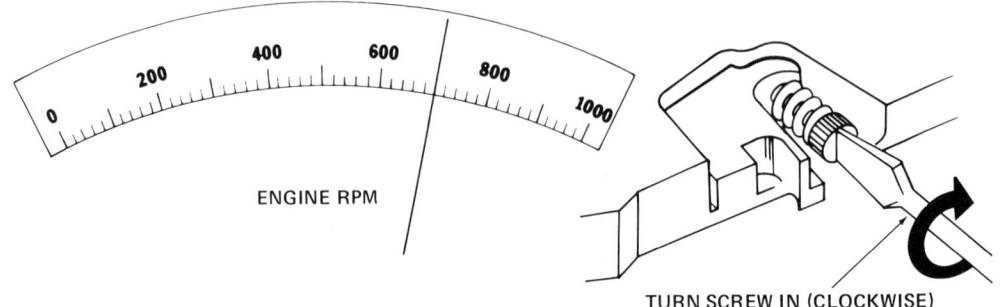

Figure 10-7. Turn the adjustment screw in until there is a definite drop in engine rpm.

change the mixture until the results appear on both meters. Also, readjust idle speed after each adjustment to maintain the specified engine curb-idle rpm.

12. Continue gradually adjusting the mixture screws in to obtain the lowest CO reading without increasing the HC level (Fig. 10-9).

13. Stop the procedure when the HC and CO readings are within manufacturer's specifications or legal limits with the engine operating smoothly. Note: If the adjustment procedure takes longer than 2 minutes, clean out the engine by running it at about 2,000 rpm for a few seconds.

14. Install new limiter caps as necessary over the idle mixture adjusting screws with their tangs resting against the full-rich stops (Fig. 10-10).

15. As necessary, turn the headlights and air conditioner off, and reconnect all hoses or lines removed and plugged under step 2.

16. Remove the analyzer test probe from the vehicle's tail pipe and disconnect the tachometer from the engine.

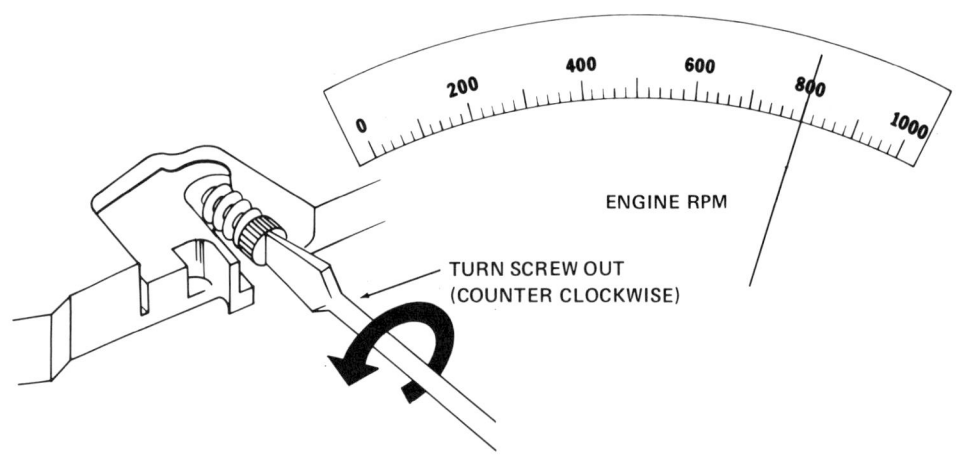

Figure 10-8. Turn the adjustment screw out until the tachometer indicates the highest steady engine reading.

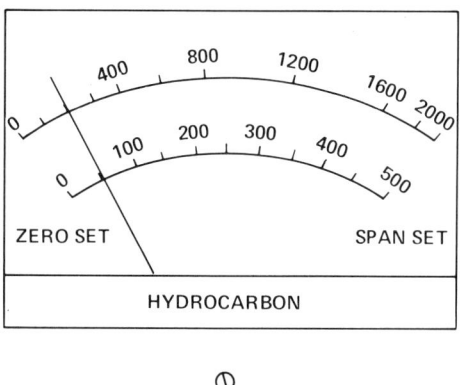
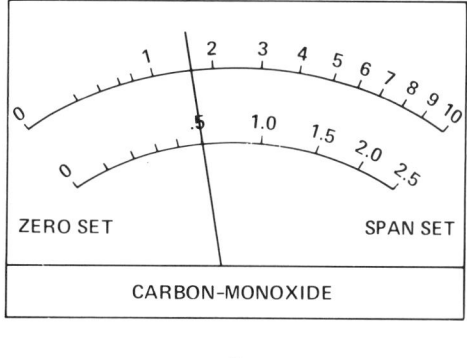

Figure 10-9. Adjust the mixture screw until the analyzer indicates the lowest CO level without increasing the HC reading.

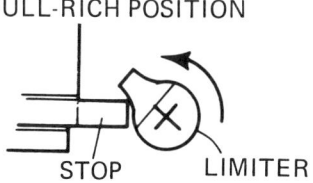

Figure 10-10. Install the limiter caps against the full-rich stops.

Carburetor Adjustments Using Propane Enrichment Equipment

As stated in Chapter 3, a large number of vehicle manufacturers now require that the mechanic adjust the carburetor idle mixture through the use of propane enrichment equipment. This action became necessary with the installation of the catalytic converter that reduces the amount of CO emissions to an almost unmeasurable amount. As a result, an infrared analyzer is not always accurate enough for carburetor mixture adjustment at idle. Consequently, the idle mixture can be set too rich, and the analyzer may not indicate the problem.

With one difference, the adjustment of the carburetor's idle mixture with propane enrichment equipment is quite similar to using the lean-drop method. The major difference between the two procedures is just how the higher rpm is reached. With propane enrichment, the technician injects a quantity of propane either through the air cleaner or directly into the intake manifold to increase the engine speed to point A of the flat portion of the curve (shown in Fig. 10-11). In this situation, the mechanic sets, with the propane flowing, the specified enriched rpm with the idle speed adjusting screw. This adjustment positions the throttle valve within the carburetor to provide the necessary rpm. The same type of results, during lean-drop procedure, is brought about by backing out the mixture screws to provide the enrichment.

When the technician shuts off the propane, the engine rpm should drop off a specified amount, which is actually the same as the rpm gain. In other words, with propane flowing into the engine, a specified rpm gain should occur. If the engine develops too much rpm gain, the mixture is too lean, and the idle mixture screws require adjustment out enough to compensate for the excess leanness. If, on the other hand,

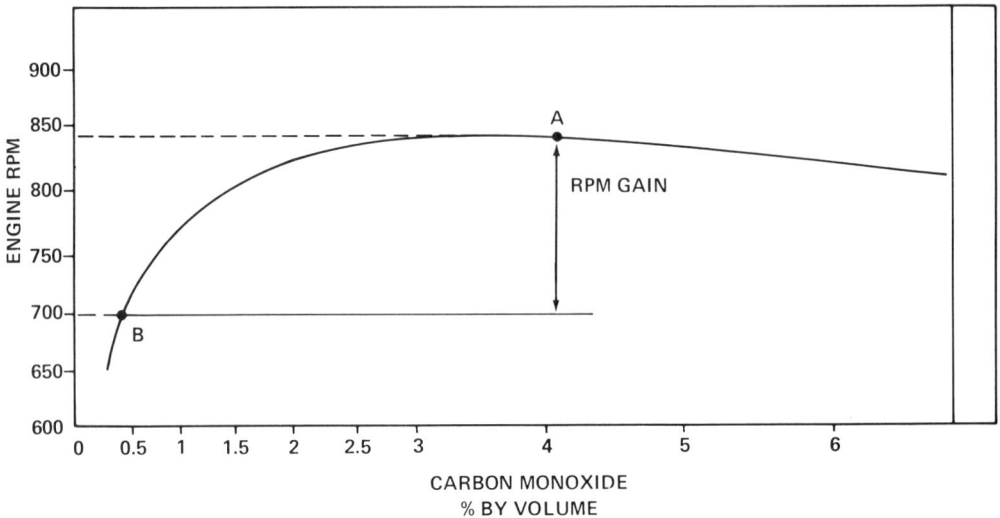

Figure 10-11. Idle gain brought about by the injection of propane into an engine.

the engine develops too little idle gain, the mixture is excessively rich, and the mixture screws require an adjustment in to lean out the mixture.

At the completion of the procedure, it may have been necessary for the technician to alter two settings of the carburetor. First, the idle adjusting screw will be properly set to provide the ideal idle air/fuel mixture, as indicated indirectly by the specified rpm gain. Second, the throttle valve position will be correct, as indicated by the engine operating at the proper rpm on the tachometer.

There are a number of reasons why it may be necessary to remove the limiter caps and adjust the air/fuel mixture using propane enrichment. For example, if a carburetor has been overhauled or its throttle body replaced, it will be necessary to readjust both the engine's idle mixture and speed. Moreover, if the vehicle fails to pass an emission inspection test by producing excessively high idle CO levels, the carburetor will require adjustment, in many cases beyond what the limiters will permit. However, in this situation, the mechanic should check and correct all other potential problems within the induction system before attempting to readjust the carburetor.

Propane Enrichment Adjustment Procedures

General Motors

As in the case of many other service techniques, each vehicle manufacturer has slight differences in the recommended propane enrichment adjustment procedures. Therefore, when using propane enrichment to adjust the air/fuel ratio on a given vehicle, always refer to the service manual or the Emission Control Information Label under the hood for the exact procedure to use.

PROPANE ENRICHMENT ADJUSTMENT PROCEDURES

There is insufficient space to explain all propane enrichment procedures used by the various manufacturers on all of their automobiles or light trucks. Consequently, we present typical procedures used by General Motors, Ford Motor Company, and Chrysler Corporation to familiarize you with the process.

To adjust idle mixture with propane enrichment on a typical General Motors product, follow these instructions:

1. Set the parking brake and block the drive wheels. On vehicles equipped with a vacuum parking brake release, disconnect and plug its vacuum hose.

2. Disconnect and plug all hoses, as directed on the Emission Control Information Label under the hood.

3. Start the engine and permit it to reach normal operating temperature. Make sure the choke valve opens and the air conditioning is off before beginning the procedure.

4. Connect an accurate tachometer to the engine. Then, adjust the engine's curb-idle speed to the specifications shown on the Emission Control Information Label (Fig. 10-12).

5. Following manufacturer's instructions, connect a timing light to the engine. Then, disconnect the vacuum advance hose and adjust ignition timing to the specifications shown on the Emission Control Information Label. Reconnect the vacuum advance hose. Note: On vehicles equipped with electronic spark timing, check the timing as directed on the label.

6. Disconnect the PCV inlet hose from the air cleaner.

7. Insert the hose from the propane enrichment tool into the inlet hose opening in the air cleaner, using the proper adapter (Fig. 10-13).

8. While holding the propane bottle in a vertical position, slowly open its control valve until the engine just reaches maximum speed, as indicated on the tachometer, with the transmission in drive or neutral for manual shift (Fig. 10-14). Note: If you apply too much propane to the engine, its rpm will begin to decrease due to an overly rich mixture. Therefore, as soon as the engine reaches maximum speed, discontinue opening the propane valve.

9. If so equipped, observe the propane flow meter to ensure the bottle is adequately full to perform the procedure.

10. With propane flowing, adjust the idle speed screw to the enriched rpm. Then, readjust the propane flow to make certain that the engine is operating at its

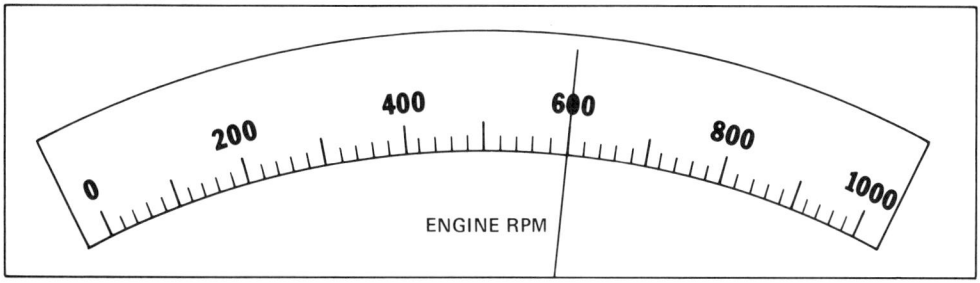

Figure 10-12. Engine's curb-idle speed before the injection of propane.

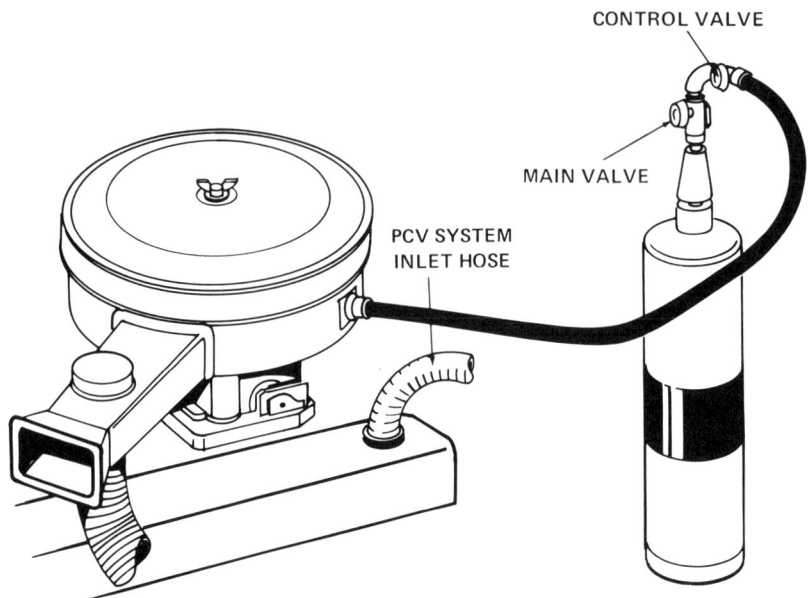

Figure 10-13. Proper connection of the propane hose into the PCV inlet tube opening in the air cleaner on a typical General Motors vehicle.

maximum speed, and readjust it again as necessary.

11. Turn off the control valve on the propane bottle. Next, place the transmission in neutral and run the engine at about 2,000 rpm for 30 seconds.

12. Place the transmission back in drive (neutral for a manual shift) and recheck the engine's curb-idle rpm. If the speed is the same as that shown on the Emission Control Information Lable, the idle mixture is correct. In this case proceed to step 17.

13. If the curb-idle speed is too low, carefully remove the limiter cap from each adjusting screw, taking care not to bend it. Then, back out the screw or screws equally (counterclockwise) one-eighth of a turn at

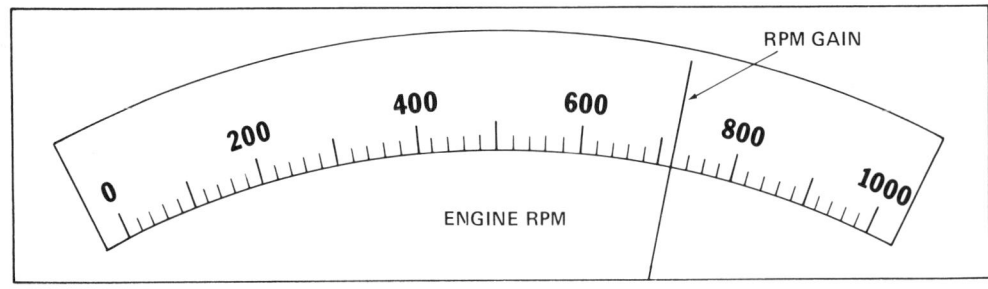

Figure 10-14. Idle gain caused by the injection of propane into the engine.

a time until the engine reaches the rpm specified on the Emission Control Information Label. If the idle speed is too high, turn the adjusting screw or screws in equally (clockwise) one-eighth of a turn at a time until the engine achieves the rpm indicated on the label.

14. Turn back on the propane control valve to check the maximum engine idle speed. If the rpm has changed from the specifications, readjust the idle speed screw to the enriched rpm with the propane flowing.

15. Turn off the propane control valve and clean out the engine at 2,000 rpm for 30 seconds in neutral. Next, place the transmission back in drive (neutral for a manual shift) and recheck the curb-idle speed. It should now be the same as that shown on the label. If not, repeat the adjustment procedure in step 13.

16. If the engine idles roughly, shut it off. Then, turn each of the mixture screws in until lightly seated. Next, back each one out equally to the average previous position; rerun the propane procedure beginning with step 8.

17. Turn off the engine and remove the propane enrichment tool. Then, connect the PCV tube to its air cleaner inlet fitting.

Ford Vehicles

To check and adjust the air/fuel mixture on a typical Ford vehicle using propane enrichment, follow this procedure:

1. Disconnect the PCV inlet hose from the air cleaner. Then, cap or plug the fitting on the housing.

2. Disconnect the canister purge hose from the air cleaner.

3. If specified on the Emission Control Information Label, disconnect the air injection supply hose from the check valve or the diverter valve outlet.

4. Connect a tachometer to the engine.

5. If the vehicle has a vacuum brake release, disconnect and plug its vacuum hose.

6. Set the parking brake and block the drive wheels.

7. Start and run the engine at fast idle until it reaches normal operating temperature.

8. As indicated on the Emission Control Information Label or in the service manual, adjust the idle speed to the normal slow idle or the recommended mixture adjusting rpm.

9. As specified, place the transmission either in neutral or drive.

10. Insert the adapter of the propane hose into the air cleaner connection from which you removed the canister purge hose (Fig. 10-15).

11. While holding the propane bottle upright, slowly open the propane control valve until the engine reaches its highest idle rpm, its maximum rpm gain. Opening this valve any further will cause the engine speed to drop as the air/fuel charge becomes excessively rich.

Results and Indications

Compare the idle gain to that listed on the Emission Control Information Label or within the service manual for the vehicle. If the rpm increase was within specifications, remove the propane tool adapter and hose from the air cleaner, and remove the tachometer. Then, remove all caps or plugs and reconnect all hoses removed under steps 1,2,3, and 5 of the test procedure.

If the rpm gain was too high, adjust the mixture as follows:

TESTING AND SERVICING ENGINE MODIFICATION SYSTEMS

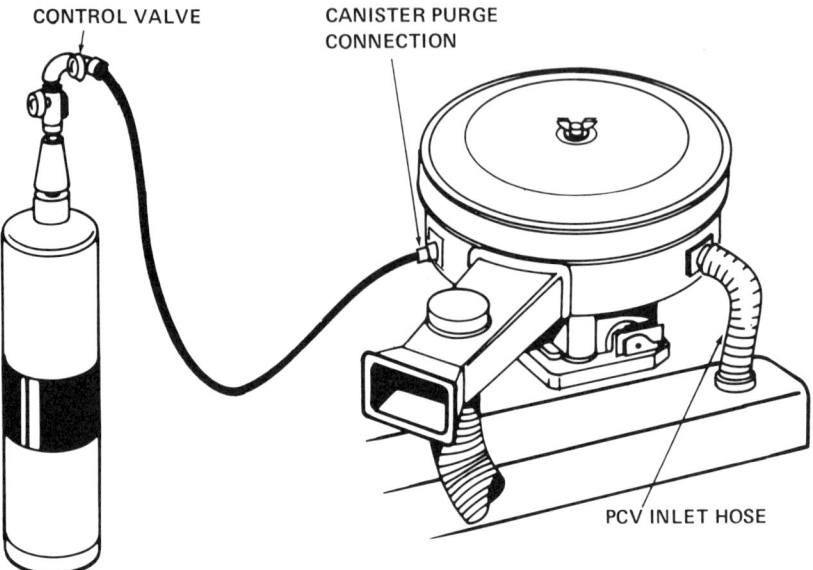

Figure 10-15. Correct connection of the propane hose into the canister purge connection on the air cleaner of a typical Ford vehicle.

1. Remove the air cleaner and plug any disconnected vacuum hoses.

2. As necessary, remove the limiter cap from each of the idle mixture adjusting screws by prying it off carefully. Be careful not to damage the adjustment screw.

3. Start the engine and turn the adjusting screw counterclockwise slowly and carefully to enrich the mixture. Continue this process until the rpm increases by the same amount that the test reading exceeded the specifications. For instance, if the idle gain was supposed to be 40 and the actual gain was 70, adjust the screw on a single-barrel carburetor until the idle speed increases 30 rpm. On two- and four-barrel units, adjust each mixture screw, one at a time, to raise engine speed a total of 15 rpm with each screw.

4. Reinstall the air cleaner and readjust the idle speed to specifications.

5. Using the propane tool, recheck the idle mixture.

6. If the mixture is satisfactory and you removed the limiter caps, install new ones with the tangs resting against the full-rich stops (see Fig. 10-10).

If the rpm gain was too low, adjust the mixture as follows:

1. Remove the air cleaner and plug any disconnected vacuum hoses.

2. As necessary, remove each mixture adjustment screw's limiter cap by prying it off carefully. Be careful not to damage the adjustment screw.

3. Start the engine and turn the mixture screw clockwise slowly and carefully to lean out the mixture. Continue the process until the rpm drops by the same amount as the original rpm gain was below the specified amount. For example, if the rpm gain

was 20 and should have been 50, adjust the mixture screw on a single-barrel carburetor until the engine speed drops by 30 rpm. On two- and four-barrel units, turn each adjusting screw, one at a time, to reduce engine speed by 15 rpm with each screw.

4. Reinstall the air cleaner and reset engine idle speed to specifications.

5. Recheck the idle mixture with the propane tool.

6. If the mixture is now correct and you removed the limiter caps, install new caps with the tangs resting on the full-lean stops (Fig. 10–16).

7. Remove the propane tool adapter and hose from the air cleaner and disconnect the tachometer from the engine.

8. Remove the caps and plugs and connect any lines or hoses disconnected under steps 1, 2, 3, and 5 of the test procedure.

Chrysler Vehicles

To perform a propane enrichment idle mixture adjustment on a typical Chrysler-built vehicle, follow these steps:

1. Place the transmission in neutral and set the parking brake.

2. Start the engine and permit it to warm up on the fast-idle cam until the thermostat is open and the radiator becomes hot. If the engine is cold (below 50 °F), it should be warmed up for at least 5 to 10 minutes on the second highest step of the fast-idle cam. If the engine is moderately warm (above 50 °F), run it on the second highest step of the cam for at least 2 to 5 minutes.

3. After the engine reaches normal operating temperature, kick off the fast-idle cam.

4. Disconnect the air cleaner vacuum supply hose from the intake manifold or carburetor. Then, connect the hose from the propane tool to the vacuum connection from which you removed the air cleaner hose (Fig. 10–17).

5. Connect a tachometer to the engine.

6. With the propane bottle held in a vertical position, slowly open its control valve until the engine just reaches maximum speed; leave the control valve open at this rpm. Note: Too much propane will begin to reduce the engine's idle speed.

7. With the propane flowing, adjust the carburetor's engine speed adjusting screw to the enriched rpm indicated on the Emission Control Information Label. As necessary, readjust the propane control valve for maximum speed.

8. Turn off the propane control valve and adjust the idle mixture screw or screws to achieve the smoothest idle at the curb-idle rpm shown on the information label. If necessary, remove the idle mixture screw limiter cap(s) to obtain the proper idle speed. Note: On some 318 CID engines with the High Altitude Package, adjust the idle mixture screw or screws to achieve the smoothest idle at 100 rpm below the curb-idle speed shown on the information label.

9. Turn on the propane control valve again and check the engine's maximum idle speed to see if it is still correct. If the speed now differs more than 25 rpm from the specified enriched speed on the information label repeat steps 6 through 9.

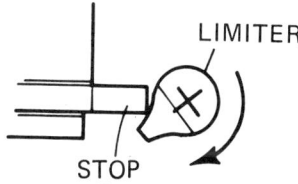

Figure 10–16. Installing a limiter cap against the full-lean stop.

TESTING AND SERVICING ENGINE MODIFICATION SYSTEMS

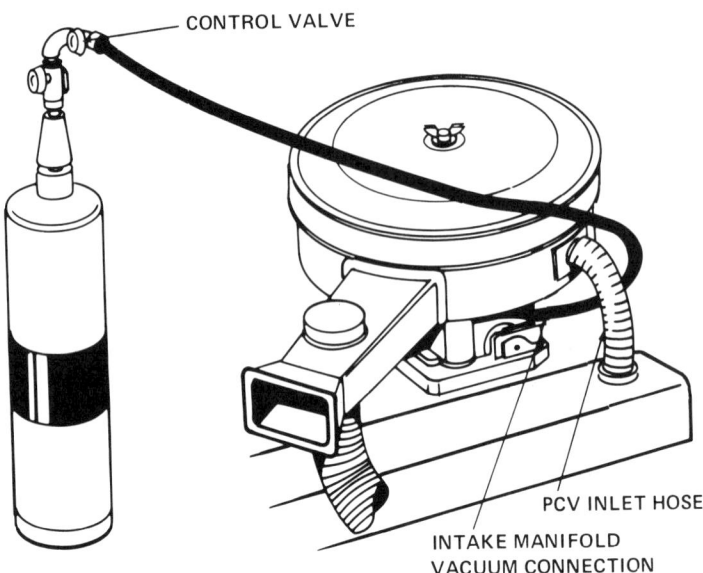

Figure 10-17. Correct connection of the propane hose into the intake manifold on a Chrysler-built vehicle.

10. Turn off the propane control valve and remove the hose from the intake manifold or carburetor. Then, reinstall the air cleaner's vacuum supply hose to its fitting.

11. On 318 CID engines with the High Altitude Package, reset the curb-idle speed to specifications by adjusting the idle solenoid.

Testing Choke-Delay Valves

As mentioned earlier, some automatic choke circuits use a delay valve to retard the action of the vacuum brake or piston to slow the opening of the choke valve during engine warm-up. If the valve does not seem to function properly, check its operation using this procedure:

1. Remove the valve from the vacuum line between the intake manifold and the choke vacuum diaphragm or piston.

2. Connect a hand-operated vacuum pump to the manifold side of the valve and a vacuum gauge to the other (Fig. 10-18).

3. With the vacuum pump apply 10 inches of vacuum to the valve. Then, note the number of seconds necessary for the second vacuum gauge to indicate the 10-inch reading.

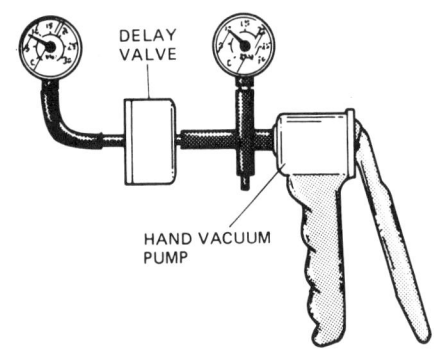

Figure 10-18. Testing a typical external choke-delay valve.

4. Compare this time delay in seconds to factory specifications. Replace the valve if the delay period is not within factory tolerances.

Dashpot Adjustment

Some carburetors come factory equipped with a dashpot. This unit retards the complete closing of the throttle in order to improve vehicle driveability or reduce its emission levels.

However, this unit cannot perform this function without the correct operating clearance. To check or adjust a typical dashpot's clearance:

1. With the engine at its normal operating temperature, adjust both the idle speed and mixture to specifications.

2. Hold the throttle valve in the curb-idle position and then depress in the dashpot plunger (Fig. 10-19).

3. Measure the clearance between the throttle lever and the tip of the dashpot plunger.

4. If the clearance is not to specifications, loosen the dashpot locknut and turn the dashpot in or out as necessary to achieve the specified measurement.

5. Tighten the locknut to secure the adjustment.

6. If removed, reinstall the air cleaner and all its attaching lines or hoses.

Adjusting an Antidiesel Solenoid

Several types of antidiesel solenoids are in use. Consequently, their adjustment procedures can vary somewhat from one manufacturer to another. However, some general steps you can use in most cases follow:

1. Check the manufacturer's specifications for both the curb-idle and shutdown-idle rpm.

2. Following the manufacturer's recommendations, disconnect and plug the hoses to the carburetor from the vapor canister, EGR valve, and the vacuum advance. Also, if the vehicle has a vacuum release parking brake, disconnect and plug its hose.

3. Connect a tachometer to the engine, following the manufacturer's instructions.

4. Set the parking brake and block the drive wheels.

5. Start the engine and permit it to reach its normal operating temperature.

6. Following the manufacturer's recommendations, place the automatic transmission in drive or neutral and turn on the air conditioner or headlights.

7. With the engine operating at curb idle, be certain that the solenoid energizes and its plunger extends fully. Note: In most

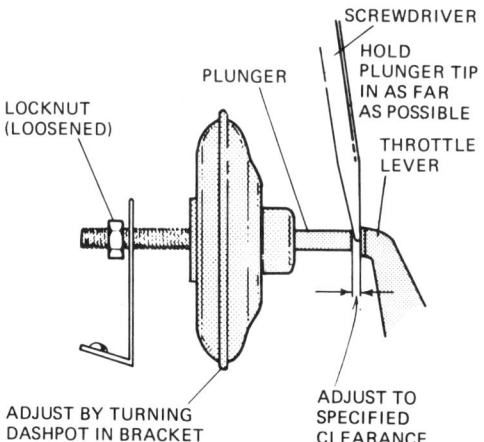

Figure 10-19. Checking and adjusting the clearance of a common-type dashpot.

TESTING AND SERVICING ENGINE MODIFICATION SYSTEMS

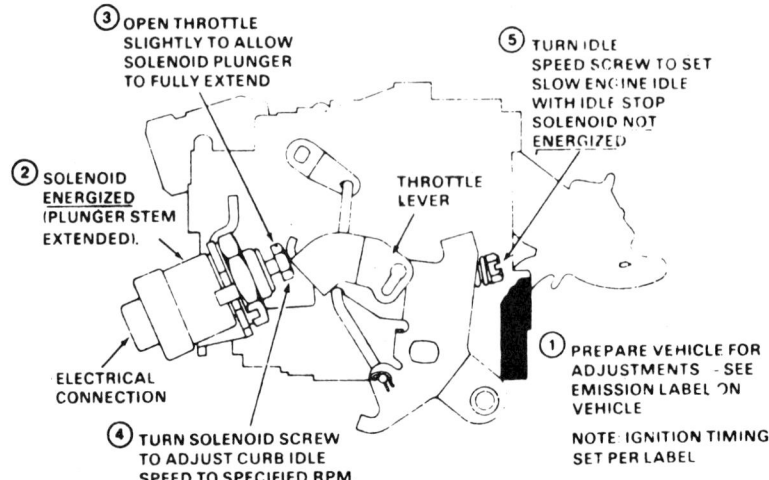

Figure 10-20. Adjusting the curb-idle speed by turning the solenoid plunger screw in or out (Courtesy of General Motors Corp.).

cases, it is necessary to open the throttle slightly off its idle position in order to permit the plunger to extend.

8. Check the idle speed on the tachometer. As necessry, adjust the curb-idle rpm by:

    a. turning the solenoid plunger adjusting screw in or out (Fig. 10-20);

    b. adjusting a hex nut on the plunger at the rear of the solenoid itself;

    c. setting a throttle linkage adjusting screw that makes contact with the solenoid plunger (Fig. 10-21);

    d. after loosening the solenoid locknut, rotating the solenoid body in its mounting bracket; or

    e. adjusting the movable bracket upon which the solenoid mounts (Fig. 10-22).

9. Disconnect the solenoid electrical lead wire to deenergize the solenoid and retract the plunger.

10. Check and then adjust as necessary the shutdown-idle rpm by turning the shutdown-idle speed adjusting screw on the throttle linkage.

11. Reconnect the solenoid lead and then return the throttle valve to the curb-idle position. Recheck the curb-idle rpm and readjust it as necessary.

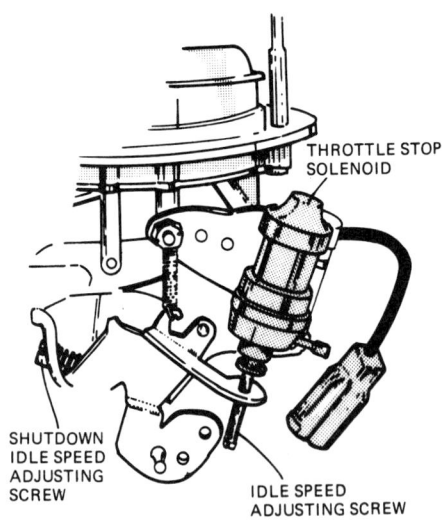

Figure 10-21. Adjusting the curb-idle rpm by turning a throttle screw in or out.

Testing and Servicing a Typical Thermostatically Controlled Air Cleaner

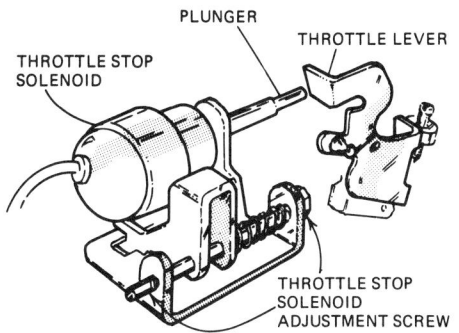

Figure 10-22. Setting the curb idle by moving the solenoid mount bracket.

12. Reconnect all hoses removed under step 2.

Adjusting an Air-Conditioning Solenoid

To check or adjust a typical air-conditioning solenoid:

1. Start the engine and permit it to reach normal operating temperature.

2. Set the air conditioner temperature control to maximum cooling, and then turn the air conditioner fan switch on.

3. Accelerate the engine slightly past the idle position to permit the air-conditioning solenoid plunger to extend. Note: This action is necessary because the solenoid does not have enough power in most cases to open the throttle valve by itself.

4. Check the idle rpm on a tachometer. If the reading is not to specification, turn the adjustment screw on the plunger in or out, until the engine speed reaches the specified amount.

5. Disconnect and remove the tachometer.

To test the operation of a thermostatically controlled air cleaner, follow these simple steps:

1. Make sure that all vacuum hoses and the heat stove-to-air cleaner flexible connections are attached and in good condition.

2. Remove the air cleaner. Then, tape a thermometer next to the air cleaner sensor (Fig. 10-23). The tape holds the thermometer alongside the sensor and prevents engine vacuum from pulling it into the air horn.

3. Replace the air cleaner cover, but do not install its wing nut.

4. With the engine shut off, observe the heat control damper door position through the snorkel opening. The snorkel passage should be open to the air cleaner element (Fig. 10-24). If not, check for binds in the door linkage.

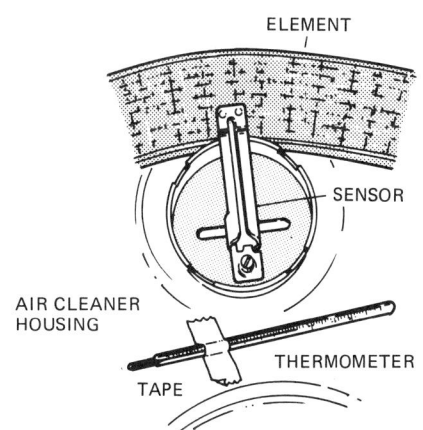

Figure 10-23. Testing the air cleaner sensor operation with a thermometer.

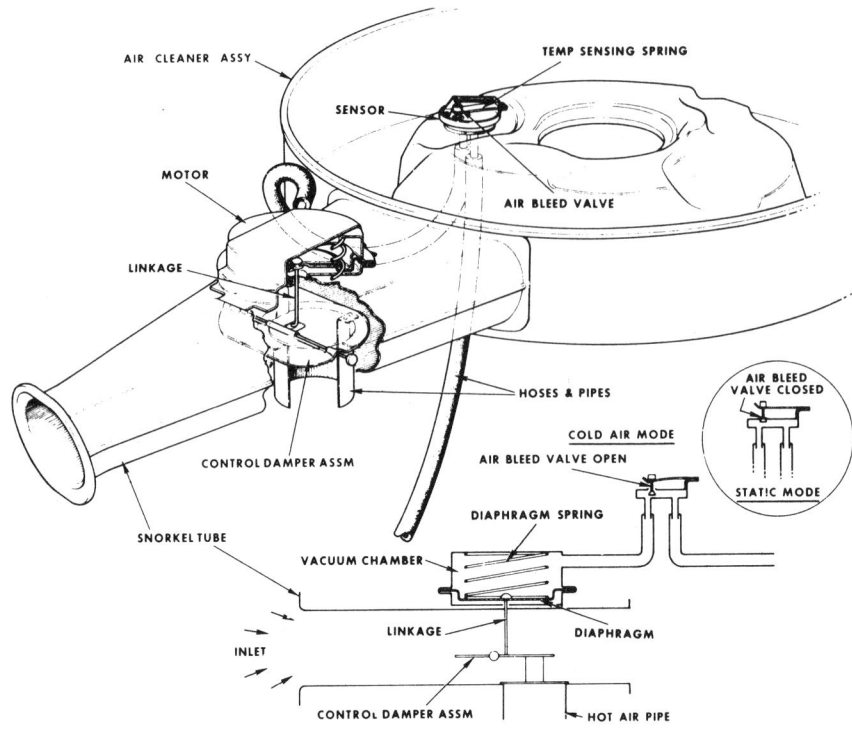

Figure 10-24. Position of the heat control damper door with the engine shut off (Courtesy of United Delco).

5. Start and idle the engine. With the air temperature below about 85 °F, the heat control door within the snorkel should be up or in the heat-on position (Fig. 10-25).

6. When the heat control door begins to move down to open the snorkel passage, remove the air cleaner cover and observe the thermometer reading. It should be between 85 to 115 °F.

Testing the Vacuum Motor Diaphragm

If the damper door does not close completely or does not open at the correct temperature, check the vacuum motor as follows:

1. Turn off the engine. Next, disconnect the vacuum line from the diaphragm.

2. Connect the hose of a portable vacuum pump to the diaphragm fitting (Fig. 10-26). Then, apply 20 inches of vacuum to the diaphragm. The diaphragm should not lose more than 10 inches of vacuum in 5 minutes.

3. Bleed off the vacuum from the hand pump. Next, apply 5 inches of vacuum to the diaphragm while observing the damper door. It should begin to lift from the bottom of the snorkel.

4. Increase the vacuum on the diaphragm to between 9 to 10 inches. The damper door should move to the full-up position.

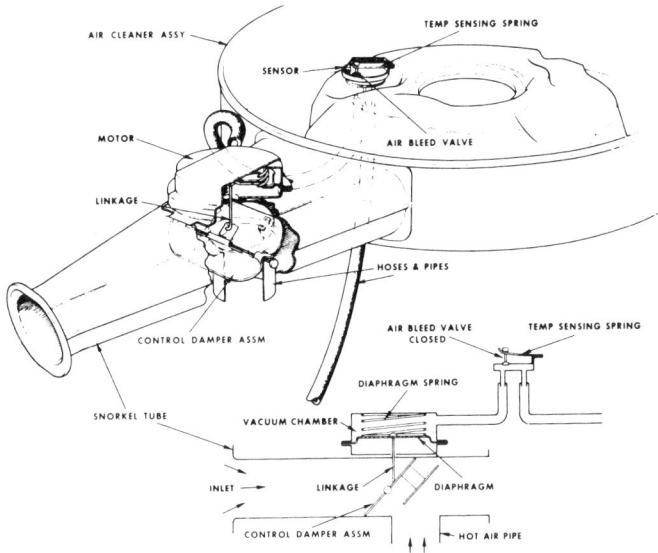

Figure 10-25. Position of the heat control damper door with an underhood temperature of less than 85 °F (Courtesy of United Delco).

5. If the vacuum diaphragm does not perform adequately, replace it following the steps outlined below and recheck the air cleaner's operation.

6. Should the vacuum diaphragm operate satisfactorily, replace the sensor (as explained below) in order to correct the damper door malfunction. Then, recheck the operation of the unit as outlined earlier in this section.

Removing and Replacing the Vacuum Motor Diaphragm

To replace a typical vacuum motor diaphragm that is defective:

1. Remove the air cleaner assembly from the engine and disassemble it.

2. Disconnect the vacuum hose from the vacuum motor diaphragm fitting. Next, tip the diaphragm slightly forward to disengage its lock and rotate counterclockwise.

3. When the diaphragm is free from the snorkel, slide the entire assembly to one side in order to disengage its operating rod from the heat control damper door (Fig. 10-27).

4. With the vacuum diaphragm assembly removed, check the damper door for freedom of operation. When the door is in the up position, it should fall freely when released. If it does not, check the door-to-snorkel sidewalls for interference by foreign material. Also, check the hinge pin for foreign matter. If interference exists, use compressed air to blow the material out of the affected areas.

5. Insert the operating rod of a new diaphragm assembly into the damper door.

6. Position the tangs of the diaphragm assembly into their respective openings in

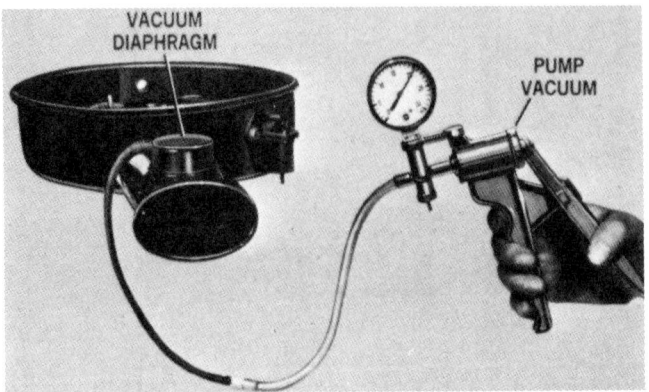

Figure 10-26. Testing the vacuum motor diaphragm with a vacuum pump (Courtesy of Chrysler Corp.).

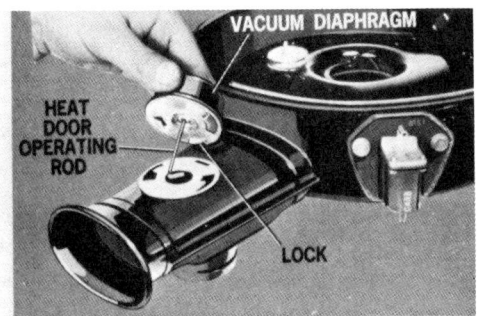

Figure 10-27. Removing the vacuum motor diaphragm assembly from the air cleaner snorkel (Courtesy of Chrysler Corp.).

the snorkel. Then, turn the diaphragm assembly clockwise until its lock engages.

7. With a vacuum pump, apply 9 to 10 inches of vacuum to the diaphragm assembly fitting (see Fig. 10-26). The damper door should move freely to the full-up (heat-on) position. Note: Do not operate the damper door manually because this can cock the rod and diaphragm, which restricts the normal movement of the door.

8. Assemble the air cleaner and install it on the engine. Then, retest the operation of the damper door to make sure that it opens at the correct temperature.

Removing and Replacing the Air Cleaner Temperature Sensor

Follow these instructions for replacing a typical temperature sensor inside an air cleaner:

1. Remove the air cleaner from the engine and disassemble it.

2. Disconnect the vacuum hoses from the sensor fitting. Then, remove the sensor's retaining clips (Fig. 10-28).

3. Remove the sensor and gasket.

4. Position a new gasket onto the air cleaner housing and install the new sensor (Fig. 10-29).

5. While supporting the sensor on its outer diameter, install two new retaining clips securely, making sure that the gasket compresses to form an airtight seal. Note: Do not support the sensor by its guard dur-

Figure 10-28. Removing the sensor retaining clips (Courtesy of Chrysler Corp.).

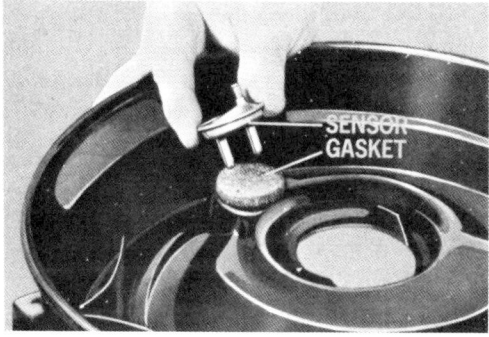

Figure 10-29. Installing a new sensor in the air cleaner (Courtesy of Chrysler Corp.).

ing this process. This can damage the bimetallic strip.

6. Install the vacuum hoses onto the sensor fitting. Then, reassemble the air cleaner.

7. Install the air cleaner onto the engine and test damper door operation.

Summary

1. A modified engine will reduce exhaust emissions internally as long as it is in good mechanical condition.

2. There are several methods used by the industry to test the mechanical condition of an engine.

3. There are a large number of devices added to the modified engine, either to further reduce exhaust emissions or improve the driveability of the vehicle, that require periodic adjustment or servicing.

4. To adjust a jet valve, follow either manufacturer's instructions or those presented in this chapter.

5. When adjusting a carburetor using an infrared analyzer, you must remember two things.

6. Adjust a carburetor's air/fuel ratio at idle with an infrared analyzer using the steps outlined in this text or manufacturer's instructions.

7. A large number of vehicle manufacturers now require that the mechanic adjust the carburetor idle mixture on their vehicles through the use of propane enrichment equipment.

8. With one difference, the adjustment of the carburetor's idle mixture with propane enrichment is quite similar to using the lean-drop method.

9. At the completion of a propane enrichment air/fuel mixture adjustment, it may have been necessary for the technician to alter two carburetor settings.

10. There are a number of reasons why it may be necessary to remove the limiter caps and adjust the air/fuel mixture, utilizing propane enrichment.

11. Use propane enrichment equipment to adjust the idle air/fuel mixture on General Motors, Ford, and Chrysler vehicles using either the procedures outlined in this chapter or factory instructions.

12. To check the operation of a choke-delay valve, follow the instructions presented in this chapter.
13. To check or adjust a typical dashpot's clearance, use the steps outlined in this text or manufacturer's procedures.
14. Adjust antidiesel solenoids utilizing manufacturer's instructions or the service tips presented in this chapter.
15. To check or adjust a typical air conditioner solenoid, follow the instructions outlined in this text.
16. To test the operation of a thermostatically controlled air cleaner, use the procedures presented in this chapter.
17. To test the vacuum motor diaphragm for serviceability, follow the steps outlined in this text.
18. If a vacuum motor is defective replace it following the procedure discussed in this chapter.
19. Follow the instructions in this chapter to replace a typical temperature sensor within an air cleaner.

Review Questions

The questions listed below will assist you in determining how well you remember the material contained in this chapter. Read each question carefully before choosing the answer. If you cannot answer the question, review the section in the chapter that covers the material.

1. When adjusting the jet valve, on a Chrysler-built engine, use a _____ -inch feeler gauge.
    a. 0.002
    b. 0.004
    c. 0.006
    d. 0.008
2. Turning a carburetor's idle mixture screw in (clockwise) _____.
    a. enriches the mixture
    b. leans out the mixture
    c. increases NOx level
    d. increases CO level
3. Propane enrichment equipment is necessary to adjust the idle air/fuel ratio on some vehicles equipped with _____.
    a. catalytic converters
    b. air injection
    c. EGR systems
    d. PCV systems
4. With the propane flowing, the engine's _____.
    a. emission levels should go down
    b. rpm should stabilize
    c. rpm should decrease
    d. rpm should increase
5. If during the propane enrichment mixture adjustment on a Ford vehicle, the idle gain was too little _____.
    a. turn the mixture screw(s) out counterclockwise
    b. turn the mixture screw(s) in clockwise
    c. leave the mixture screw(s) alone
    d. seat the mixture screw(s)
6. On a Ford vehicle, the propane test hose connects into the _____.
    a. air cleaner's canister hose connection
    b. air cleaner's PCV inlet hose connection

c. intake manifold
d. carburetor
7. On a Chrysler vehicle, the propane test hose connects into the _____.
   a. purge hose connection on the air cleaner
   b. PCV hose connection on the air cleaner
   c. intake manifold
   d. EGR valve hose connection
8. A _____ is necessary to check a choke-delay valve.
   a. vacuum pump
   b. engine analyzer
   c. infrared analyzer
   d. thermometer
9. To test the operation of a thermostatically controlled air cleaner, a (an) _____ is necessary.
   a. vacuum gauge
   b. engine analyzer
   c. infrared analyzer
   d. thermometer
10. To test the serviceability of a vacuum motor diaphragm within the air cleaner, use a (an) _____.
    a. engine analyzer
    b. infrared analyzer
    c. vacuum pump
    d. vacuum gauge

For the answers to these review questions, turn to the Appendix.

Chapter 11

Spark-Timing Control Systems

Scientists and automotive engineers proved early on that the amount of spark advance directly affected HC, CO, and NOx emission levels. This led to the development of many types of spark-timing control devices. Manufacturers began installing these devices in the late 1960s and continued their use until 1975, when the catalytic converter came into use.

After the introduction of the catalytic converter, many types of spark-timing control devices were no longer necessary. The converter itself so reduced HC and CO emissions that it made the spark-timing control devices obsolete. In addition, the exhaust gas recirculation (EGR) system was found to better control NOx emissions, eliminating another need for the control of ignition timing in most cases.

Spark Timing Effects on Emissions

To understand how these devices actually curb harmful emissions, let us review the effect of spark timing on HC, CO, and NOx levels. Before the advent of exhaust emission control devices, engines operated on advanced basic ignition timing and used the vacuum and centrifugal advance mechanisms to move the timing ahead to promote maximum engine performance and fuel economy. This advanced timing provided an arc at the spark plug before the piston reached TDC on its compression stroke. This action permitted combustion to occur before TDC and end shortly after TDC.

This arrangement allows the various pistons to liberate the maximum amount of heat energy in the fuel during the combustion process. The advantages of operating an engine on advanced basic ignition timing, along with an adequate advance curve, are an increase in fuel economy and engine performance.

However, an engine with advanced ignition timing produces excessive HC emissions at idle. In this situation, the pistons convert more of the heat energy within the fuel into useful work too early in the engine's cycle. This lowers the temperature of the exhaust gases. As these temperatures decrease, the oxidation of hydrocarbons in the exhaust manifold also lowers. As a result, an engine with advanced basic ignition timing emits more hydrocarbons out of its exhaust and into the atmosphere.

By retarding the basic timing, there is a slightly higher compression of the unburned air/fuel mixture prior to ignition but a delayed burn time (Fig. 11-1). This delayed burn time keeps the exhaust gas temperatures higher, permitting the burning or oxidation of excess unburned hydrocarbons to continue right out into the exhaust manifold and system. This lowers the amount of hydrocarbons emitted into the atmosphere but does reduce engine performance and increases fuel consumption.

Note in Fig. 11-2 that by advancing the ignition timing, CO emissions decrease. However, by doing so, the HC levels increase (see Fig. 11-1). In other words, if a mechanic advances the timing as much as 2.5° over the recommended setting, CO levels decrease but the HC emissions increase. It should be very clear then that advancing the basic ignition timing creates a larger problem, namely, the increase in HC emissions.

Another advantage of retarded timing is an increase in air flow through the carburetor. When any engine operates on retarded basic ignition timing, its idle speed is naturally lower than with advanced timing. This decrease in rpm is due to less fuel heat energy being converted into work by the pistons, and more heat supplied to the oxidation process within the exhaust manifold.

In order to raise the rpm up to the specified amount, the throttle valve must be in a

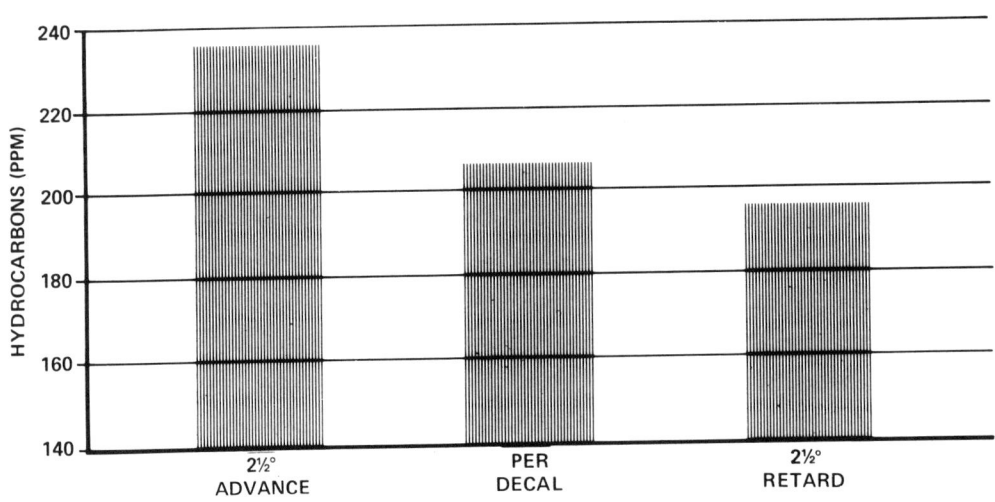

Figure 11-1. Effect on HC emissions with changes in basic ignition timing.

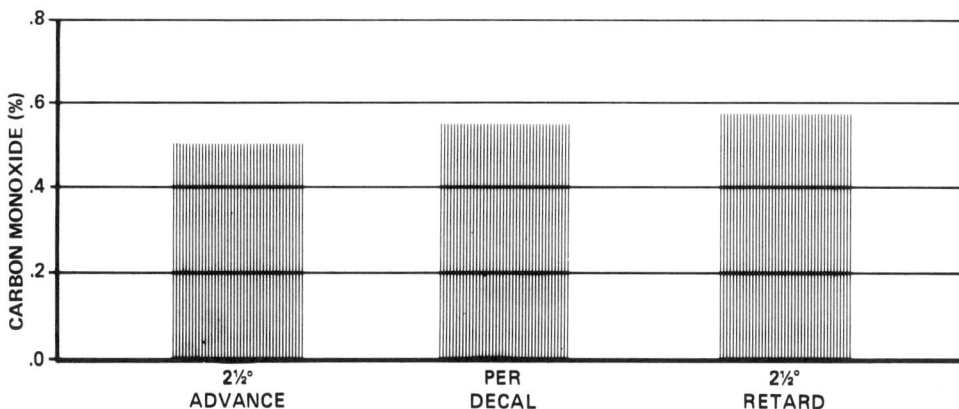

Figure 11-2. Effect on CO levels with changes in basic ignition timing.

somewhat wider open position. This new position permits additional air to pass through the carburetor and into the engine's combustion chambers. The additional air dilutes the amount of residual exhaust gases within the cylinder while supplying more oxygen for combustion. As a result, the engine idles faster with a leaner, more combustible mixture. Finally, the increased air flow provides the oxygen necessary for the oxidation of both hydrocarbons and carbon monoxide, and their emission levels therefore decrease.

Advanced timing also has a negative effect on NOx emissions. As stated before, advanced timing creates high combustion chamber temperatures and pressure—factors which increase NOx emissions. However, the timing has to be advanced somewhat during engine acceleration in order for the engine to develop sufficient power with reasonable fuel economy.

NOx emissions are not a large problem during engine idle: it is only when the engine accelerates between idle and cruise that it emits increased levels of NOx. The reason is, of course, that during these periods, the engine is operating with an enriched mixture, a higher combustion chamber temperature, and cylinder pressure.

During cruise conditions, the engine is running with a light load. Consequently, its mixture is leaner and its combustion chamber temperature and cylinder pressure are lower. Thus, the ignition timing can be advanced for increased fuel economy during cruise without affecting NOx levels to a great extent.

Types of Spark-Timing Control Devices

A variety of spark-timing control devices have been in use since the late 1960s. The type of device used on a given vehicle depends on its manufacturer, the type and size of engine, and the design of other attached exhaust emission control systems. Therefore, a separate discussion of the various systems used by the many automobile manufacturers is beyond the scope of this chapter. However, an overview of some of the main approaches utilized to control distributor spark advance will be

Dual-Diaphragm Vacuum Advance

Figure 11-3 shows a Ford distributor with a dual-action vacuum advance. This mechanism consists basically of two diaphragms: a vacuum retard and a vacuum advance. The purpose of the retard diaphragm is to retard the ignition timing from about 4° to 10° during engine idle and closed-throttle deceleration. This action reduces HC emissions during these phases of engine operation.

The retard diaphragm receives a vacuum signal directly from the intake manifold. Therefore, full manifold vacuum acts on the diaphragm, which is sufficient to overcome its spring tension and pull the retard diaphragm to the right. This action retards ignition timing.

The advance diaphragm, on the other hand, receives its vacuum signal from a vacuum fitting above the carburetor's throttle valves. In other words, the advance diaphragm receives a ported vacuum signal.

In operation, the ported vacuum signal applies itself to the advance diaphragm as the throttle valves open above idle. As a result, this signal, along with the diaphragm spring, moves both diaphragm assemblies to the left, thus overriding the retard side. This action provides the engine with a normal vacuum advance during moderate acceleration and cruise operation.

However, once the throttle valves close to the idle position, full manifold vacuum applies itself once again to the retard diaphragm. This causes the retard diaphragm to retard the timing, which reduces HC emissions during engine deceleration and idle.

Temperature-Sensitive Vacuum Control Valves

If an engine with a dual-action vacuum advance idles for some time in hot weather with retarded timing, it tends to overheat. In this situation, it is necessary to override a retard-advance system to cool the engine off. Manufacturers accomplish this through a temperature-sensitive vacuum control valve (Fig. 11-4).

This device is nothing more than a thermostatically operated valve. The valve, in turn, controls three port openings. One of the openings connects via a hose or line to the carburetor's ported vacuum signal (Fig. 11-5). The second port connects by means of a hose to the vacuum advance side of the dual-action advance mechanism. The last port of the valve attaches into a hose or line, which has full manifold vacuum applied to it.

When the engine is operating at normal temperature, the valve assumes a position as shown in Fig. 11-5. In this position, full manifold vacuum reaches the retard side of the vacuum advance, and carburetor-ported vacuum passes to the advance side of the assembly. Consequently, the engine idles with retarded timing, but the vacuum advance begins to function as soon as the engine begins to accelerate.

Figure 11-6 illustrates what occurs within the valve when the engine overheats. In this case, the ball valve moves upward, uncovering the vacuum passage from the distributor and switches off the carburetor-ported vacuum. As a result, full

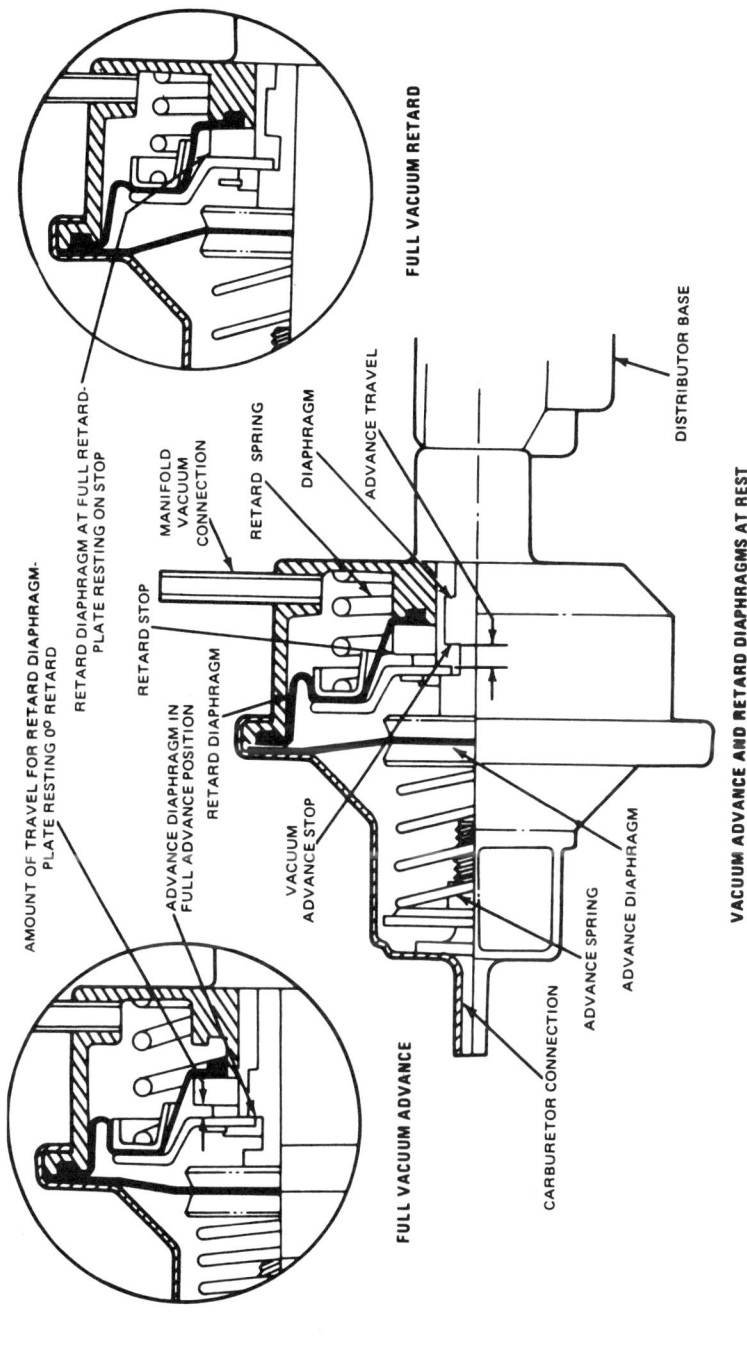

Figure 11-3. Typical dual-diaphragm advance-retard mechanism (Courtesy of Ford Motor Co. of Canada Ltd.).

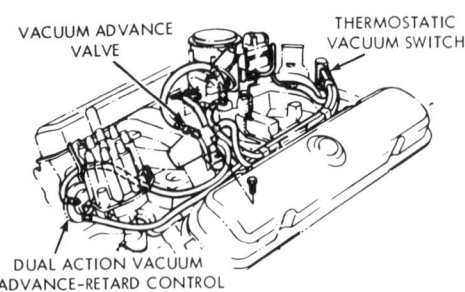

Figure 11-4. Temperature-sensitive vacuum switch is necessary to override a retard-advance system if the engine overheats (Courtesy of United Delco).

manifold vacuum applies itself to both sides of the dual-action assembly. The vacuum advance diaphragm therefore overrides the retard side, and the timing advances between about 4° to 10°. This action makes the engine idle much higher, which causes faster coolant circulation to reduce engine temperature. However, as the engine cools down, the thermostatic valve reverts back to the position shown in Fig. 11-5.

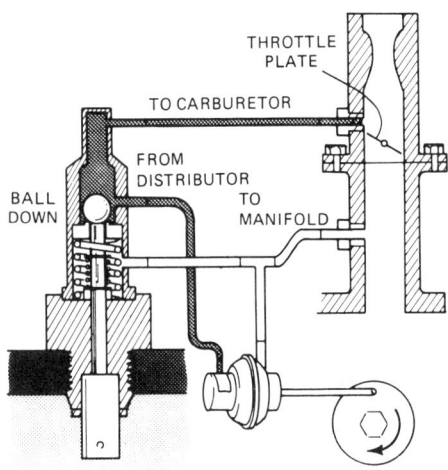

Figure 11-5. Design and operation of a typical temperature-sensitive vacuum switch at normal operating temperature.

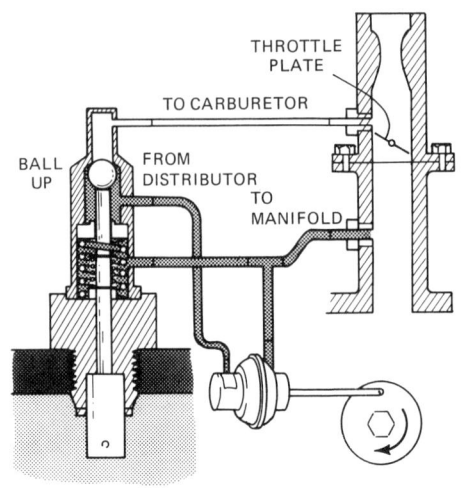

Figure 11-6. Operation of a temperature-sensitive vacuum switch when the engine overheats.

Spark-Delay Valves

As mentioned above, during periods of moderate acceleration with either a dual-diaphragm or standard vacuum advance mechanism on the distributor, engine combustion temperatures rise to create excessive NOx emissions. Since a given amount of advance is necessary for proper engine performance during this time, the manufacturer can only retard the timing enough to curb high emissions. This is usually done by controlling the amount of vacuum advance permitted during light-load acceleration.

One of the methods used by manufacturers to retard the action of the vacuum advance is the use of a spark-delay valve. This valve delays or slows down full vacuum advancement from the distributor by restricting the amount of vacuum that acts on the advance diaphragm.

Figure 11-7 shows the installation and design of a typical Ford spark-delay valve. The valve itself fits into the vacuum line between the ported spark carburetor fit-

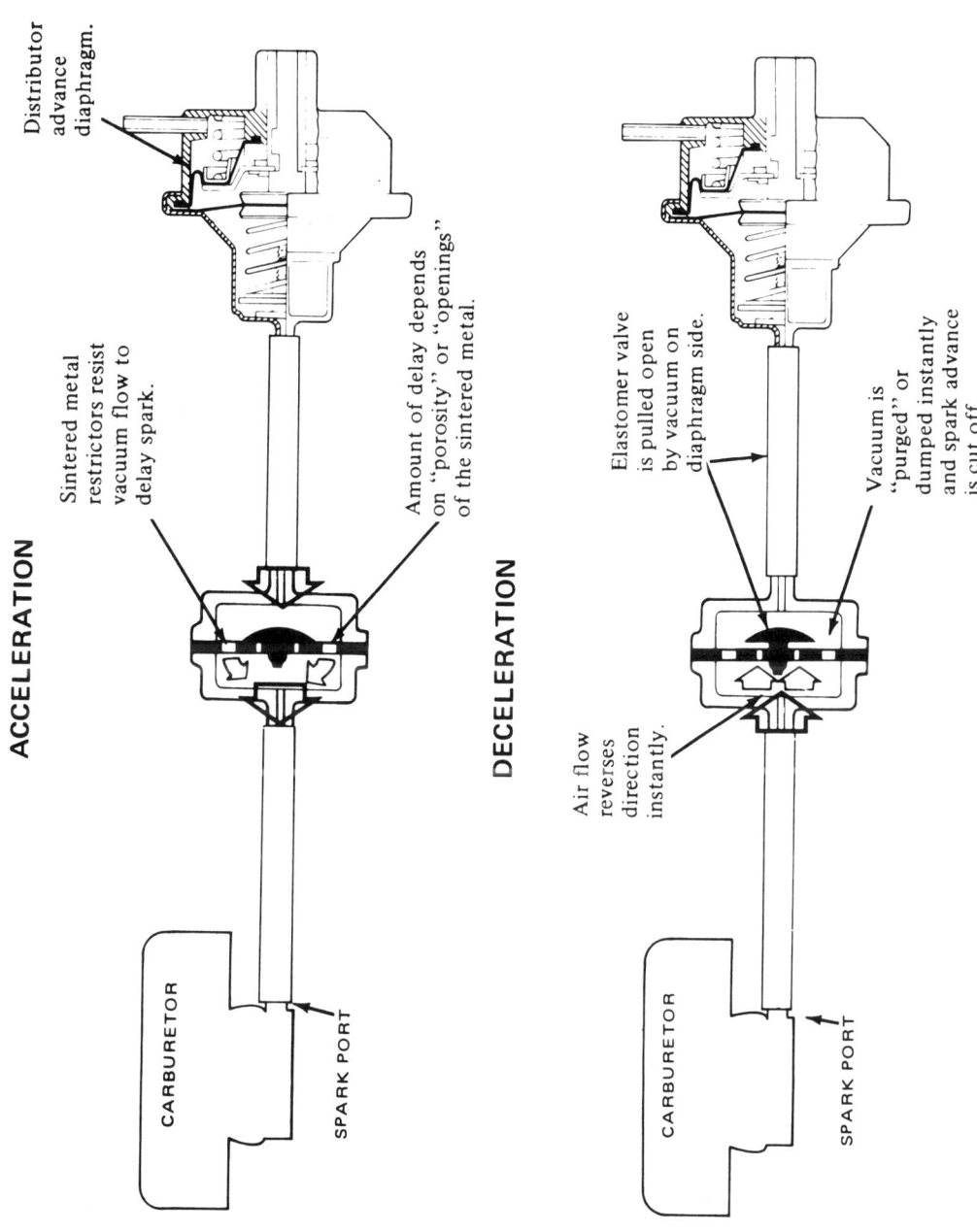

Figure 11-7. Installation and design of a typical spark-delay valve (Courtesy of Ford Motor Co. of Canada Ltd.).

ting and the advance diaphragm. Consequently, the ported vacuum signal to the advance unit must pass through the valve on its way to the diaphragm.

The valve contains a sintered metal restrictor, which resists flow to delay spark advance. The amount of this delay depends on the porosity of the openings within the sintered metal. In addition, the valve has an elastomer, which opens by vacuum on the diaphragm side of the valve. This action occurs when the ported vacuum signal drops to zero as the throttle closes. Finally, in most cases, the valve has color coding that designates the delay time provided by the valve. Consequently, when replacing this valve, always use a new one with the correct color coding.

OSAC Valves

The *orifice spark advance control* (OSAC) valve is a device used by Chrysler Corporation to control NOx and to some degree HC emissions (Fig. 11-8). This valve also delays spark advance during vehicle acceleration by restricting the signal from the carburetor's ported vacuum fitting to the vacuum advance unit. This delay period may be from 10 to 27 seconds when the throttle moves from idle to the part-throttle position.

The restriction within this unit is in one direction only. Therefore, during closed-throttle deceleration or wide-open throttle acceleration, there is no delay in retarding the spark. The retarding of the spark in these situations is brought about by a one-way valve, which bypasses the timing-delay orifice.

Some OSAC valves also contain a thermo control device to retain reliable driveability during engine warm-up. This portion of the device mounts into the air cleaner assembly and senses carburetor inlet air temperature (Fig. 11-9). Consequently, the design of the OSAC system makes it operational when ambient temperature is higher than about 58 °F.

The thermo control device consists of a bimetallic disc that opens the one-way valve whenever it senses temperature below 58 °F. With the one-way valve open, the vacuum signal is unrestricted during all phases of engine operation. When the air temperature within the air cleaner

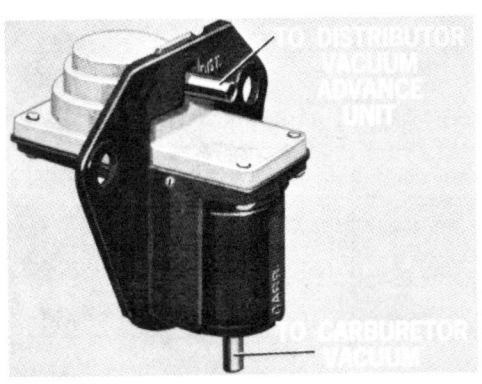

Figure 11-8. An OSAC valve (Courtesy of Chrysler Corp.).

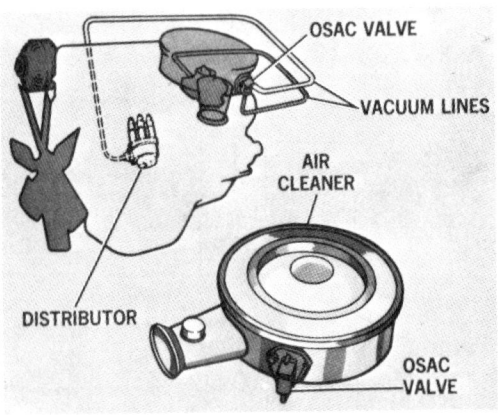

Figure 11-9. Thermostatic control portion of an OSAC valve mounted in the air cleaner assembly (Courtesy of Chrysler Corp.).

reaches about 60 °F, the bimetallic disc again closes the one-way valve, and the OSAC valve delays normal vacuum advance.

One some OSAC systems, a thermo ignition control (TIC) valve senses engine temperature and switches full manifold vacuum directly to the vacuum advance. This action occurs if the engine should overheat. The valve itself has a similar design and operates in much the same manner as the temperature-sensitive vacuum control valve, mentioned earlier in this chapter.

Vacuum Control Valves

Beginning in the middle to the late 1960s, all domestic automobile manufacturers used some form of vacuum control (deceleration) valve on their vehicles with manual transmission. This device prevented an overretard condition during engine deceleration or gear shifting that created high CO emissions in some engines. However, the use of the vacuum control valve was discontinued in the early 1970s because it was not effective against HC or NOx emissions.

To understand why a vehicle with a manual transmission produces high CO levels during deceleration or gear shifting, let us examine what occurs within the power plant during these phases of operation. For instance, when an engine is decelerating, the throttle valves are in the idle position, but the rear wheels attempt to drive the engine at road speed, which is higher than idle rpm.

During this time each intake stroke of the pistons works like a vacuum pump and tries to pull in a full charge of air/fuel mixture into the cylinders. However, the nearly closed throttle valves limit the amount of mixture drawn into the various cylinders. As a result, the density of the mixture is greatly reduced, and the fuel particles separate sufficiently to prevent ease of ignition and uniform and complete combustion.

To make matters worse, the intake valves open before the completion of the exhaust strokes. This permits the entrance of some exhaust gas into the cylinders on their intake strokes. This action dilutes the mixture sufficiently to interfere with its complete combustion. As a result, partly burned or completely unburned fuel passes out of the exhaust to produce CO and HC emissions.

A vacuum control valve (Fig. 11–10) reduces CO emissions during these periods by advancing ignition timing. In order to perform this function, the valve connects to the distributor vacuum advance unit, the carburetor's ported vacuum fitting, and the intake manifold.

Inside the valve assembly there are two chambers, a valve, and a calibrated spring. The chamber on one side of the valve connects to both the carburetor's ported vacuum fitting and the distributor advance unit by means of hoses. The chamber on the other side of the valve connects via a hose into the intake manifold.

The valve controls which source of vacuum will reach and operate the vacuum advance. For example, when the valve is in the down position, as shown in Fig. 11–10, the carburetor's ported vacuum signal acts on the vacuum advance. However, if the valve is in the up position, intake manifold vacuum applies itself to the vacuum advance diaphragm.

A diaphragm connected to the valve and the calibrated spring control the valve's action. For instance, the spring normally holds the valve in the closed position. However, a strong manifold vacuum

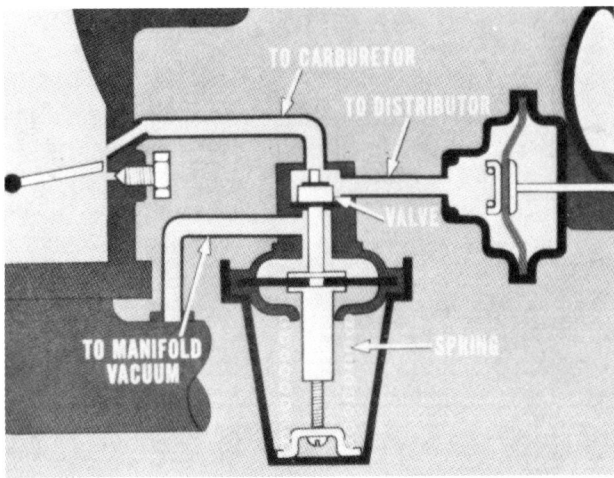

Figure 11-10. The vacuum control valve reduces CO emissions during deceleration or gear shifting (Courtesy of Chrysler Corp.).

signal, acting on the valve's diaphragm, can overcome the spring's force and open the valve.

Operation of the Vacuum Control Valve

During the engine idle, the vacuum valve (Fig. 11-11) does not affect timing because intake manifold vacuum is not strong enough to open the valve against spring tension. The only vacuum force that can act on the advance diaphragm is from the carburetor's ported vacuum fitting. However, at idle, the vacuum from this fitting is not great enough to move the distributor diaphragm and advance the ignition timing.

During moderate acceleration and at cruising speeds, the carburetor's throttle valves are open, and intake manifold vacuum is still too low to open the vacuum control valve. Therefore, only vacuum from the carburetor's ported vacuum fitting applies itself on the distributor's vacuum advance diaphragm. Now since the ported vacuum signal increases with the opening of the throttle valves, there is an increase in ignition advancement due to the action of the distributor diaphragm reacting to the vacuum signal.

Anytime the engine decelerates, the throttle valves move toward their closed position. Now because the carburetor's ported vacuum fitting is above the throttle valves, the vacuum signal from this source is insufficient to provide any movement of the distributor diaphragm, with a corresponding advancement of the timing.

However, intake manifold vacuum is now at or near maximum, so it opens the vacuum valve (Fig. 11-12). When the valve opens, intake manifold vacuum acts on the distributor's diaphragm. This action causes movement of the diaphragm, which provides maximum vacuum advancement of the timing during deceleration (Fig. 11-13). The timing remains advanced for a few seconds until the vacuum in the intake manifold drops below a predetermined amount. When this happens, the vacuum control valve closes, and the timing retards.

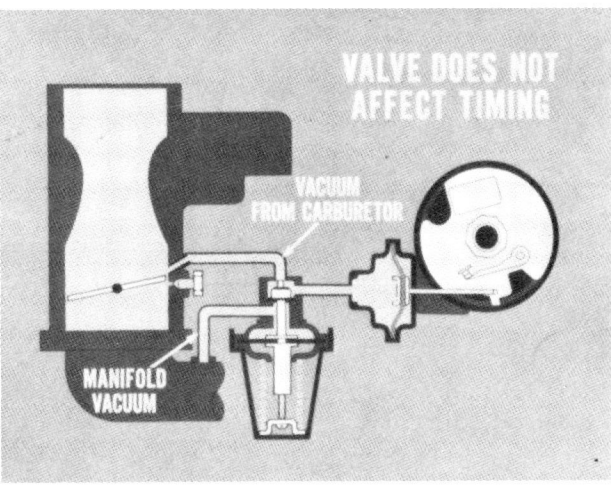

Figure 11-11. Operation of a vacuum control valve during engine idle (Courtesy of Chrysler Corp.).

Because the diaphragm greatly advances the ignition during engine deceleration, combustion starts much earlier, permitting more time for the complete burning of the air/fuel mixture. As a result, CO emissions are reduced to an acceptable level. But as pointed out earlier, the engine still can produce excessive HC emissions.

Transmission-Controlled Spark

Beginning in 1970, some domestic vehicles came from the factory with a transmission-controlled spark (TCS) system. The purpose of this system was to prevent any distributor vacuum advance when the vehicle was in the lower gear ranges or tra-

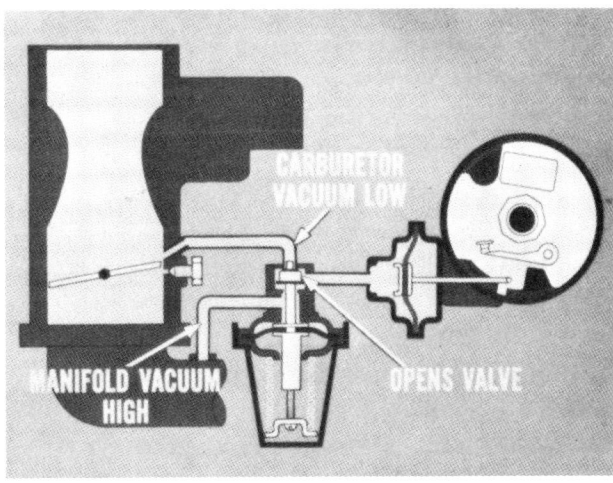

Figure 11-12. Operation of a vacuum control valve during deceleration or gear shifting (Courtesy of Chrysler Corp.).

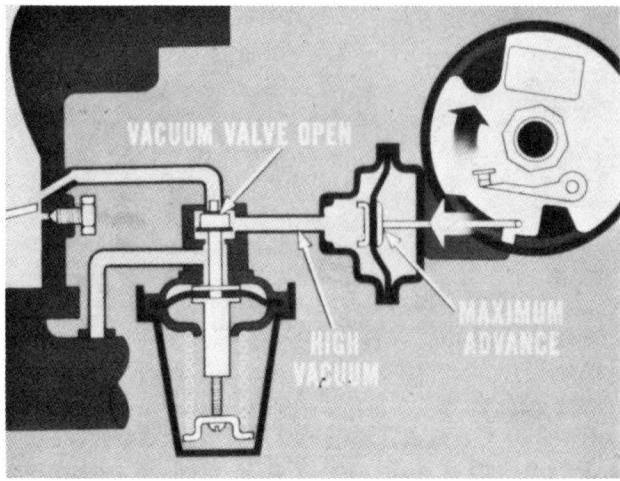

Figure 11-13. With the control valve open, the advance diaphragm provides maximum ignition advancement (Courtesy of Chrysler Corp.).

veling slowly. This action lowered peak cylinder operating temperatures and thereby reduced NOx emissions.

Figure 11-14 illustrates a typical TCS system, which consists of a solenoid-operated vacuum advance switch, transmission switch, temperature switch, and time relay. The solenoid-operated switch, which energizes by the grounding of a normally open transmission switch, controls vacuum to the advance mechanism within the distributor. For instance, when the solenoid is in the nonenergized position, it cuts off vacuum to the vacuum advance and vents it to the atmosphere through a filter at the opposite end of the solenoid. This venting action prevents the vacuum advance from becoming locked in an ad-

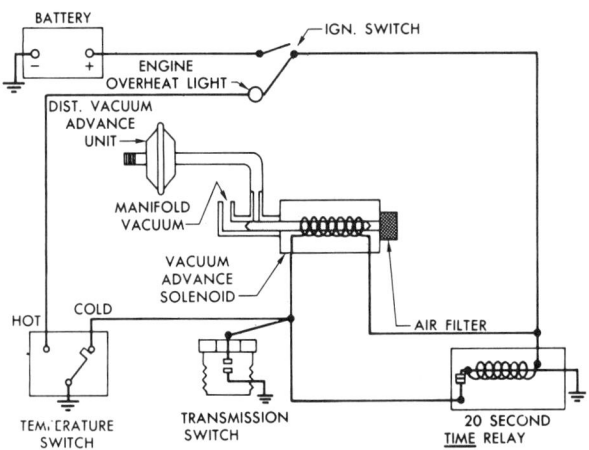

Figure 11-14. Components of a TCS system.

vanced position. As the solenoid energizes, the plunger uncovers the vacuum port and shuts off the air vent. This permits vacuum to act on the vacuum advance diaphragm.

Two switches and a time-delay relay control the operation of the vacuum advance solenoid. The transmission switch used with a manual transmission senses shift-lever position, while the switch used with an automatic transmission operates from hydraulic fluid pressure from the governor circuit. In either case, the switch provides a ground circuit for the solenoid-operated vacuum advance switch when the vehicle is in high gear; but it breaks the circuit when the vehicle is in reverse, neutral, or the lower forward gear ratios.

The thermostatic coolant temperature switch provides a cold override of the TCS system below about 90 °F. In operation, the relay opens the vacuum solenoid switch during starting and for a few seconds after the engine is operating. This provides normal vacuum advance for starting and warm-up.

Some TCS systems have a dual-temperature override switch. In addition to the cold-override provision, the dual switch also has a hot-override terminal. With this design, the thermostatic switch provides a ground to energize the vacuum solenoid when coolant temperature exceeds about 232 °F.

The time-delay relay is an electrically operated on-off type of switch. When its coil energizes, it begins to heat the bimetallic strip in order to open the normally closed relay points in about 20 seconds. This action occurs after the driver turns on the ignition switch. During this 20-second period, the vacuum advance solenoid energizes through the closed relay points.

If for some reason the engine has not started within this 20-second period and the time relay has completed its cycle, it cuts off vacuum advance operation, via the solenoid, until the relay cools off. In other words, once the relay has passed through a cycle with the ignition switch on, it must cool off before reactivating the solenoid. It should be obvious then that the time-delay relay opens the solenoid vacuum switch during starting and for the first few seconds after the engine starts, whether the engine is warm or cold. This provides normal vacuum advance to facilitate starting the engine.

TCS System Operation

Figure 11-14 shows the system components at their at-rest position with the engine off and cold. In this situation, the temperature switch cold-terminal contact points are in the closed position. Also, the time-delay relay points are closed, and the transmission switch points are open. As a result, the vacuum solenoid deenergizes, and its plunger shuts off the port to the vacuum advance unit.

When the driver turns on the ignition switch, a circuit is completed from the ignition switch, through the vacuum solenoid, and to ground through the temperature switch (Fig. 11-15). At the same time, another circuit is completed from the ignition switch, through the time-delay coil, and to ground. Also, as long as the relay points remain closed, they provide a path to ground for the vacuum solenoid. With either the temperature switch or time-delay contacts providing a ground circuit, the vacuum advance solenoid energizes, permitting vacuum to the distributor.

In low and intermediate forward gear operation, with the engine temperature above about 93 °F, the temperature cold-override and transmission switch points are open (Fig. 11-16). Also, if 20 seconds

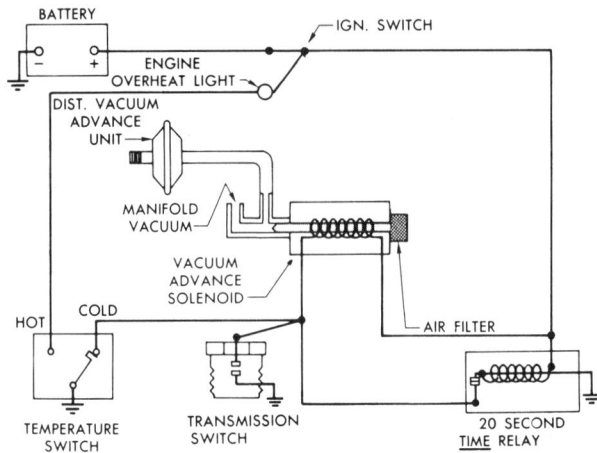

Figure 11-15. TCS system operation with the cold-override energized.

have elapsed, the time-delay points are open. This switch action breaks all the normal grounds that had energized the solenoid. Therefore, the solenoid deenergizes, permitting its plunger to block vacuum to the advance mechanism and open the vacuum advance port to the atmosphere.

As the transmission shifts into high gear, the transmission switch points close (Fig. 11-17). This completes the circuit from the ignition switch, through the solenoid switch, and to ground. As a result, the vacuum advance solenoid energizes, permitting vacuum to pass through the valve to the advance unit.

If the coolant temperature of a system

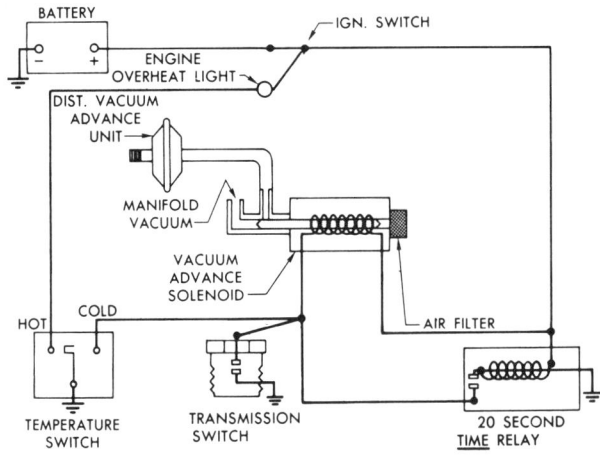

Figure 11-16. TCS system operation in the lower forward gear ratios with the engine at normal operating temperature.

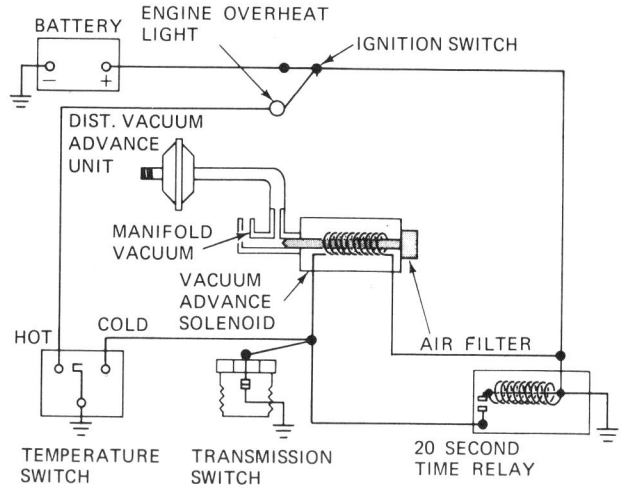

Figure 11-17. TCS system operation in high gear with the engine warmed up.

with a heat-override feature reaches about 232 °F, the hot-override points close (Fig. 11-18). This action completes the electrical circuit of the vacuum solenoid to ground no matter what gear range the transmission is in. Consequently, the solenoid energizes and provides vacuum to the advance mechanism.

Electronically Controlled Timing

Function

Up to this point, this text has described several ways of controlling ignition timing by slightly altering the normal operation of the vacuum advance mechanism. However, in many cases, the vacuum advance

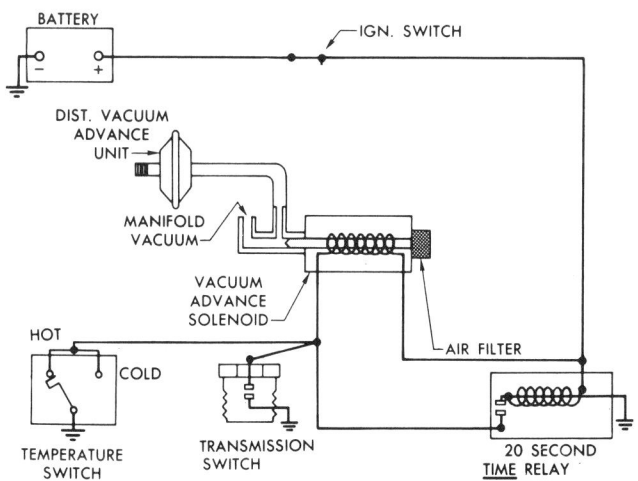

Figure 11-18. TCS system operation with engine temperature at or above 232 °F.

as well as the centrifugal advance mechanisms often cannot react fast enough to changes in engine operating conditions to meet emission control standards and fuel-mileage requirements. As a result, manufacturers have developed computer-controlled ignition-timing systems to provide more precise timing advance during the various phases of engine operation. These systems work in conjunction with electronic ignition systems.

Design of a Typical Electronic Spark Control System

Figure 11-19 is a schematic of a Chrysler electronic lean-burn system, which uses an electronic ignition system and a spark control computer to regulate ignition timing. The computer adjusts ignition timing automatically, according to signals received from the various system sensors.

The spark control computer (Fig. 11-20) mounts onto the air cleaner. The early-model computer for this system had two printed circuit boards inside: the ignition schedule module and the ignition-timing control module. However, the computer used on recently built vehicles has only one circuit board that performs both functions.

The ignition schedule module receives signals from various engine sensors; and after computing them, it determines the exact spark-timing requirement of the engine. Then, the ignition schedule module directs the timing module to advance or retard the timing accordingly.

Depending on system design, it may have six to seven sensors (Fig. 11-21) that feed information to the computer. These include a start pickup coil, run pickup coil, coolant temperature sensor, air temperature sensor, carburetor switch sensor, throttle position transducer, and vacuum transducer. The start pickup coil fits inside the distributor. Its function is to provide a fixed amount of advance during engine cranking.

The run pickup coil is also inside the distributor. This coil supplies a basic timing signal and permits the computer to determine engine rpm. In later-model systems, both the start and run pickup coils are combined into one coil to provide all timing signals to the computer. This coil is shown in Fig. 11-21 as the pickup coil.

The coolant temperature sensor mounts into the water-pump housing. This sensor signals the computer when the coolant temperature is low, so that it can regulate the timing accordingly, in order to improve the driveability of the vehicle during the warm-up period.

The air temperature sensor fits inside the computer. This device is a thermistor, which provides a varying amount of resistance with changing air temperature. For example, as air temperature increases, resistance decreases; as temperature decreases resistance increases.

The carburetor switch senses two operating phases of the engine. That is, the sensor informs the computer if the engine is at idle or at off-idle. This signal, of course, is necessary so that the computer can regulate the ignition timing during these periods of engine operation.

The throttle position transducer connects to the throttle valve lever. A *transducer* is a device that changes mechanical movement into an electrical signal. The transducer itself has a coil and a movable metal core. The small amount of voltage applied to the coil varies in strength as the core moves within the coil. This varying voltage signal is interpreted by the computer. In the case of the throttle position transducer, this core movement informs

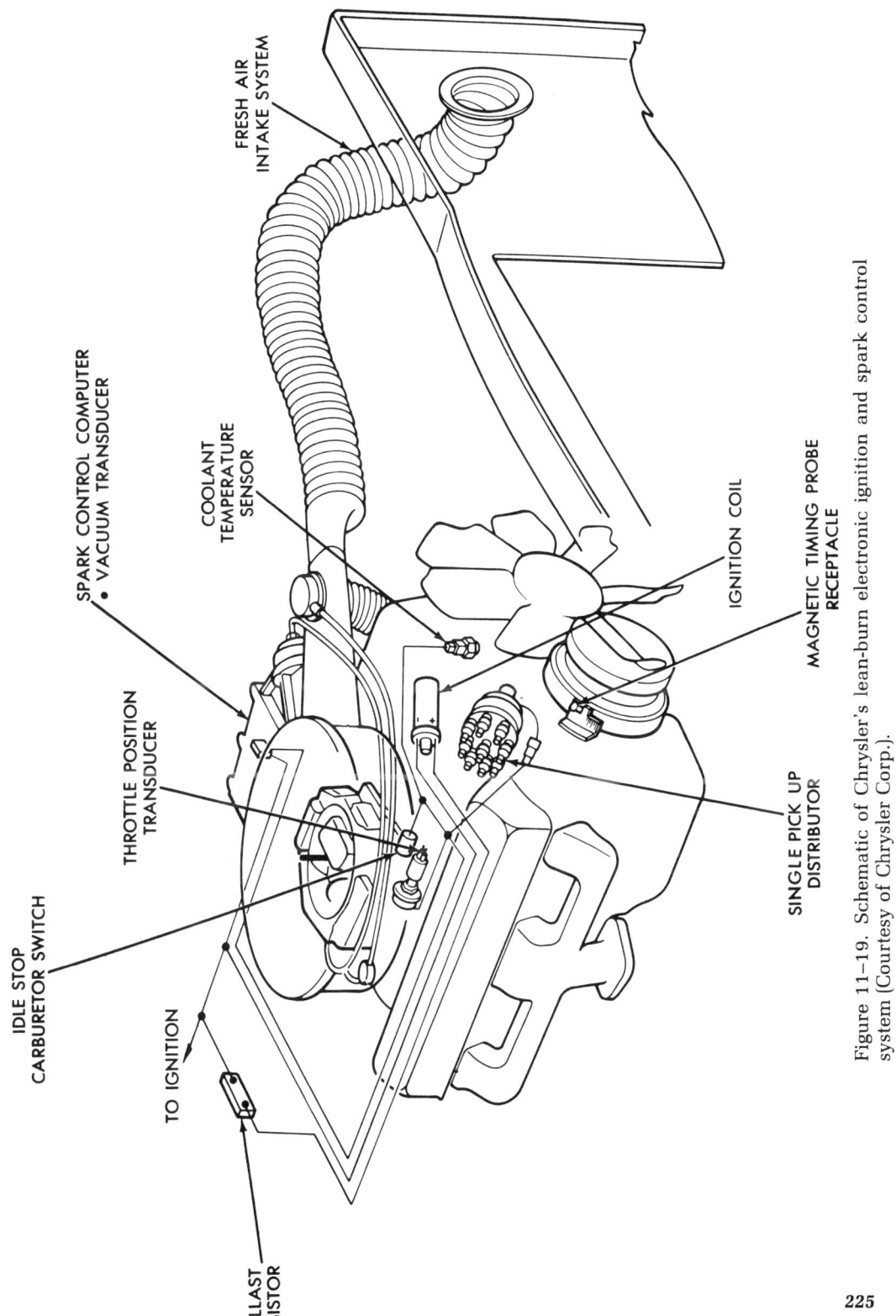

Figure 11-19. Schematic of Chrysler's lean-burn electronic ignition and spark control system (Courtesy of Chrysler Corp.).

226   SPARK–TIMING CONTROL SYSTEMS

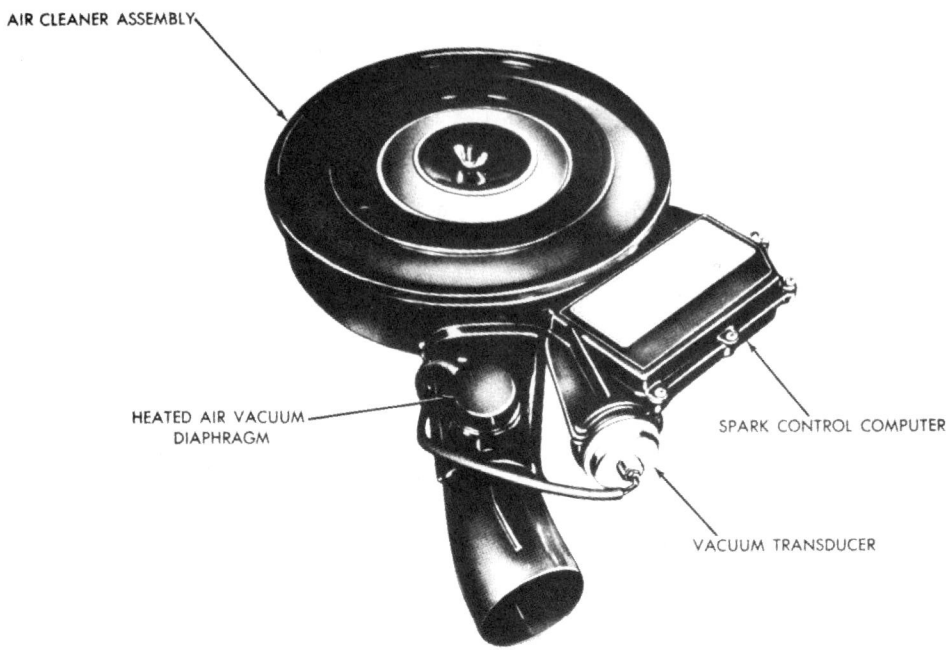

Figure 11-20. Location of the spark control computer (Courtesy of Chrysler Corp.).

TYPES OF SPARK–TIMING CONTROL DEVICES

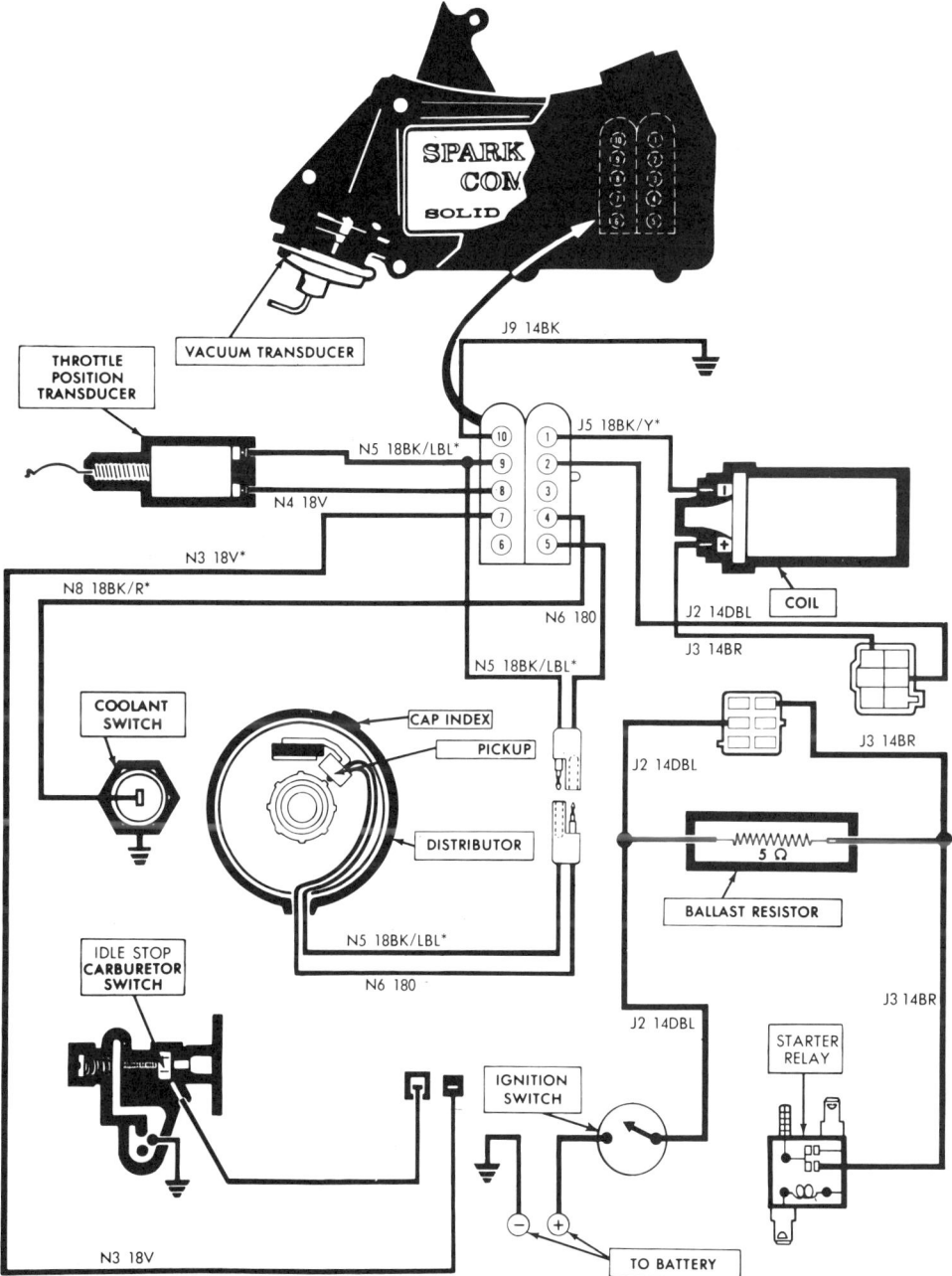

Figure 11–21. Wiring schematic showing the various sensors in the lean-burn system (Courtesy of Chrysler Corp.).

the computer of the position, and the rate of change in the position, of the throttle valves.

The vacuum transducer also mounts onto the air cleaner and spark control computer. This device contains a diaphragm, activated by engine vacuum. Movement of the diaphragm with changes in engine vacuum moves the core within the transducer coil to signal the computer of the various changes in engine vacuum. As a result, the computer alters ignition timing according to engine load.

In this system, the computer depends on distributor shaft rotation to direct a crankshaft position signal to the control module. In other types of systems on the market, the computer receives crankshaft position information from a sensor mounted on the front of the engine and from a disc attached to the crankshaft (Fig. 11-22).

In this situation, the signals taken from the crankshaft are more accurate than those from the distributor shaft. This is mainly because the gears or chain driving the camshaft along with the gear rotating the distributor shaft have operating tolerances or a given amount of looseness. While these tolerances are small, they can combine to cause a significant difference between crankshaft position and ignition timing.

Figure 11-23 shows the location of the magnetic timing probe receptacle at the front of a typical engine. This receptacle accommodates a magentic probe used in conjunction with special testing equipment to check timing on some electronic timing systems.

Summary

1. Scientists and automotive engineers proved early on that the amount of spark advance directly affected HC, CO, and NOx emission levels.

2. After the introduction of the catalytic converter, many types of spark-timing control devices were no longer necessary.

3. To increase fuel economy and engine performance, most vehicles, before

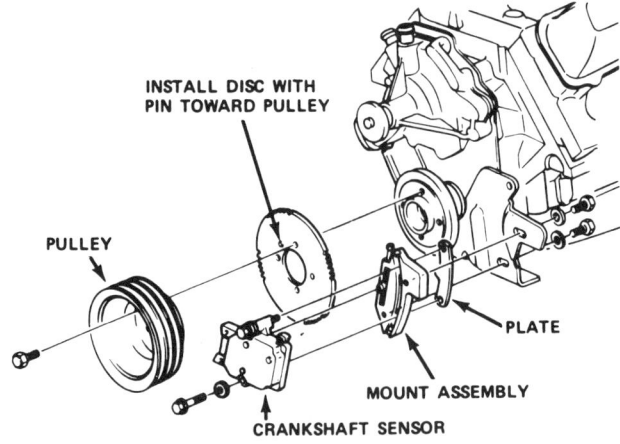

Figure 11-22. Crankshaft position sensor and rotating disc (Courtesy of General Motors Corp.).

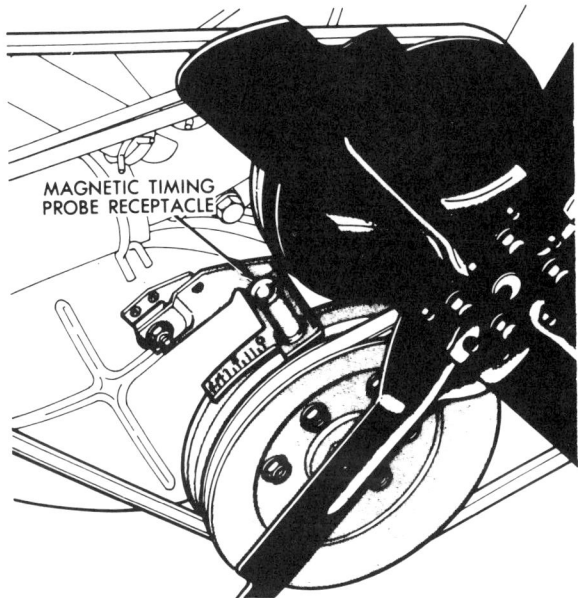

Figure 11–23. Location of a typical timing probe receptacle. (Courtesy of Chrysler Corp.).

the advent of exhaust emission control devices, operated on advanced basic ignition timing.

4. An engine with advanced ignition timing produces excessive HC emissions at idle.

5. Retarding an engine's basic ignition timing creates a delayed burn time.

6. By advancing ignition timing, CO emissions decrease.

7. Another advantage to retarded timing is an increase in the air flow through the carburetor.

8. Advanced ignition timing has a negative effect on NOx emissions.

9. NOx levels are higher during engine acceleration but are lower during cruise operation.

10. The type of spark-timing control device used on a given vehicle depends on its manufacturer, the type and size of engine, and the type of other attached exhaust emission control systems.

11. A dual-action vacuum advance consists of two diaphragms: a vacuum retard and a vacuum advance.

12. The retard diaphragm of a dual-action assembly retards ignition timing during idle and closed-throttle engine deceleration.

13. If an engine overheats, the temperature-sensitive, vacuum control valve advances the ignition timing on units equipped with a dual-action vacuum advance.

14. A temperature-sensitive vacuum control valve has three port openings.

15. One of the methods used by manufacturers to slow down the action of the vacuum advance to reduce NOx emis-

sions is the installation of a spark-delay valve.

16. The spark-delay valve contains a sintered metal restrictor and a one-way valve.
17. The OSAC valve is used by Chrysler Corporation to control NOx and to some degree HC emissions.
18. An OSAC valve delays spark advance from 10 to 27 seconds through the use of a restriction.
19. Some OSAC valves contain a thermo control device.
20. A vacuum control (deceleration) valve was necessary on some vehicles with manual transmissions to prevent an overretard condition during engine deceleration.
21. A vacuum control valve reduces CO emissions during deceleration by advancing ignition timing.
22. A vacuum control valve consists of two chambers, a valve, and a calibrated spring.
23. A vacuum control valve opens during engine acceleration due to high intake manifold vacuum.
24. A TCS system prevents any distributor vacuum advance when a vehicle is in a lower gear range or traveling slowly.
25. A typical TCS system contains a solenoid-operated, vacuum-operated switch, transmission switch, temperature switch, and time relay.
26. Manufacturers have developed computer-controlled ignition timing systems to provide more precise timing advance during the various phases of engine operation.
27. A Chrysler electronic lean-burn system uses an electronic ignition system and a spark control computer to regulate ignition timing.
28. The Chrysler lean-burn system consists of a spark control computer that receives signals from six to seven sensors.
29. The sensors used in the lean-burn system include a start pickup coil, run pickup coil, coolant temperature sensor, air temperature sensor, carburetor switch, throttle position transducer, and a vacuum transducer.
30. In some engines, the computer receives crankshaft position information from a sensor mounted on the front of the engine and a disc attached to the crankshaft.
31. Many vehicles with electronic spark control systems have a magnetic timing probe receptacle at the front of the engine.

Review Questions

The questions listed below will assist you in determining how well you remember the material contained in this chapter. Read each question carefully before choosing the answer. If you cannot answer the question, review the section in the chapter that covers the material.

1. Spark-timing devices were not necessary in most vehicles after _____.
   a. 1974
   b. 1975
   c. 1976
   d. 1977
2. Retarded ignition timing _____ _____ emissions.
   a. decreases HC
   b. decreases CO

c. increases HC
d. increases NOx
3. Advanced ignition timing _____ _____ emissions.
   a. has no effect on CO
   b. decreases HC
   c. increases NOx
   d. increases CO
4. At idle, a dual-action vacuum advance unit retards ignition timing about _____.
   a. 2° to 8°
   b. 4° to 10°
   c. 6° to 12°
   d. 8° to 14°
5. A spark-delay valve _____.
   a. reduces NOx emissions
   b. delays spark advancement
   c. reduces CO emissions
   d. both a and b
6. An OSAC valve delays ignition advancement by the vacuum advance for _____ seconds.
   a. 10 to 27
   b. 12 to 29
   c. 14 to 31
   d. 16 to 33
7. TCS systems began to appear on automobiles beginning in _____.
   a. 1966
   b. 1968
   c. 1970
   d. 1972
8. The early type computers in a Chrysler lean-burn system has _____ printed circuit board module(s).
   a. one
   b. two
   c. three
   d. four
9. In a Chrysler lean-burn system, the air temperature sensor is in the _____.
   a. carburetor
   b. air cleaner
   c. water jacket
   d. computer
10. The vacuum transducer of a Chrysler lean-burn system is in or on the _____.
    a. air cleaner
    b. water jacket
    c. intake manifold
    d. carburetor

For the answers to these review questions, turn to the Appendix.

Chapter 12

Testing Spark-Timing Control Systems

As mentioned in the last chapter, spark-timing control devices reduce HC, CO, and NOx emissions during various phases of engine operation while still retaining reasonable engine performance and fuel economy. If one of these devices malfunctions, there can be a number of negative effects. These include a lack of engine power and an increase in fuel consumption and exhaust emissions.

There have been a large variety of spark control devices used on automobiles and light- to medium-size trucks since their introduction in 1970. The type and design of the device itself depends largely on the particular manufacturer, the vehicle's engine size and type, along with the number of other exhaust emission control devices utilized. Consequently, when testing a particular spark-timing control device, it will be necessary in most cases to use factory instructions and specifications.

For this reason, this chapter only presents samples of typical procedures used by the industry to test spark control devices. The procedures include the testing of dual-diaphragm distributors, thermostatic vacuum switches, spark-delay valves, OSAC valves, vacuum control valves, TCS systems, and electronic spark controls. The design and operation of these particular components or systems were described in the last chapter.

Testing Dual-Diaphragm Distributors

A dual-diaphragm vacuum advance distributor retards basic ignition timing 4° to 10° at idle or on deceleration to reduce HC emissions, but permits normal vacuum advance during moderate engine acceleration and cruise. To test the advance and retard portions of this unit, follow these instructions:

1. Following the manufacturer's recommended procedures, connect both a timing light and a tachometer to the engine.

2. Disconnect and plug both vacuum hoses to the distributor.

3. Clean off the timing marks on both the crankshaft pulley and timing flange. Then, go over the specified timing marks with white chalk to brighten them.

4. Operate the engine at the recommended timing rpm, and observe the initial (basic) timing with the light (Fig. 12–1).

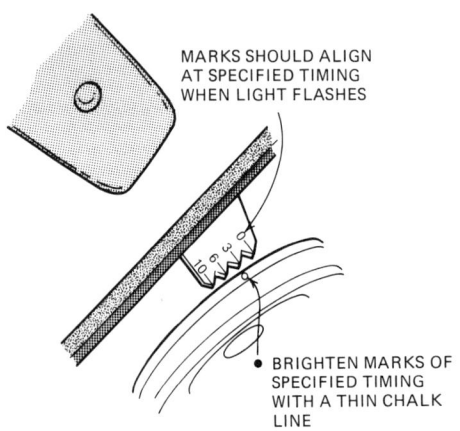

Figure 12–1. Checking initial timing with a timing light.

As necessary, readjust the timing to manufacturer's specifications.

5. Unplug and connect the intake manifold hose to the rear (inner) vacuum chamber fitting, and then observe the timing setting with the timing light. The timing should immediately retard about 4° to 10° from the basic timing specifications (Fig. 12–2). If the rear diaphragm doesn't retard the number of degrees specified by the manufacturer, the distributor vacuum unit is defective and requires replacement. Note: You can verify a leaky diaphragm, at this point, with a vacuum pump before replacing the unit.

6. Adjust engine rpm to 2,000.

7. While checking the timing with the light, unplug and connect the carburetor vacuum hose to the outer advance diaphragm. The timing should now immediately advance (Fig. 12–3). If it does not, a problem exists either in the vacuum advance diaphragm or within the distributor. In this situation, use a vacuum pump to check the diaphragm for leakage and the distributor for a stuck breaker plate.

8. Return the engine speed to the normal curb-idle setting and check the timing

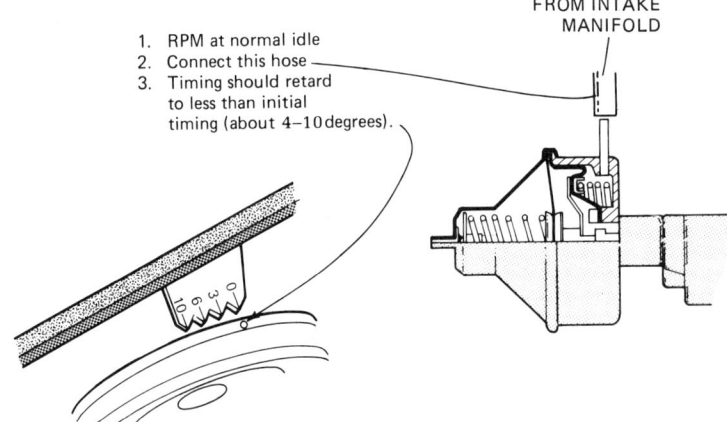

Figure 12–2. Testing the vacuum retard of a dual-diaphragm distributor.

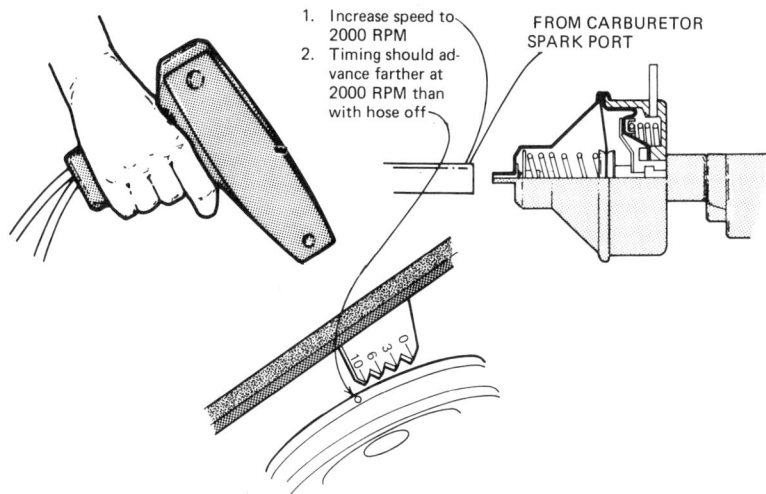

Figure 12-3. Checking the advance portion of a dual-diaphragm distributor.

with the light. The timing should retard in the same manner as in step 5.

9. At the conclusion of the test, disconnect and remove the timing light and tachometer from the engine.

Testing Temperature-Sensitive Vacuum Control Switches

As mentioned in the last chapter, many spark control systems utilize a temperature-sensitive vacuum control switch. Although manufacturers use similar devices with other exhaust emission control systems, a thermostatic vacuum switch, when used with a spark control system, usually permits full vacuum advancement if the engine overheats.

Although thermostatic vacuum switches come in various shapes, sizes, and have different names such as thermo vacuum switch (TVS), ported vacuum switch (PVS), thermo ignition control (TIC), and thermo vacuum valve (TVV), they all operate on the same principle. That is, the switches activate by coolant or air temperature to either cut off or permit a vacuum signal to a given operating unit.

Because manufacturers use so many types of these devices to perform various functions, devices that have different numbers of port openings and activate at different temperatures, it is impossible in the space provided to cover the test procedures for them all. Consequently, when testing one of these devices, always refer to the appropriate shop manual for each type and follow its instructions and specifications. However, this chapter will present the procedures for checking both two- and three-port switches.

To check a typical two-port unit, follow these instructions (Fig. 12-4):

1. Disconnect both vacuum hoses from the valve.

2. Connect a vacuum gauge hose to one port and the hose from a vacuum pump to the other.

TESTING SPARK-TIMING CONTROL SYSTEMS

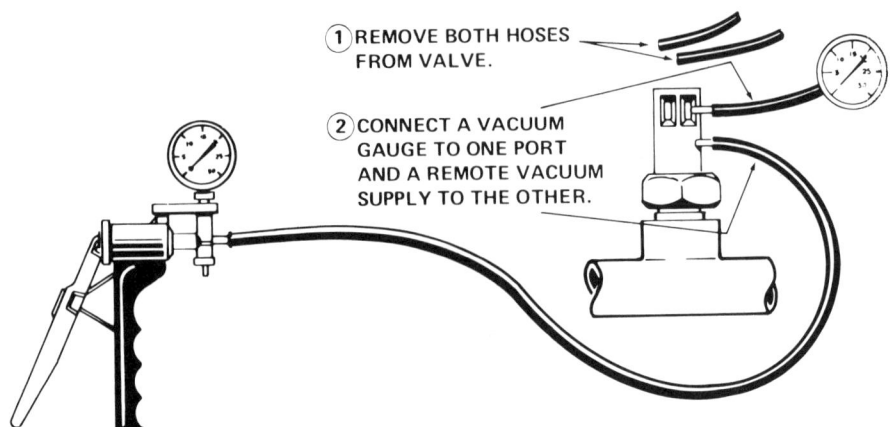

Figure 12-4. Testing a typical PVS assembly.

3. With the pump, apply 10 to 15 inches of vacuum to the valve.

4. Note the reading on the vacuum gauge. The reading should be zero with the engine at normal operating temperature. If the gauge indicates any reading, the valve is defective and requires replacement.

5. With the engine shut off, slowly and carefully remove the radiator pressure cap. Then insert a thermometer into the upper radiator tank.

6. Position a piece of cardboard in front of the radiator core, blocking the normal air flow through its tubes. Next, start the engine and permit it to operate until the coolant reaches the opening temperature of the ported vacuum valve.

7. Check the reading on the vacuum gauge. It should now be the same as on the hand-operated pump gauge. If not, the valve is defective and requires replacement.

To test a typical thermo vacuum valve, use these steps (Fig. 12-5):

1. Check all the hoses leading to the thermo vacuum valve to make certain their routing is correct. The lower hose should run from the valve to the intake manifold. The center hose must connect to the vacuum advance fitting, and the upper hose should run to the carburetor's ported vacuum fitting.

2. Start the engine and permit it to reach normal operating temperature.

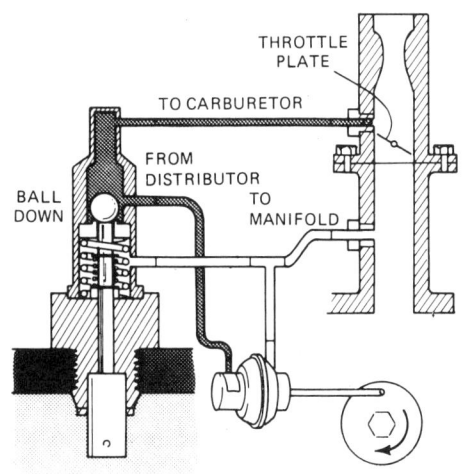

Figure 12-5. Typical TVV vacuum hose connections to the intake manifold, carburetor, and vacuum advance.

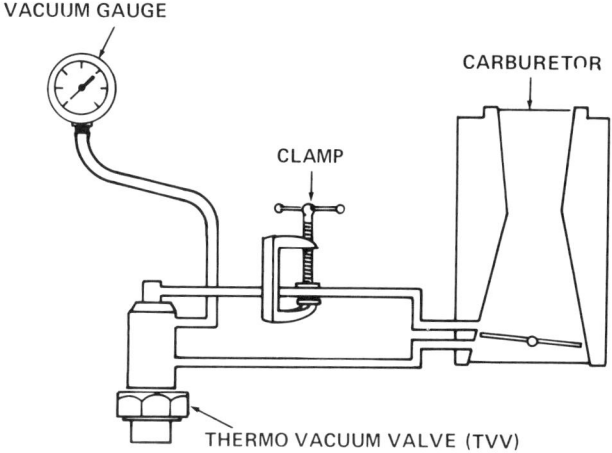

Figure 12-6. Testing a TVV for internal leaks.

Next, with a timing light check and, as necessary, adjust basic ignition timing to specifications.

3. Clamp off the upper vacuum hose from the valve (Fig. 12-6).

4. Remove the center hose from the TVV, and attach a hose from a vacuum gauge to the open fitting.

5. Note the reading on the vacuum gauge. The gauge should read zero.

6. Remove the vacuum hose and clamp from the valve's upper fitting.

7. Remove the lower hose from the TVV. Then, connect both the center and lower hoses together with a plastic fitting (Fig. 12-7).

8. With a light, check ignition timing. It should have advanced about 14° to 20°. If it does, the vacuum advance is functioning properly.

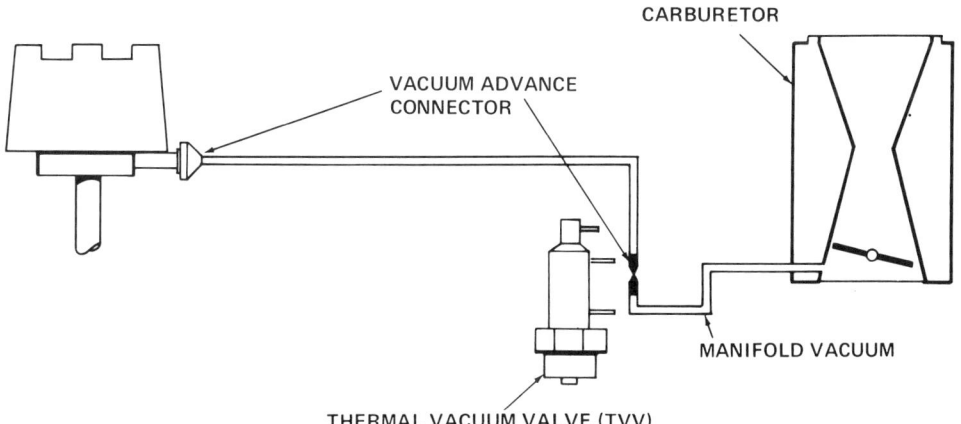

Figure 12-7. Testing the action of the vacuum advance.

9. Reconnect all hoses properly to the TVV (see Fig. 12-5).

10. With the engine shut off, slowly and carefully remove the radiator pressure cap. Then, insert a thermometer into the upper radiator tank.

11. With a piece of cardboard, block off the normal air flow through the radiator. This will elevate engine temperature.

12. Start the engine and operate it until the thermometer indicates the opening temperature of the TVV. Then, recheck ignition timing with the light. The timing should now have advanced 14° to 20°, and the engine's speed at idle should be higher. If not, the thermo vacuum valve is defective and requires replacement.

13. Remove the cardboard from in front of the radiator. The engine should now cool down, the timing should return to the basic setting, and the engine speed should decrease down to the normal curb-idle rpm. If this does not occur, the TVV is sticking in the open position and must be replaced.

14. Disconnect and remove the timing light from the engine.

Testing and Servicing Spark-Delay Valves

Some manufacturers recommend that the spark-delay valve be replaced at given intervals such as every 12,000 vehicle miles or 12 months. In addition, since the spark delay must vary with different engine applications, the valve is color coded for identification. Therefore, when replacing this valve, it is important that the new unit have the same color as the defective valve.

If you suspect a delay valve is malfunctioning, test it following this procedure:

1. Remove the carburetor vacuum hose from the valve. Then, connect the hose from a vacuum pump to this fitting.

2. Remove the hose from the opposite side of the valve, and connect the hose from a vacuum gauge to this fitting (Fig. 12-8).

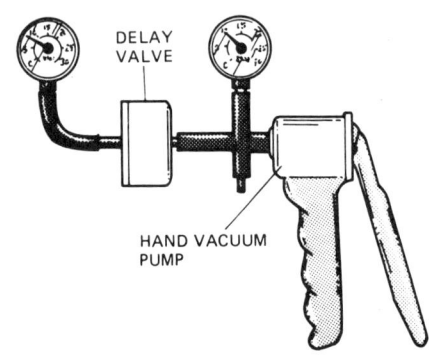

Figure 12-8. Testing a spark-delay valve.

3. Apply 15 inches of vacuum to the valve with the hand pump.

4. Note the reading on the vacuum gauge on the opposite end of the valve from the pump. Depending upon specifications, this gauge needle should rise 15 inches of vacuum in a given number of seconds. If it does not, the valve is defective and requires replacement.

5. Reverse the gauge and pump connections on the valve in order to check the valve's release operation.

6. Apply 15 inches of vacuum to the valve while noting the reading on the other vacuum gauge. Its needle should rise immediately to 15 inches of vacuum. If not, the valve is defective and requires replacement.

Testing an OSAC Valve

The OSAC valve, as found on a Chrysler automobile, may be found on the firewall or on the air cleaner. In any case, to test this valve for serviceability on the vehicle, follow these steps (Fig. 12-9):

1. Remove the distributor vacuum hose from the OSAC valve. Then, connect a hose from a vacuum gauge over the DIST. fitting on the valve.

2. Start the engine and permit it to reach normal operating temperature.

3. Set the parking brake and then operate the engine at 2,000 rpm in neutral.

4. Observe the needle on the vacuum gauge. The valve is functioning properly if you notice a gradual rise in vacuum on the gauge. It should take as long as 20 seconds for the gauge to indicate a stabilized level, depending on the engine and vehicle.

Results and Indications

If the needle immediately rises to indicate intake manifold vacuum, the OSAC valve is defective and requires replacement. If the valve has a temperature override feature, it should permit an immediate vacuum increase on the gauge below 60°F. Therefore, heat and cool the temperature-sensitive portion of the valve in order to check its operation above and below 60°F before condemning the valve.

If no indication occurs on the gauge at any time, the valve is defective and requires replacement.

Testing and Adjusting a Vacuum Control (Deceleration) Valve

One test hookup is all that is necessary to both check and adjust the calibration of a vacuum control (deceleration) valve. To prepare for and perform the actual test and adjustment of a vacuum control valve, follow these directions:

1. Connect an accurate tachometer to the engine.

2. Install a T fitting along with a vacuum gauge into the hose between the control valve and the distributor's vacuum advance (Fig. 12-10).

3. Clamp off the hose between the vacuum control valve and the intake manifold.

4. Start the engine and permit it to reach normal operating temperature. Note: At this point, with the idle speed and mixture set properly, the vacuum gauge should indicate less than 6 inches of vacuum. If not, adjust the engine's idle speed to obtain this reading.

5. Unclamp the hose to the vacuum control valve.

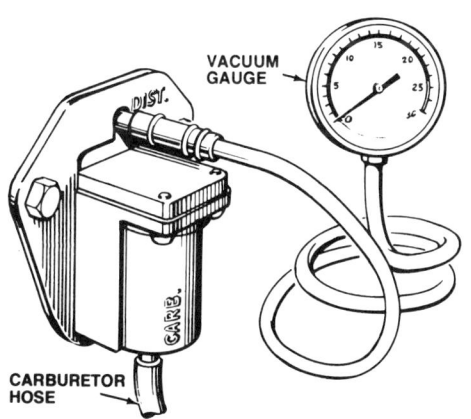

Figure 12-9. Testing an OSAC valve.

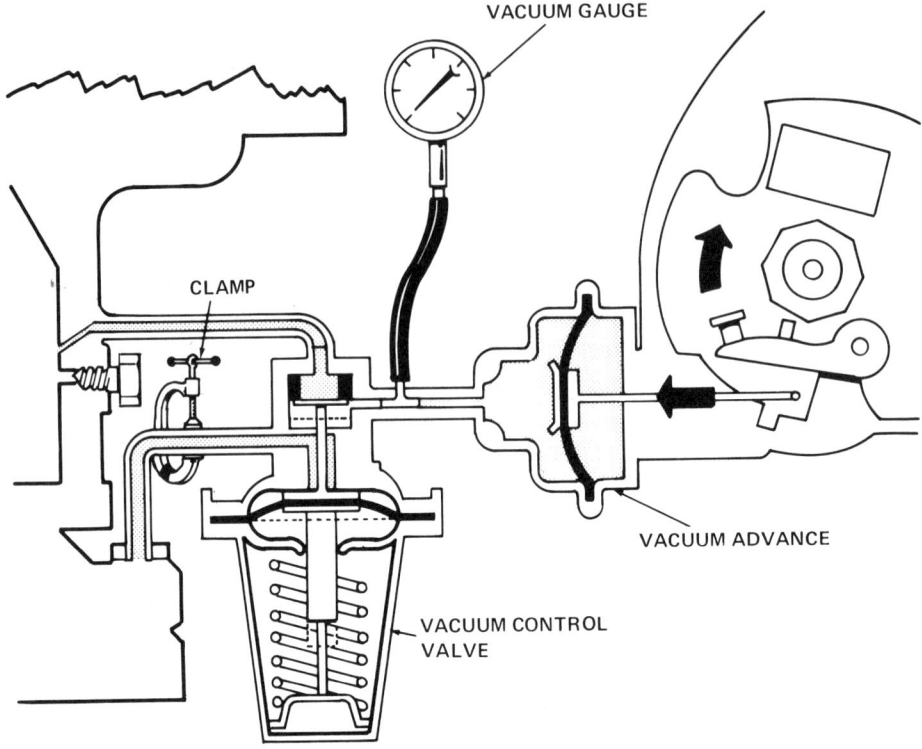

Figure 12-10. Testing a vacuum control valve.

6. Accelerate the engine to about 2,000 rpm, and hold at this speed for approximately 5 seconds. Then, release the throttle.

7. After releasing the throttle, note the reading on the vacuum gauge. The gauge should read above 15 inches of vacuum for a minimum of 1 second. Next, the gauge reading should fall below 6 inches of vacuum in 3 seconds.

Results and Indications

If the vacuum drops below 6 inches in less than 1 second or takes more than 3 seconds to drop below this figure, the valve requires adjustment. To adjust a typical valve:

1. Snap off the plastic cover on the lower part of the valve.

2. To increase the amount of time the valve is open, turn the adjusting screw counterclockwise. This reduces the effective closing pressure of the spring and permits the valve to stay open longer.

3. To decrease the time the valve is open, turn the adjusting screw clockwise. Note: One turn of the adjusting screw changes the valve setting approximately one-half inch of vacuum. For instance, if at the end of 3 seconds the vacuum reading has only dropped back to 7 inches, two clockwise turns of the screw should bring the reading down to 6 inches of vacuum in 3 seconds. If the adjustment does not bring the timing to within limits, the vacuum

control valve is defective and requires replacement.

4. After completing the adjustment procedure, retest the valve's closing time.

Testing a Typical TCS System

Test the transmission-controlled spark (TCS) system mentioned in Chapter 11, following this procedure:

1. Check all vacuum hoses and electrical wiring for proper routing, connection to their respective units, and damage. Repair or replace any damaged units.

2. Tee a vacuum gauge into the vacuum hose to the distributor vacuum advance (Fig. 12–11).

3. With the engine cool, carefully remove the radiator's pressure cap. Then install a thermometer into the upper radiator tank.

4. With the coolant temperature below about 82 °F, start the engine and observe the vacuum gauge. The gauge should read intake manifold vacuum.

5. Continue to operate the engine for at least 3 minutes or at least until the coolant temperature is 90 °F or above. Now observe the vacuum gauge; it should read zero. Stop the engine.

6. After waiting for a few minutes restart the engine and watch the vacuum gauge. The gauge should read intake manifold vacuum for about 20 seconds.

7. Raise the back wheels and install safety stands under the vehicle or position it on a dynamometer.

8. Start the engine and permit it to run at slightly above idle at normal operating temperature with the transmission in neutral or park.

9. Observe the reading on the vacuum gauge. A zero reading indicates the TCS system is functioning properly at this point.

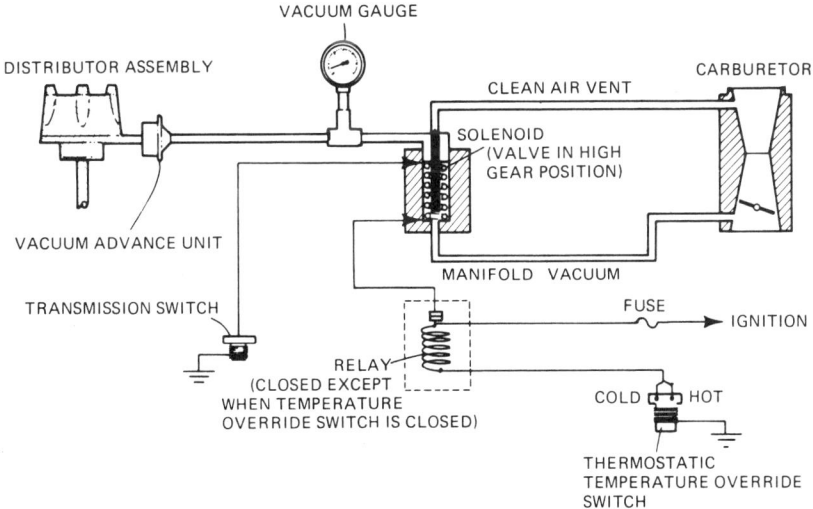

Figure 12–11. A vacuum gauge installed into the hose to the vacuum advance in preparation for testing the TCS system.

10. While watching the vacuum gauge, place the transmission into gear. Next, accelerate the engine until the automatic transmission shifts into high gear, or you can shift the manual transmission into fourth gear.

11. Observe the indication on the vacuum gauge. It should now read full manifold vacuum. Note: Some manual transmissions use transmission switches that activate the TCS system in both third and fourth ratios. In these situations, the gauge will show a full vacuum reading in both ratios.

Results and Indications

1. If the solenoid does not activate and the vacuum gauge shows a reading in step 4 of the test, check for electrical power to the solenoid with a test light or voltmeter.

2. If the solenoid has power but only energizes for 20 seconds during step 4, test the temperature override switch. To test the switch:
   a. Remove its wiring connection.
   b. With a jumper lead, ground the wire or wires in the harness.
   c. Check the operation of the solenoid. It should now energize. If it does not, the solenoid is defective and requires replacement.

3. If the solenoid did not energize during step 6, check the operation of the time-delay relay. To check the relay:
   a. Disconnect and ground the single wire from the solenoid to the relay.
   b. Check the solenoid for operation. If it now energizes, the relay is defective and requires replacement.

4. If the solenoid has power and still does not energize after you check both the temperature switch and time-delay relay, check the solenoid itself for serviceability. To perform this check:
   a. Remove the wiring connector from the solenoid.
   b. Connect a jumper lead from a 12-volt source to one solenoid terminal and ground the other terminal, using a second jumper lead.
   c. Check the operation of the solenoid. It should now energize. If it does not, replace the solenoid because it is defective.

5. The vacuum gauge reading should be zero during step 9 but rise to indicate full manifold vacuum after step 10. If this does not occur, remove the single wire from the transmission switch and ground it with a jumper lead. If the solenoid now energizes, the transmission switch is either defective or in some cases, out of adjustment. Replace or adjust the switch as necessary.

Testing A Typical Electronic Spark Control System

In many cases, the mechanic can still use a powered timing light to test the operation of an electronic spark control system. This text described the powered timing light and how to use it in Chapter 4.

However, many electronic spark control systems require the use of a magnetic timing unit to check its operation. The Sun™ 1115 engine analyzer shown in Fig. 12–12 also has a MAG-81™ magnetic timing unit. This unit works in conjunction with the analyzer to accurately measure the basic ignition timing and advancement

TESTING A TYPICAL ELECTRONIC SPARK CONTROL SYSTEM

Figure 12–12. A typical engine analyzer.

the crankshaft, which indicates TDC position of the number-one piston. The magnetic timing unit then uses this information to provide a digital readout of exact engine timing at any moment and at any engine speed.

The readout window displays timing information in digital form. In other words, the readout window shows a series of numbers representing basic as well as any advancement in ignition timing due to the action of mechanical or electronic spark advancement mechanisms. Finally, numbers within the readout window also verify the correct adjustment for the offset angle of the magnetic probe receptacle.

Between the readout window and the offset angle knob is a two-position switch. The technician uses the manual position of this switch for testing magnetic timing of current model vehicles. The manufacturer provided by an engine with a magnetic timing receptacle and an electronic spark control system. The MAG-81 magnetic timing unit, unlike the standard or powered timing light, displays initial and timing advance information digitally in the readout window on the unit.

The MAG-81 basically consists of a magnetic probe and a front panel that has a readout window, a control switch, and an offset angle knob. The magnetic probe connects via an electrical harness to the left side of the panel and has a probe that fits into a receptacle on or near the engine's timing bracket (Fig. 12–13).

The probe itself picks up a signal from

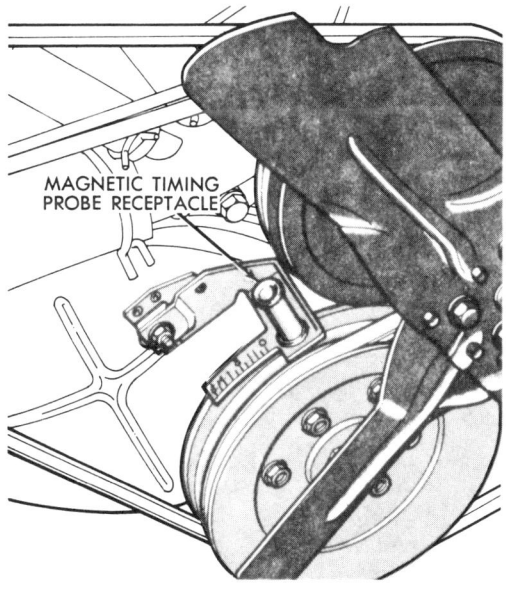

Figure 12–13. A typical magnetic timing receptacle (Courtesy of Chrysler Corp.).

of this unit has set aside the other position of the function switch for future applications; therefore, it is not utilized at this time.

The offset angle knob is necessary to compensate for the offset angle of the timing receptacle. In other words, the probe receptacle in most cases is offset a number of degrees from the zero mark on timing bracket (Fig. 12-14). The offset angle may range from a $-9.5°$ on AMC and General Motors vehicles to as much as a $-135°$ on Ford-built vehicles. In any case, before actually performing the magnetic timing check, the mechanic adjusts this knob one way or the other until the correct number of degrees of offset angle appear in the readout window.

To check basic ignition timing and the amount of advancement provided by a typical electronic spark control system using the MAG-81 timing unit, follow these steps:

1. As necessary, insert the proper probe adapter into its slot in the timing bracket (Fig. 12-14).

2. Insert the magnetic pickup probe into the opening in the adapter or receptacle so that the probe tip is in contact with the engine's vibration damper. Note: In some installations, the lock ring on the probe contacts the adapter before its tip bottoms against the vibration damper. This design permits the probe, in this situation, to enter the adapter to the correct depth for test purposes.

3. Make certain that the probe and lead assembly are clear of the fan, pulleys, etc.

4. Turn the oscilloscope ac power on. This action lights up the MAG-81.

5. Prepare the scope tester and make all the analyzer connections as shown in Fig. 12-15, except for the red trigger pickup that fits around number-one spark plug lead.

6. Connect the probe harness to the receptacle on the left side of the MAG-81 control panel (Fig. 12-16).

7. Start the engine and adjust the offset angle knob until the correct setting appears in the readout window.

8. Connect the red trigger pickup around the number-one spark plug cable.

9. With the engine at curb idle, read the initial timing setting in the readout window.

10. Increase engine speed to 2,500 rpm, or the speed recommended by the manufacturer.

11. Note the amount of total advancement appearing in the readout window.

Results and Indications

If the initial timing or the total advancement curve provided by the elecronic spark control system are incorrect, follow

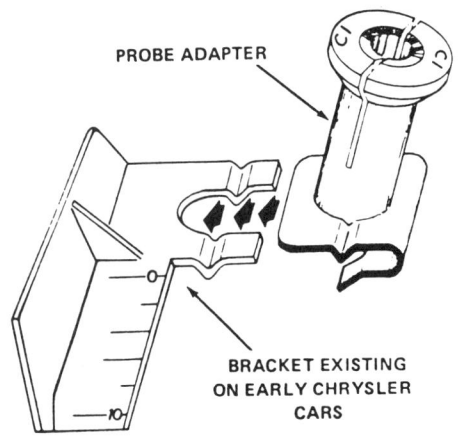

Figure 12-14. The timing receptacle is usually offset a number of degrees from the zero or TDC mark on the timing bracket (Courtesy of Sun Electric Corp.).

TESTING A TYPICAL ELECTRONIC SPARK CONTROL SYSTEM

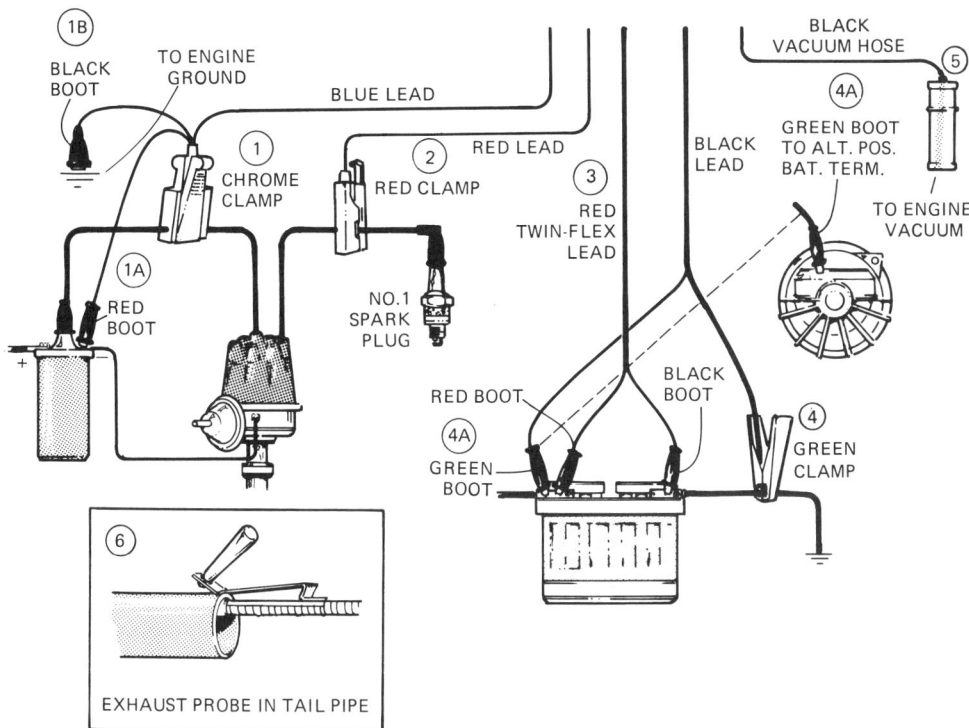

Figure 12–15. Engine analyzer's lead connections to the vehicle.

Figure 12–16. Control panel of a Sun MAG-81.

the manufacturer's instructions for adjustment of basic timing or troubleshooting the system for malfunctions.

Summary

1. If a spark control device malfunctions, there will be a lack of engine power and an increase in fuel consumption and exhaust emissions.
2. When testing a particular spark-timing control device, it will be necessary in most cases to use factory instructions and specifications.
3. To test the advance and retard portions of a dual-diaphragm distributor, follow manufacturer's instructions or those presented in this text.
4. A thermo vacuum switch, used with a spark control system, usually permits full manifold vacuum if the engine overheats.
5. All thermostatic vacuum switches operate upon the same principle.
6. When testing a thermo vacuum switch, always refer to the appropriate shop manual for instructions and specifications.
7. To test a typical two- and three-port thermo vacuum switch, follow the instructions set forth in this chapter or within the vehicle service manual.
8. Some manufacturers recommend that a spark-delay valve be replaced at given intervals.
9. If you suspect a delay valve is malfunctioning, test it using the procedure outlined in this chapter or manufacturer's instructions.
10. To test an OSAC valve, as found on a typical Chrysler automobile, use the manufacturer's instructions or those found in this text.
11. To prepare for and perform the actual test and adjustment of a vacuum control valve, use the procedures outlined either in this chapter or those found in the appropriate service manual.
12. To test the serviceability of a TCS system, follow the typical procedures outlined in this text or the vehicle service manual.
13. Many electronic spark control systems require the use of a magnetic timing unit to check its operation.
14. The MAG-81 basically consists of a magnetic probe and a front control panel that has a readout window, control switch, and an offset angle knob.
15. To check basic ignition timing and the amount of advancement provided by a typical electronic spark control system using the MAG-81 timing unit, follow the steps presented in this chapter.

Review Questions

The questions listed below will assist you in determining how well you remember the material contained in this chapter. Read each question carefully before choosing the answer. If you cannot answer the question, review the section in the chapter that covers the material.

1. With the rear chamber of a dual-diaphragm distributor connected to the intake manifold via a hose, the timing should retard _____.
   a. 3° to 8°
   b. 4° to 10°
   c. 7° to 12°
   d. 9° to 15°

2. To raise engine temperature in order to check the operation of a thermo vacuum valve _____.
   a. block the radiator core
   b. race the engine
   c. disconnect the fan belt
   d. remove the thermostat

3. When a thermo vacuum valve opens, ignition timing should advance _____.
   a. 8° to 14°
   b. 10° to 16°
   c. 12° to 18°
   d. 14° to 20°

4. Manufacturers recommend the replacement of some spark-delay valves every 12 months or _____ vehicle miles.
   a. 8,000
   b. 10,000
   c. 12,000
   d. 14,000

5. An OSAC valve should delay the application of full manifold vacuum to the distributor for _____ seconds.
   a. 20
   b. 18
   c. 16
   d. 14

6. A vacuum control valve, on engine deceleration, should maintain 15 inches of vacuum on the distributor for _____ second(s).
   a. 3
   b. 2
   c. 1
   d. 0

7. On a typical TCS system the engine has normal vacuum advance below _____ °F.
   a. 72
   b. 82
   c. 92
   d. 102

8. In order to test many electronic spark control systems, you must use a _____ assembly.
   a. vacuum gauge
   b. engine analyzer
   c. infrared analyzer
   d. magnetic timing

For the answers to these review questions, turn to the Appendix.

Chapter 13

Air Injection Systems

Since the passage of the first exhaust emission control standards, manufacturers have extensively used the air injection system. One of the main reasons for its popularity is that this system is adaptable to, and functions well on, most gasoline engine designs. In other words, the device is an external add-on unit utilized with or without the engine modification system mentioned earlier in this text.

Different vehicle manufacturers call their air injection systems by different names. For instance, American Motors calls it "Air Guard." Chrysler Corporation refers to it "Air Injection." While Ford Motor Company names its unit "Thermactor." Lastly, General Motors calls its version of the basic system "Air Injection Reactor" or "AIR."

Regardless of the name, its function is the same no matter which vehicle it is on. That is, the air injection system introduces fresh air, with its oxygen content, into the exhaust of an operating engine. This process causes further oxidation (burning) of the hydrocarbons and carbon monoxide left in the hot exhaust gases. In other words, the oxygen unites with the carbon monoxide to form carbon dioxide, a harmless gas. The oxygen also combines with the hydrocarbons to produce water, usually in vapor form. As a result, the air injection system is very effective in lowering both HC and CO emissions from any automotive-type gasoline engine.

In some vehicles, the air injection system directs the air into the base of the exhaust manifold to assist the oxidation process in this area. However, other systems inject the air through the cylinder head, at the exhaust ports, causing the oxidation process to begin within this area.

The Design of the Air Injection System

Air Pump

A typical air injection system consists of a belt-driven air pump, combination diverter and air pressure relief valve, check valve, and air injection manifolds (Fig. 13-1). The belt-driven air pump mounts onto the front of the engine. The crankshaft pulley drives this vane-type pump, which in turn supplies a high volume of air at low pressure to the injection system (Fig. 13-2).

The Saginaw Division of General Motors manufactures the air pumps used on domestic vehicles. The early Saginaw pumps (1966-67) were three-vane units that could be rebuilt in the field if necessary. However, a Saginaw two-vane design replaced the original style on 1968 and later models. This unit is usually not rebuildable in the field; therefore, it requires replacement when it fails. Finally, when most of the older three-vane pumps fail, most mechanics replace them with the newer two-vane units.

The main operating difference between the two pumps styles is the manner in which they filter the incoming air. For example, the three-vane pump draws its fresh air supply through a separate air filter (Fig. 13-3) or from the clean side of the air cleaner assembly. But the two-vane unit utilizes an impeller-type centrifugal filter fan.

The centrifugal filter fan mounts on one end of the air pump's rotor shaft and therefore turns at pump rpm. This unit is not a true filter but it does clean the air as it enters the inlet port of the pump, which eliminates the need for separate air filter and hose connections.

The centrifugal filter itself is nothing more than a vaned wheel that, when rotating, actually opposes the normal air entry into the pump's inlet; this opposition is not sufficient to hamper the flow substantially. However, the opposing force causes the discharge of any foreign particles within the air before they actually enter the pump.

In operation, the air enters the pump inlet by passing between the vanes of the centrifugal filter fan (Fig. 13-4). The vanes of the fan assembly are rotating at a relatively high rpm. Consequently, the turning vanes strike any heavier-than-air particles attempting to enter the pump's inlet with the incoming air and rebounds them out and away from the pump. In other words, the fan vanes force the foreign particles to move out of the filter in a direction opposite to the flow of air coming into the pump's inlet.

Along with the filter fan, later air pump designs also have a set of two vanes mounted on a rotor inside the pump housing (Fig. 13-5). These vanes fit into slits within the rotor and are 180° apart. Moreover, a set of seals (two per vane) provide a seal between the vanes and the rotor.

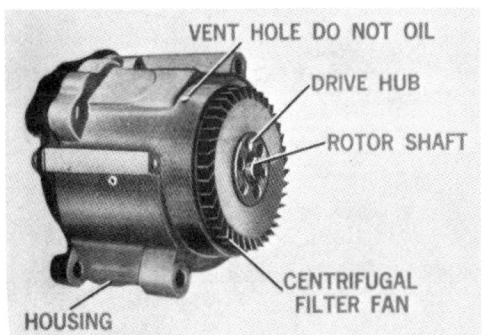

Figure 13-2. Typical air injection pump (Courtesy of Chrysler Corp.).

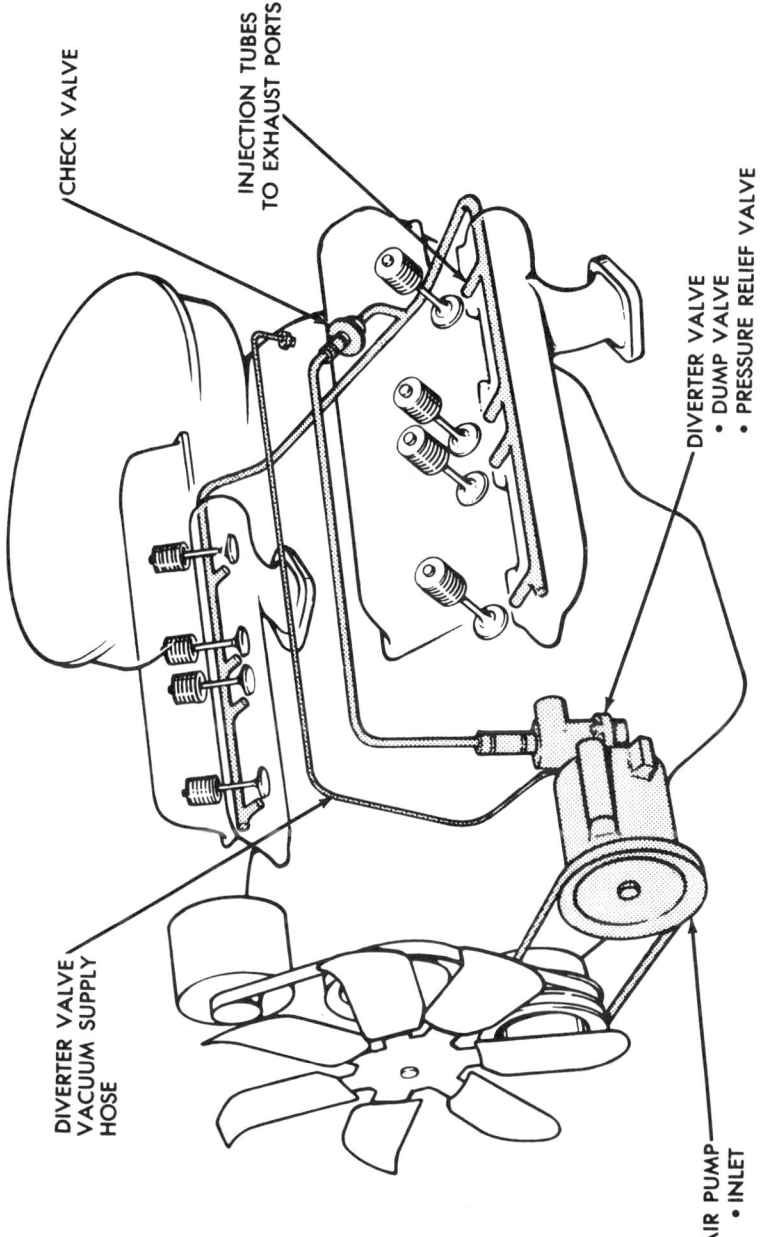

Figure 13–1. Design of a typical air injection system (Courtesy of Chrysler Corp.).

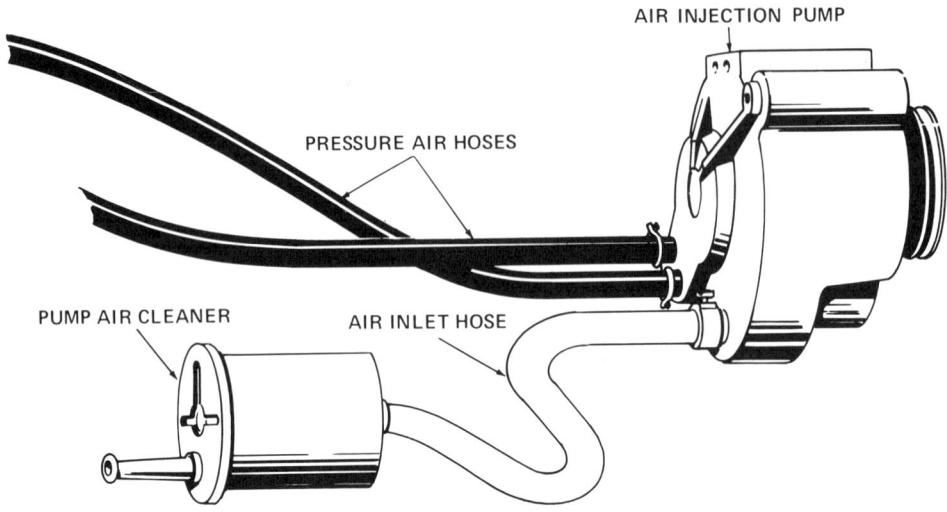

The pump pulley, through a shaft, drives the rotor that turns on an axis different from the center line of the pump bore. The vanes, on the other hand, turn about the center line of the housing bore.

The vanes themselves are in constant contact (or rather nearly constant contact) with the pump housing bore as the rotor turns. Therefore, the vanes must constantly slide back and forth in the sealed rotor slits during pump operation. This action is necessary because the vanes and rotor operate on different centers.

Air Pump Operation

As the pump begins to operate after the engine starts, a vane moves past the inlet port. This action increases the volume in the pump chamber near the inlet port, which has the effect of producing a vac-

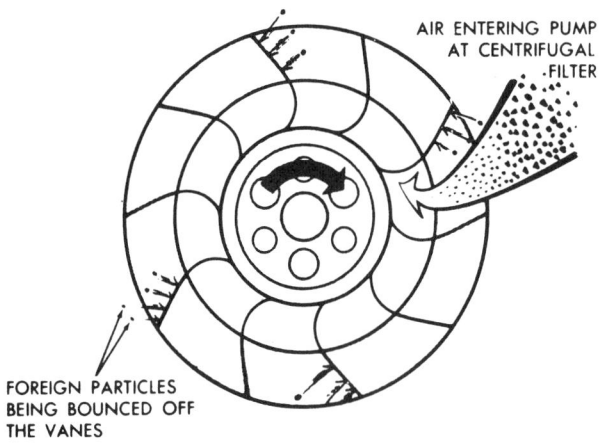

Figure 13-4. Operation of a centrifugal filter (Courtesy of United Delco).

THE DESIGN OF THE AIR INJECTION SYSTEM

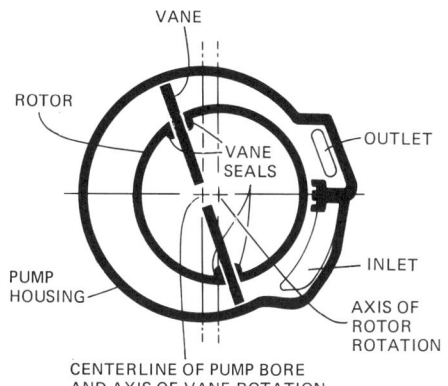

Figure 13-5. Schematic of the construction of a two-vane air pump.

uum. This vacuum causes atmospheric pressure to force air through the centrifugal filter and into the pump chamber (Fig. 13-6a).

As the first vane continues to rotate within the housing bore, the second vane passes the inlet port (Fig. 13-6b). At this point the air that previously entered the pump is trapped between the two vanes. As the vanes continue to turn, they trap and carry this air into a smaller area of the chamber. This action compresses the air.

Continued rotation of the first vane takes it past the outlet port (Fig. 13-6c). Once the vane passes the outlet port, the compressed air exhausts out of the port and into the remainder of the injection system. Although this discussion has concentrated on only one cycle, note that two cycles are made by the pair of vanes during one complete revolution of the rotor.

Diverter Valve

The diverter valve can be a separate unit connected to the pump by a hose or bolted to the pump itself. The purpose of the valve is to prevent a backfire in the exhaust system during engine deceleration. When an engine decelerates, there is a strong vacuum produced within the intake manifold, which applies itself just beneath the carburetor throttle valves. However, since the throttle valves are now in the closed position, they prevent sufficient air from filling the cylinders on their respective intake strokes. Under these conditions, the mixture is too rich to completely burn within the combustion chambers.

Since the mixture does not completely burn during the power strokes, a quantity of charge exits the cylinders via the exhaust strokes. If at this same time the injection pump continued to supply air to the manifold or ports, this air would combine with the mixture and a backfire would occur. However, the diverter valve prevents the backfire by momentarily exhausting the air pump's output, so that it does not reach the exhaust ports or manifolds during the initial stages of engine deceleration.

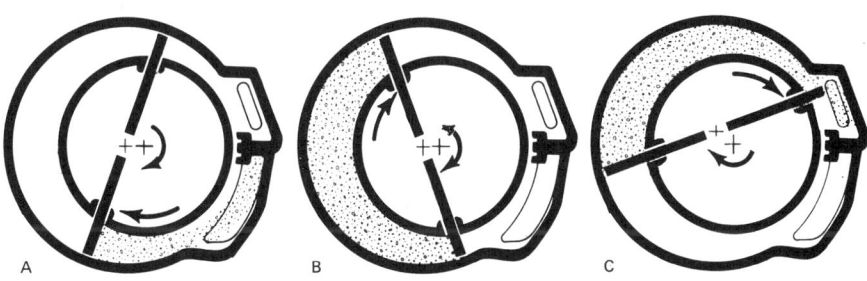

Figure 13-6. Operation of a two-vane air injection pump.

The diverter valve illustrated in Fig. 13-7 consists of a diaphragm and spring that control the operation of a double-acting metering valve. For example, the sudden rise in intake manifold vacuum during deceleration creates a strong vacuum condition under the diaphragm on its spring side, since this portion of the unit connects to the manifold via a hose. The vacuum signal causes the diaphragm and its attached metering valve to move upward against the return-spring tension.

This valve position (closed) causes two reactions within the system. First, the upper portion of the valve seats to cut off injection air to the exhaust ports or manifolds. Second, the lower portion of the metering valve unseats to momentarily bypass or divert the pump's output through the silencer material to the atmosphere.

However, this diversion of pump air flow only lasts for about 1 to 3 seconds. This is due to the orifice hole in the diaphragm assembly, which soon equalizes the vacuum on both sides of the unit. Consequently, the return spring quickly brings the diaphragm and metering valve back down to the normal operating (open) position in a matter of seconds. In other words, the diverter valve, in effect, turns off the air supply suddenly in order to prevent a backfire and then turns it back on gradually for the purpose of starting the oxidation process in the exhaust ports or manifolds.

Relief Valve

The diverter valve shown in Fig. 13-7 also has a built-in pressure relief valve. This valve controls pressure within the system by exhausting excessive pump output at higher engine rpm to the atmosphere through the silencer.

The relief valve assembly consists of a valve body, which encloses a preload spring, movable valve, and valve seat. When the pressure of the pump builds up to a predetermined amount, it forces the valve off its seat, compressing the preload spring. As a result, excess pump pressure exhausts through the silencer to the atmosphere. As the air pressure drops below the tension of the preload spring, it closes the valve, which stops the further exhausting of pump pressure. In other words, the pre-

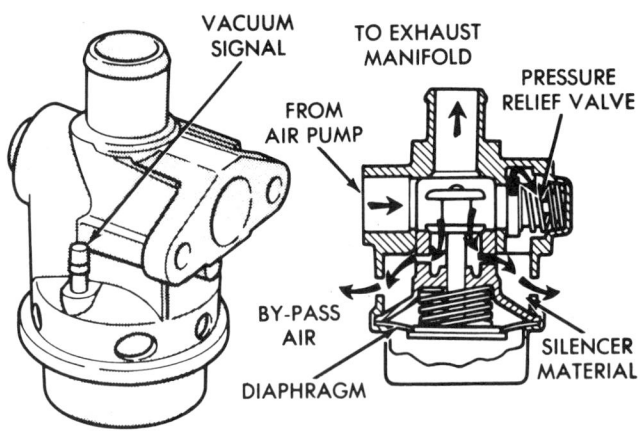

Figure 13-7. Design of a diverter valve (Courtesy of Chrysler Corp.).

load tension of the spring determines at what air pressure the valve opens and therefore the maximum amount of system pressure during all phases of pump operation.

Check Valves

A check valve (Fig. 13-8) fits into the air injection manifold. This valve has a one-way diaphragm, which allows pump pressure to enter the manifold but prevents hot exhaust gases from backing up into the hose and pump. In other words, the valve protects the hose and pump from damage from corrosive gases in the event of drive belt failure, abnormally high exhaust system pressure, or air hose rupture.

Injection Manifolds and Tubes

The air injection system shown in Fig. 13-1 has two injection manifolds, which incorporate a number of stainless steel tubes—one for each engine cylinder. The manifolds distribute the air to the tubes while they direct the injection air into the exhaust ports close to the exhaust valves, near the manifold side of the port, or into the exhaust manifold itself.

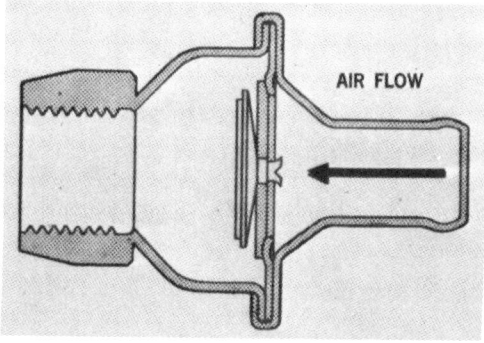

Figure 13-8. Typical injection system check valve (Courtesy of Chrysler Corp.).

Injection System Operation

When the engine is operating, the air pump displaces a large volume of low-pressure air. This air passes through a hose to the diverter valve, which is normally open.

After the air flows through the diverter valve, it moves through a hose or pipe to the check valve. The check valve is open anytime the pressure in the air injection system is higher than the pressure within the exhuast system.

Once the air passes through the check valve, it enters the injection manifolds. The manifolds then distribute the air to each of the tubes within the exhaust ports or manifolds. If at any time the air pressure within the injection manifolds exceeds the preload on the relief-valve spring, the valve opens and reduces manifold air pressure.

During engine deceleration, the diverter valve closes; this action momentarily exhausts air pressure into the atmosphere. This prevents a backfire in the exhaust system. However, within 1 to 3 seconds, the diverter valve once again opens to permit normal operation of the injection system.

Air Injection Systems Used With Catalytic Converters

An air injection system used on a vehicle with a catalytic converter does about the same task as the basic system just described. In other words, the system assists in oxidizing the hydrocarbons and carbon monoxide within the exhaust gases through the addition of fresh air. Consequently, the newer systems utilize many of the same components as those found on vehicles without a converter.

But the newer systems do require additional controls for effective use of the injection air with the catalytic converter. The following sections explain some of the basic alterations made to the Chrysler, Ford, and General Motors systems to convert them for use with catalytic converters.

Chrysler Catalytic Air Injection System

In 1977 Chrysler modified its basic air injection system by the addition of an air-switching valve and coolant-controlled engine vacuum switch (CCEV). These two units (Fig. 13-9) control the air flow from the injection pump during and after engine warm-up to assist in oxidizing HC and CO emissions within the catalytic converter without interfering with the control of NOx by the EGR system. This text will cover the EGR system in a later chapter.

In other words, the switching valve and vacuum switch direct the injection air where it does the best job of burning up the pollutants. For instance, when an engine is cold, the injection air oxidizes the pollutants more efficiently right at the engine's exhaust ports, where there is enough heat. Moreover, this additional air helps to warm up the mini converter used on these engines. However, after the engine is at normal operating temperature, more air is needed within the main underfloor converter in order for it to function properly in reducing HC and CO levels. Finally, by redirecting the injection air away from the

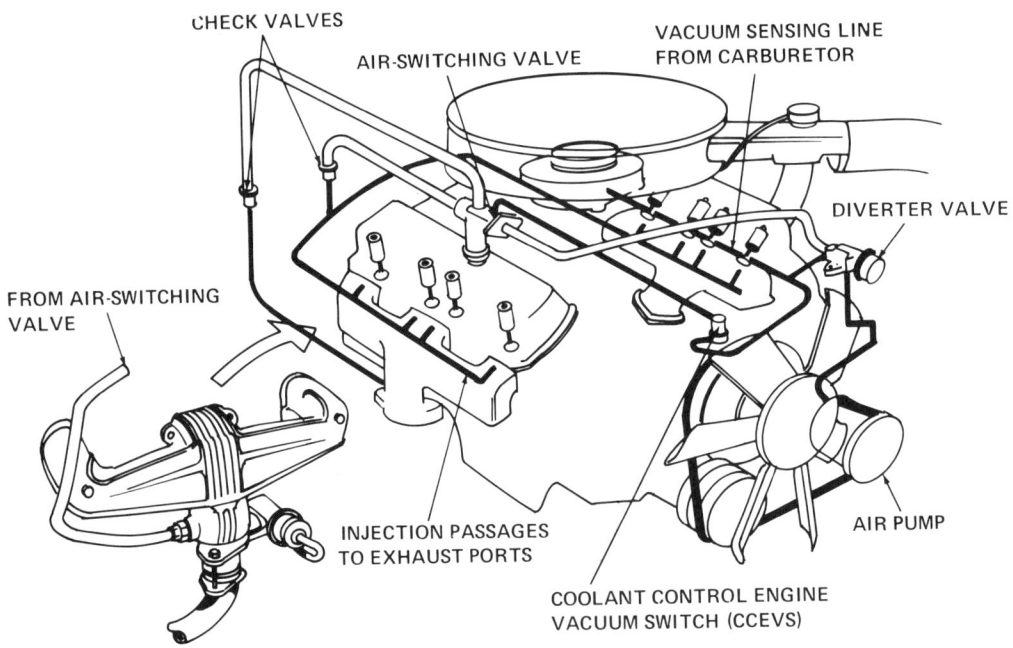

Figure 13-9. Chrysler added this switching valve and CCEV switch to its basic air injection system in 1977.

AIR INJECTION SYSTEMS USED WITH CATALYTIC CONVERTERS

exhaust ports after the engine is warm, it is prevented from recirculating with the exhaust gases into the EGR system, where it would excessively lean out the mixture.

As in the past, air output from the injection pump goes to the diverter valve, but it then proceeds to the air-switching valve. This valve can send the air either to the exhaust ports or to a single entry point downstream.

The air-switching valve (Fig. 13-10) operates by intake manifold vacuum, regulated by a CCEV switch. This switch is similar in design and operation to the thermo vacuum valve described earlier. When, for instance, the engine coolant temperature is below 125 °F on the 318 V-8 or below 98 °F on any other engine, the CCEV switch is open, permitting manifold vacuum to operate the air-switching valve. This action directs the air to the exhaust ports in the same manner as in a normal air injection system.

However, when the coolant temperature goes above 125 °F on the 318 V-8 or above 98 °F on any other engine, the CCEV switch shuts off the vacuum to the switching valve. In this situation, the spring within the valve opens the tube to the downstream connection. As a result, 85 percent of the injection air enters the system downstream while about 15 percent continues to flow into the exhaust ports.

The actual downstream connection point differs between engine styles. For example, on V-8 engines, the entry point is on the right side manifold, above the heat control valve (see Fig. 13-9). On 6-cylinder engines, on the other hand, the downstream entry point is within the exhaust pipe between the mini converter and the main converter.

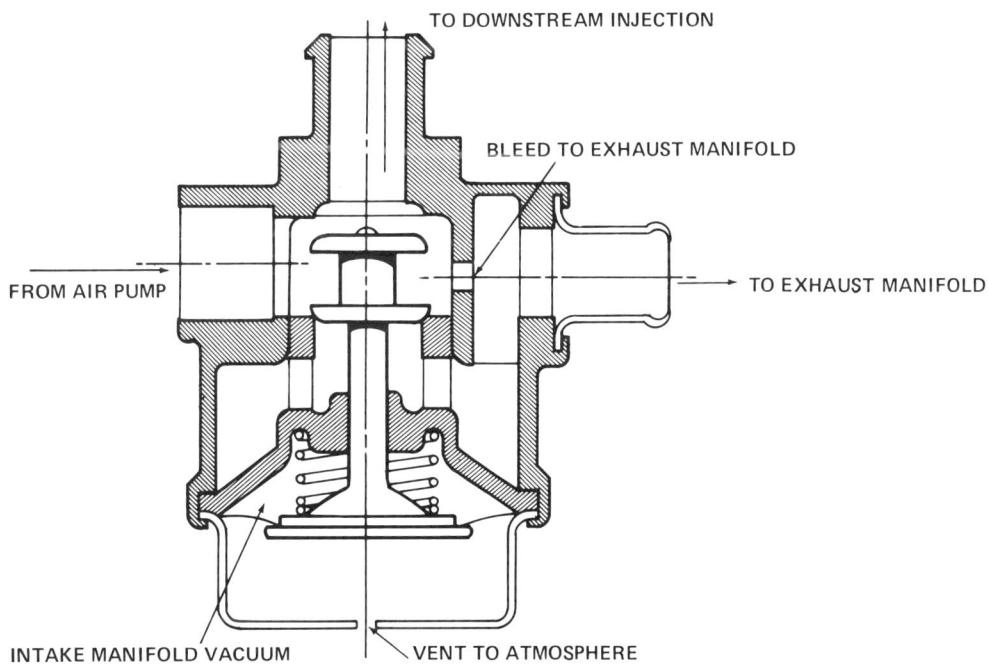

Figure 13-10. Design of a typical air-switching valve.

Ford Catalytic Air Injection System

Since 1975, the air injection system Ford uses on vehicles with catalytic converters has become quite complicated. This system complexity results from, in this case, the installation of air injection along with a catalytic converter on so many different engine-vehicle configurations. These various engines all do not produce the same amount of power or negative exhaust emissions nor operate efficiently at the same temperature. Consequently, a variety of controls are necessary to coordinate the actions of the air injection system to the catalytic converter without damaging the latter.

This chapter could not possibly cover all the variations. However, this section will describe the major modifications made to the pump, the design and operation of the different diverter valves used, and the air control valve used with dual-converter systems.

Thermactor Air Pumps

With the addition of the catalytic converter in 1975, Ford altered its basic air injection pump in several ways. First, it was necessary to increase the capacity of the pump so that it could provide additional air when necessary to the converter. Second, Ford discontinued installing the relief valve on the newer pumps. Instead, the valve is now built into the diverter valve.

Ford Diverter (Timed Air ByPass Valve)

Before the installation of catalytic converters on its vehicles, Ford used a diverter valve similar to the one discussed earlier in this chapter. But since 1975, Ford has introduced several different types now known as timed air bypass valves.

Figure 13–11 illustrates one type of air bypass valve design. This valve looks quite similar to the diverter valve used before 1975 that was made by Carter Carburetor. However, the newer design has a small hose connection on the end, instead of on the side.

The main difference between the older diverter valve and this timed air bypass unit is what occurs when the valve is in its normal operating position. The old diverter valve, for example, was always in the open position, which, except during engine deceleraion, permitted air flow from the pump to the exhaust ports. But the new unit will operate normally in the closed position. In other words, the old-style valve permitted the air to pass through it to the engine exhaust ports whether its small vacuum sensing line or hose was connected or not.

On the other hand, the timed air bypass valve, being normally in the closed or dump position, requires that the small sensing line be connected to manifold vacuum in order to open the valve from the dump position to the normal running position. This is opposite from the older diverter valve.

When operating, the timed air bypass valve shown in Fig. 13–11 receives its vacuum signal directly from the intake manifold. This signal holds the valve open, its normal run position. The valve then permits normal air delivery to the exhaust ports or manifolds.

The vacuum signal is available to the bypass valve during engine idle, acceleration, and cruise conditions. However, during deceleration, long periods of idle, and cold engine operation, the vacuum signal

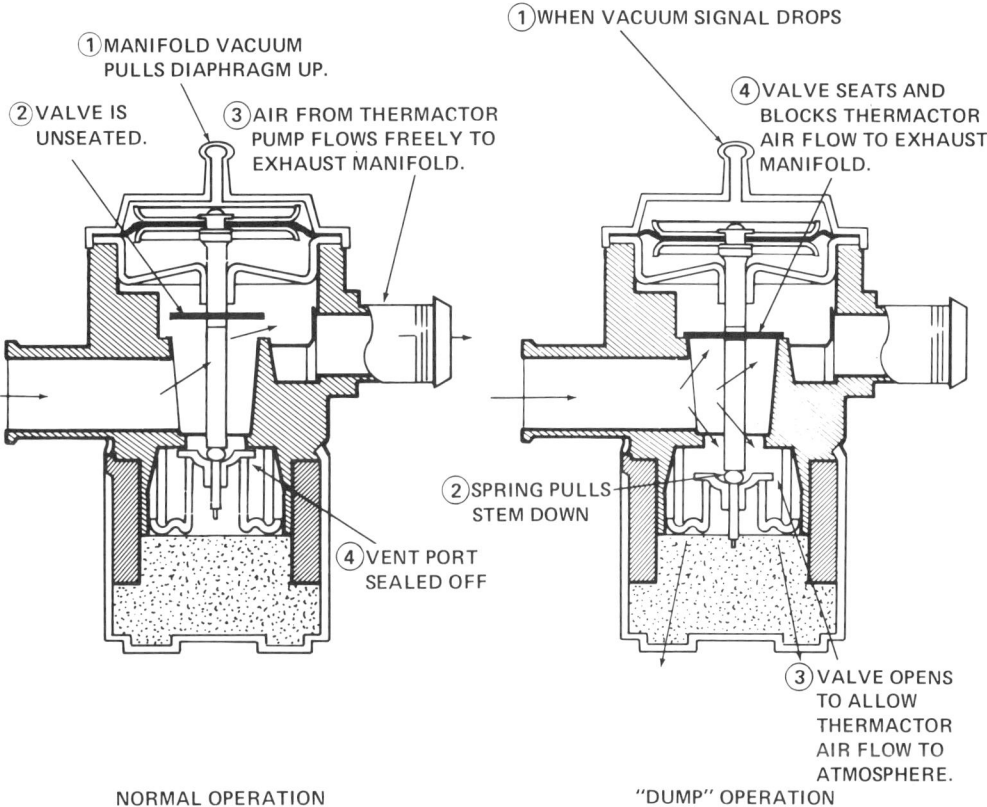

Figure 13-11. This timed air bypass valve uses vacuum to hold the air passage open to the exhaust manifold.

is cut off to the valve. This causes the valve to close and permit the pump pressure to dump into the atmosphere.

In order to cut off the signal to the bypass valve during engine deceleration, a vacuum differential valve (Fig. 13-12) is necessary. The vacuum differential valve (VDV) connects into the vacuum signal line between the intake manifold and the bypass valve.

The design of the differential valve is such that intake manifold normally passes through it and to the bypass valve. But when manifold vacuum increases during deceleration, the VDV bleeds off the vacuum in the signal line. This action eliminates the signal to the bypass valve, and the valve closes to dump pump pressure into the atmosphere.

A later refinement of this system connected the signal line to a ported vacuum source instead of one from the intake manifold (Fig. 13-13). This installation eliminated the VDV because now the bypass valve only receives its vacuum signal above idle. Consequently, at engine idle, the diaphragm spring pulls the stem and valve down to the closed position, which allows pump pressure to dump to the atmosphere. This valve also assumes this po-

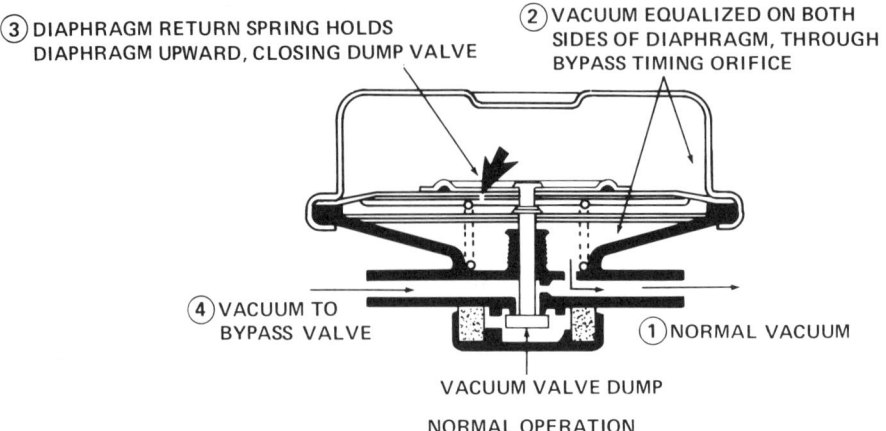

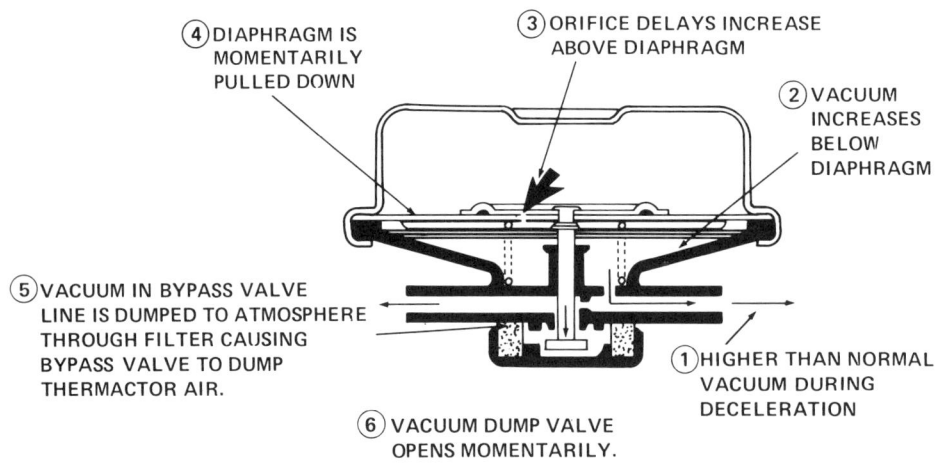

Figure 13–12. Differential valve closes off the sensing vacuum line to the bypass valve during engine deceleration.

sition during engine deceleration because the carburetor's throttle valves at this time are closed or in the idle position. Therefore, there is no vacuum signal to the bypass valve's signal line.

Some of these systems also have a delay valve that is similar to a spark-delay valve. This valve is in the signal line, and it delays for a few seconds the drop in vacuum to the bypass valve when the throttle valves close. Therefore, air pump pressure does not dump to the atmosphere every time the

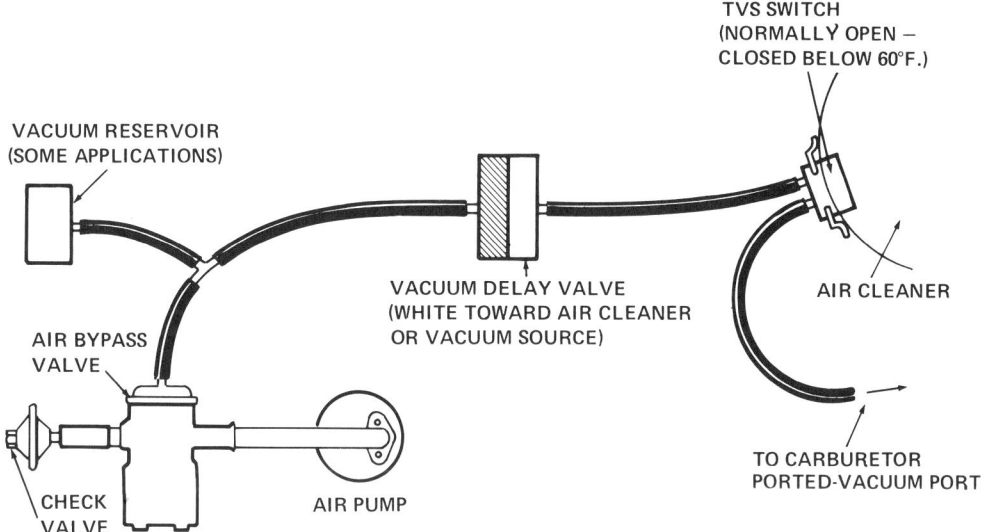

Figure 13-13. Bypass valve connected to a ported vacuum signal.

driver takes his foot off the accelerator pedal during heavy traffic conditions.

The manufacturer also installs temperature controls into the signal line. This is usually in the form of a thermo valve that shuts of the vacuum signal when the engine is cold. This action prevents the pump from supplying air to the engine exhaust ports or manifolds until it warms up.

There may be another sensor installed under the vehicle above the catalytic converter. This device is in the system to shut off a solenoid if the converter overheats. The solenoid in turn cuts off the vacuum to the bypass valve, and it moves into the dump position. This stops injection air flow to the exhaust ports or manifolds until the catalytic converter cools down.

The later style of air bypass valve is shown in Fig. 13-14. This valve has two small signal hoses connected to it. Both of these hose fittings index into chambers, one on each side of the diaphragm within the valve assembly. The hose from the intake manifold connects to the lower of the two fittings while a hose from a separate on-off vacuum valve or solenoid connects to the fitting closer to the end of the unit.

Stretched across between the two chambers and attached to the poppet valve is a special type of diaphragm. This device has a small hole so that the vacuum or pressure on each side of the diaphragm will equalize at all times.

The separate vacuum switch or solenoid controls the movement of the diaphragm and its attached poppet valve. For example, a closed valve or solenoid seals the upper end chamber from the atmosphere. Now, since there is a bleed hole within the diaphragm, a vacuum exists on both sides of it. As a result, there is no force on the diaphragm other than the spring, which moves the poppet valve to the open posi-

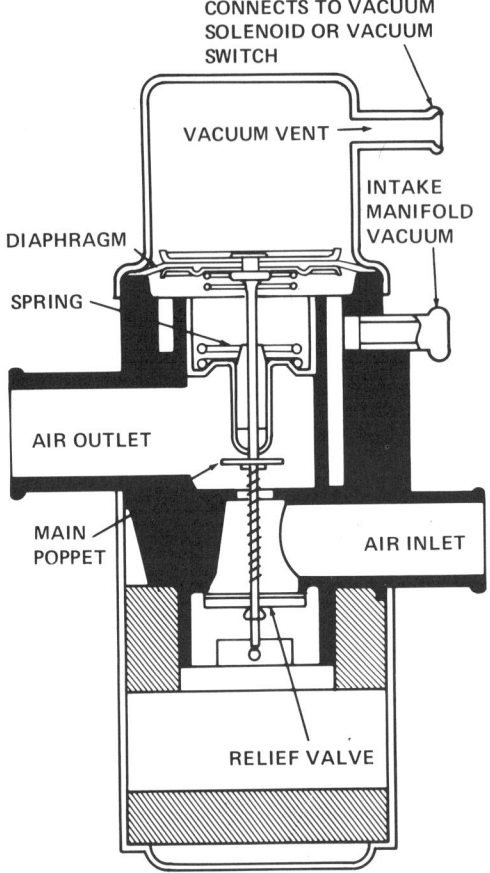

Figure 13-14. Bypass valve with an integral vacuum differential function.

tion, permitting pump air to flow into the exhaust ports or manifolds.

However, when the separate valve opens, it admits atmospheric pressure to the upper side of the diaphragm. In this situation, with manifold vacuum on the bottom side and atmospheric pressure on the other, the diaphragm and poppet valve move down to the closed position. As a result, pump air pressure dumps into the atmosphere.

In the air injection system shown in Fig. 13-15, the separate valve is an idle vacuum valve. This valve controls the venting to the atmosphere of the upper side of the air bypass valve diaphragm. The valve itself is nothing more than a vacuum-operated vent. For example, with a vacuum signal applied to the valve, the vent closes. But when there is no signal on the valve, it opens.

The vent portion of the valve connects to the upper chamber of the bypass valve. Therefore, vacuum actually never passes through the idle vacuum valve at any time. The vacuum signal directed at this valve only opens or closes the vent. Finally, due to the fact that the idle vacuum valve connects to a ported vacuum signal, the valve keeps the vent closed above idle rpm but opens it during engine idle and deceleration.

As mentioned earlier, venting of the bypass valve is only necessary during prolonged idle operation to prevent the catalytic converter from overheating. To prevent the dumping of pump air for one-half to a full minute, a vacuum-delay valve is in the hose near the idle vacuum valve. This valve traps the vacuum, and it takes one-half to one full minute for it to bleed down. This action ensures that the bypass valve only goes into the dump position during a prolonged idle period.

The system illustrated in Fig. 13-15 also has a thermo vacuum switch (TVS). This unit is on the air cleaner to shut off vacuum to the idle vacuum valve below about 60 °F. With no vacuum applied to the valve, its vent opens. This permits the bypass valve to move into the dump position.

On some engines, a vacuum reservoir connects into the hose between the idle vacuum valve and the TVS. This compo-

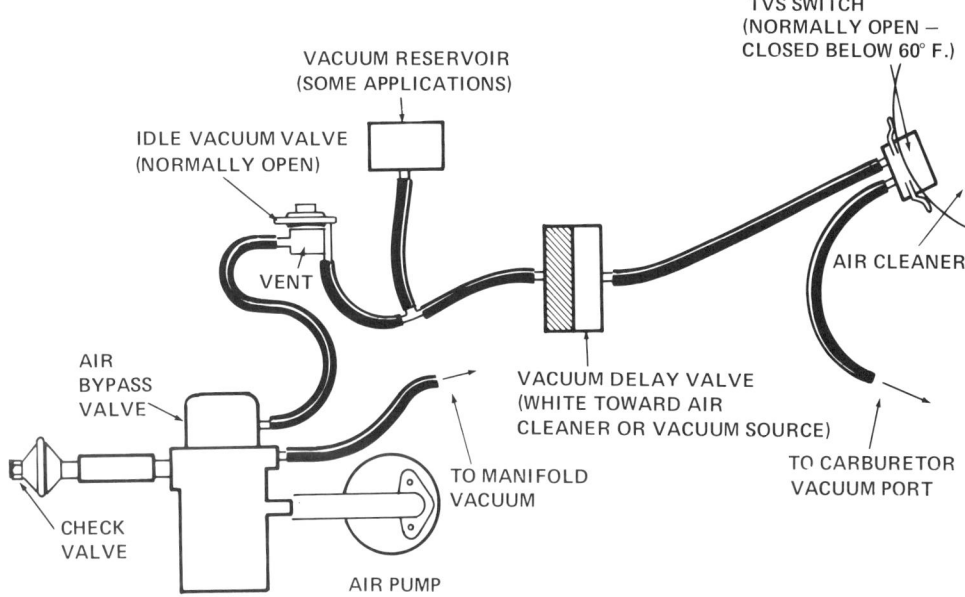

Figure 13-15. Air pump system using a timed air bypass valve with a vacuum vent and TVS switch.

nent increases the amount of vacuum available to the delay valve, so it takes longer for the valve to bleed down. This ensures that it will be a full minute before pump air dumps to the atmosphere during periods of idle operation.

Air Injection System Used With Dual Catalytic Converters

On many Ford compact models, there is a special air injection system used in conjunction with a dual-bed catalytic converter (Fig. 13-16). In this system, the pump and bypass valve are about the same as in other later systems, but there is a secondary air control valve and ported vacuum switch (PVS) added to the system. The secondary air control valve receives the air under pressure from the pump and bypass valve, and directs it either to the exhaust ports, as before, or directly into the midpoint in the catalytic converter.

The switching action of the secondary air control valve is done by vacuum from the PVS assembly. This switch, by the way, is the same one that directs a vacuum signal to the thermo vacuum switch (TVS) mounted in the air cleaner. For example, when the engine is cold, the PVS directs vacuum to the secondary air control valve, and it directs the air flow from the pump to the engine exhaust ports. But when the engine warms up, the PVS shuts off the vacuum, and the secondary air control valve sends the pump air to the rear half of the converter.

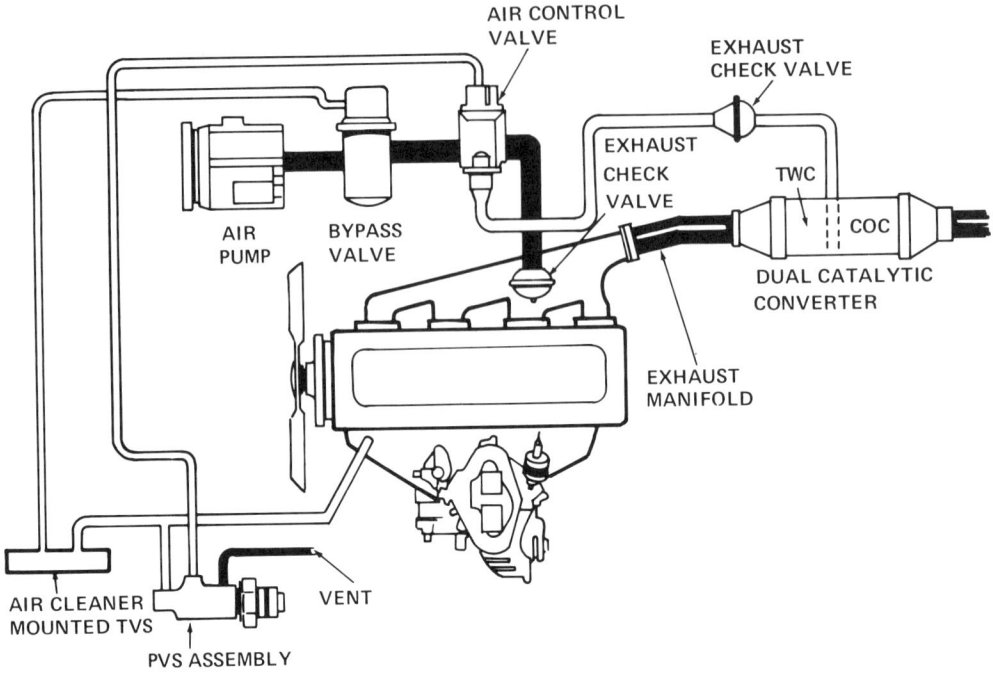

Figure 13-16. Thermactor system used in conjunction with a dual-bed catalytic converter.

If the converter was warm when the vehicle was first started cold, there would be no need to switch the pump air from the converter to the exhaust ports. However, when the engine starts cold, the only warm area where oxidation of HC and CO emissions can take place is at the exhaust ports. Therefore, the system must pump the air into this area so that HC and CO emissions can be kept at reduced levels during warm-up (Fig. 13-17).

The additional air introduced into the front of the converter reduces its efficiency in reducing NOx emissions. However, this is not a big problem during engine warm-up because NOx levels are not very high in a cold engine.

But after the engine warms up, the pumping of injection air into the exhaust port area in front of the converter would make it difficult for the front section of this device to reduce NOx emissions. So the control valve switches the pump air from the exhaust port area to the rear half of the converter. This air flow then finishes the oxidation process of any remaining HC and CO compounds within the exhaust gases.

General Motors Catalytic Air Injection Systems

As previously mentioned, the original designation for a General Motors injection system was Air Injection Reactor (AIR) or in some cases, Manifold Air Injection

AIR INJECTION SYSTEM USED WITH DUAL CATALYTIC CONVERTERS

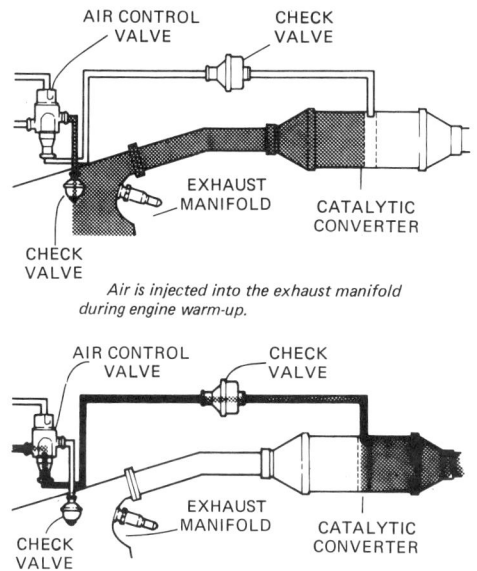

Figure 13-17. Operating phases of an air injection system used in conjunction with a dual-bed catalytic converter.

(MAI). But when General Motors modified the original system for use on some converter-equipped vehicles, the newer system became known as the Converter Air Injector Reactor (CAIR) or Converter Air Injection (CAI).

Within the CAIR or CAI systems, the air pump, diverter valve, and check valve are about the same as in the original system. However, in the newer systems, the air from the pump injects into the exhaust pipe downstream from the exhaust manifold (Fig. 13-18). This type of injection delivers pump air directly to the catalytic converter to assist in the oxidation of HC and CO emissions within this unit. Furthermore, by preventing the air from reaching the exhaust ports, the combustion chamber temperature is lower because the EGR system admits less air; this lowers NOx emissions.

However, on later-model General Motors vehicles, mainly those without cata-

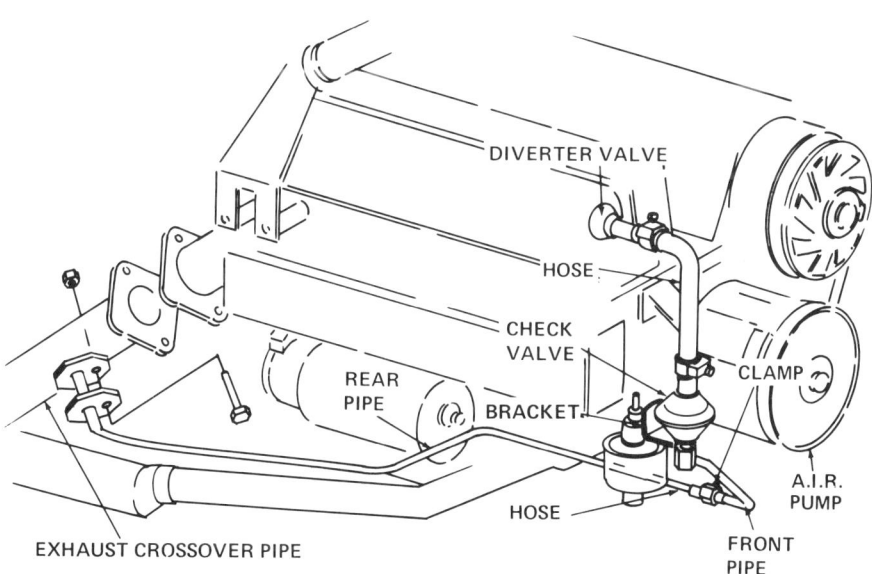

Figure 13-18. Schematic of a typical air injection system used by General Motors on a vehicle with a catalytic converter.

lytic converters, the system still injects the air into the exhaust ports. This is the hottest place on these particular engines to oxidize the HC and CO emissions.

Aspirator-Type Air Injection Systems

Function and Design

Since 1975, General Motors, Chrysler, and Ford have used a much simpler type of air injection system on some of their vehicles. This device is known as either the aspirator-air, pulse-air, or suction-air injection system.

In any case, the system does not use an air pump to force air into the air manifolds. Instead, the system uses exhaust pressure pulsations to draw air into the exhaust system. Each time an exhaust valve closes, there is a period when the pressure within the manifold drops below that of the atmosphere. During these low-pressure (vacuum) pulses, air from the clean side of the air cleaner (Fig. 13-19) moves into the exhaust manifolds. This air in turn provides the oxygen necessary to oxidize the HC and CO compounds remaining in the exhaust manifolds as well as supply needed oxygen to the catalytic converter, if so equipped.

The aspirator system shown in Fig. 13-19 consists basically of a steel tube, one-way aspirator valve, and a length of hose from the valve to the air cleaner. The steel tube flanges into the exhaust manifold. This tube carries the air from the aspirator valve to the exhaust manifold.

Threaded into the open end of the steel tube is the aspirator valve. This unit (Fig. 13-20) is a one-way check valve that per-

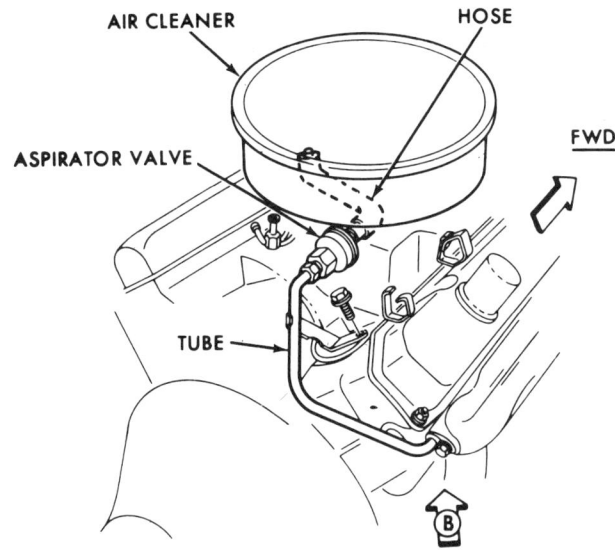

Figure 13-19. Design of an aspirator-type air injection system (Courtesy of Chrysler Corp.).

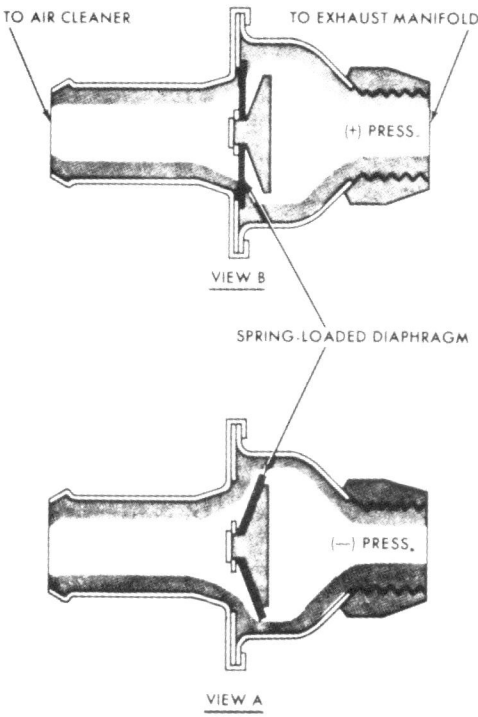

Figure 13-20. Design and operation of an aspirator valve (Courtesy of Chrysler Corp.).

mits air to flow from the clean side of the air cleaner, through the steel tube, and into the exhaust manifold. However, the same valve prevents exhaust gases from backing up the tube, hose, and into the air cleaner.

The one-way aspirator valve may contain a spring-loaded diaphragm (Fig. 13-20) or have a metal reed valve. In either case, the valve operates in much the same manner and connects via a hose to the clean side of the air cleaner.

System Operation

When an engine with an aspirator system is idling or operating at low speeds, the aspirator-air valve opens due to the pulsating negative pressure within the exhaust manifold. In other words, the valve opens each time there are low-pressure (vacuum) pulses, which occur each time an exhaust valve closes. When the diaphragm or reed valve opens (Fig. 13-20a), atmospheric pressure forces fresh air from the clean side of the air cleaner, through the aspirator valve, steel tube, and into the exhaust manifold.

At higher engine speeds when the exhaust gas pressure is above that of the atmosphere, the spring closes the diaphragm or reed valve (Fig. 13-20b). Now, the assembly acts as a one-way check valve to prevent exhaust gases from backing up into the air cleaner. This usually occurs more at higher engine speeds because the vacuum pulses follow each other too quickly for the valve to respond by opening. Consequently, the valve's internal spring just keeps the diaphragm or reed valve closed. As a result, no fresh air enters the exhaust manifolds during these periods, and the exhaust gases cannot flow back to the air cleaner.

While the system just discussed only uses one aspirator valve and one steel tube, other systems will have more of both. For example, the General Motors six-cylinder engine illustrated in Fig. 13-21 has a system that consists of four pipes of equal length and four check valves. Each of the pipes secures to a check valve, which fits into a grommet in the pod. The pod is a plenum welded to the top of the rocker-arm cover. The other end of the pipe routes to and secures into an exhaust manifold passage within the cylinder head assembly.

A separate pipe fits between the pods and acts as a fresh air supply source (Fig. 13-21a). The pipe has a tee onto which fits a hose that leads to a nipple on the clean side of the air cleaner.

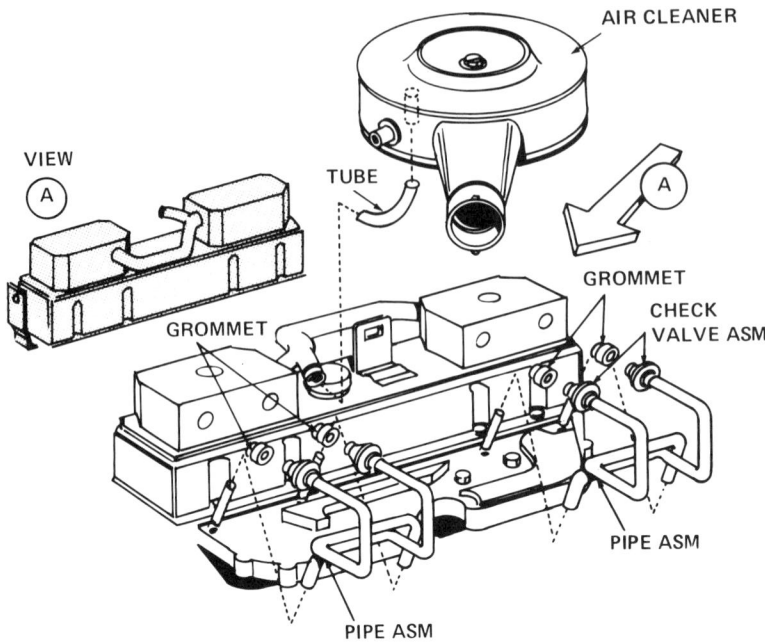

Figure 13-21. Schematic of a GM six-cylinder engine with a pulse-air system.

Pulse Air Injection Used with Multiple Catalytic Converters

Figure 13-22 illustrates a unique pulse-air system utilized by Chrysler Corporation on some of its compact automobiles. This particular system supplies secondary air into the exhaust system between the front and rear catalytic converters. This additional air promotes the oxidation of HC and CO emissions within the rear converter.

Chrysler calls the system shown Fig. 13-22 the Pulse Air Feeder (PAF), which consists of a seal cover, pulse-air feeder, and hoses and lines that connect the feeder to the air cleaner, exhaust system, and engine.

The seal cover is inside the engine. Its function is to seal the crankcase of number-three cylinder. However, the cover does have a little hole for discharging engine oil and accumulated blow-by gases from the crankcase.

The pulse-air feeder assembly controls the air flow into the exhaust system during all phases of engine operation. To accomplish this task, the assembly has two valves: the main-reed and the sub-reed valves. The main-reed valve activates in response to the movement of the diaphragm below it. The diaphragm itself moves in response to pressure pulsations generated by the reciprocating motion of the number-three piston in its sealed crankcase.

The sub-reed valve operates in much the same manner as the aspirator valve mentioned earlier in this chapter. That is, the sub-reed valve activates to permit air flow due to exhaust system vacuum pulses gen-

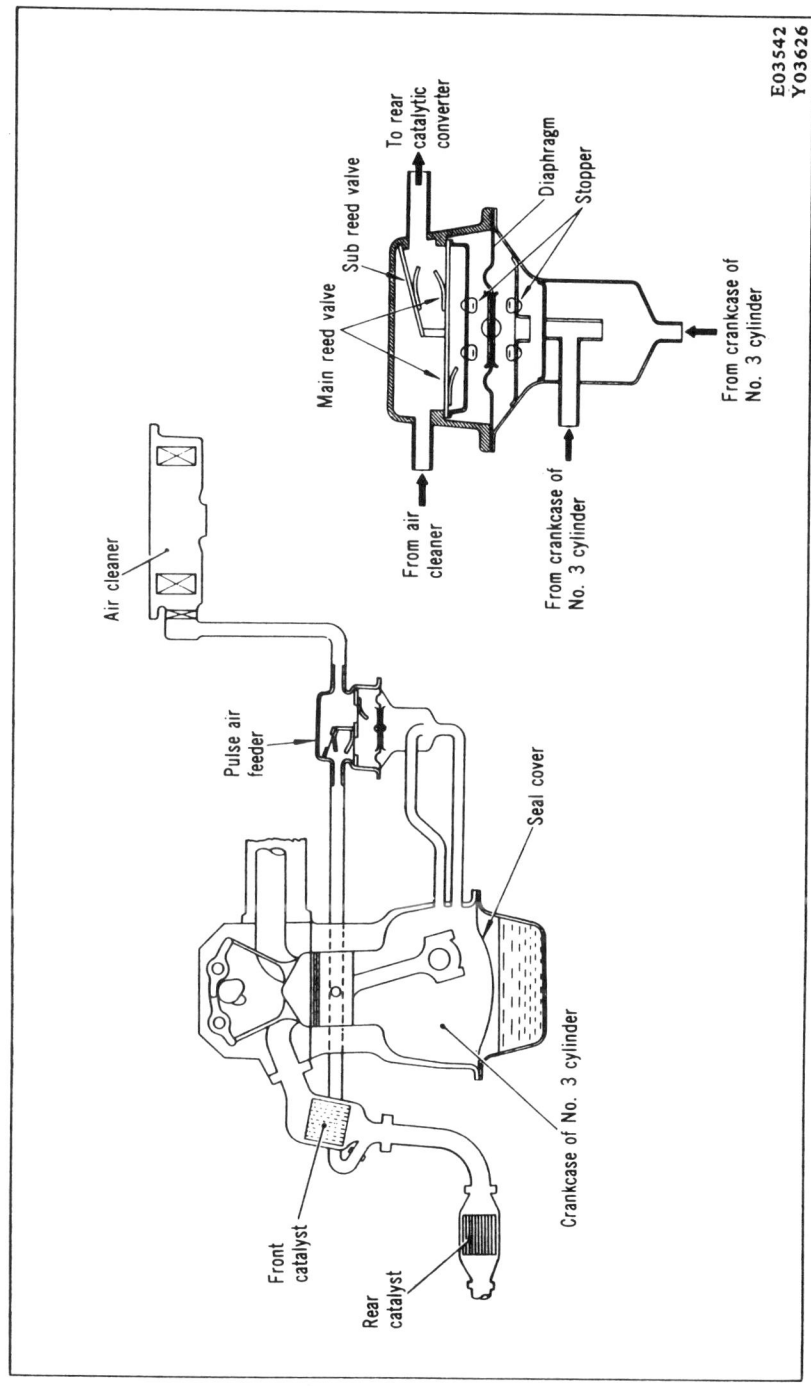

Figure 13-22. Pulse-air feeder system (Courtesy of Chrysler Corp.).

erated in the system between the front and rear catalytic converters.

Summary

1. Since the passage of the first exhaust emission control standards, manufacturers have extensively used the air injection system.
2. The most common names for the early air injection systems are Air Guard, Air Injection, Thermactor, and Air Injection Reactor.
3. The air injection system introduces fresh air into the exhaust ports or manifolds of an operating engine.
4. A typical air injection system consists of an air pump, combination diverter and air pressure relief valve, check valve, and manifolds.
5. Before 1968, air pumps had three vanes.
6. The main operating difference between the two- and three-vane pumps is the manner in which they filter the incoming air.
7. A centrifugal filter fan mounts on the end of later-type air pumps and cleans the air before it enters the inlet port.
8. A centrifugal filter fan cleans the air by deflecting away from the pump's inlet any heavier-than-air particles.
9. Along with the filter fan, a later-type air pump has two vanes, vane seals, and a rotor and shaft, which operate inside the pump housing.
10. The pair of vanes within an air pump complete two cycles of air delivery during one complete revolution of the rotor.
11. The diverter valve prevents a backfire in the exhaust system during engine deceleration.
12. A typical diverter valve consists of a diaphragm and spring that control the operation of a double-acting metering valve.
13. The diversion of pump air flow during deceleration only lasts for about 1 to 3 seconds.
14. The relief valve controls the maximum pressure within the system by exhausting excessive pump output at higher engine rpm to the atmosphere.
15. Inside the relief valve body is a preload spring, movable valve, and valve seat.
16. The check valve allows pump pressure to enter the manifolds but stops hot exhaust gases from backing up into the hose and pump.
17. The air injection system used on vehicles with catalytic converters performs the same task as it did in earlier systems but requires additional controls.
18. The Chrysler catalytic air injection system has an air-switching valve and coolant-controlled engine vacuum switch.
19. A Ford catalytic air injection system is complex because of its installation on so many different engine-vehicle configurations.
20. Ford uses several different types of diverter (timed air bypass) valves on its versions of catalytic air injection systems.
21. The early version of the air bypass valve required a vacuum differential valve to control its operation during deceleration.

22. In a Ford catalytic air injection system, there may also be a delay valve, thermo valve, and a converter overheat sensor.
23. On the later Ford air bypass valve, a separate vacuum switch or solenoid controls the movement of its diaphragm and the attached poppet valve.
24. The idle vacuum valve vents to the atmosphere the upper side of the air bypass valve's diaphragm.
25. The delay valve in a Ford catalytic air injection system prevents the dumping of pump air for one-half to one minute during engine idle.
26. The Ford air injection system with the later-style air bypass valve also has a thermo switch in the air cleaner and a vacuum reservoir.
27. On many newer Ford compact models, there is a special air injection system used in conjunction with a dual-bed catalytic converter.
28. The switching action of the secondary air control valve is done by vacuum from the PVS assembly.
29. The secondary air control valve switches the air injection from the exhaust ports to the rear half of the converter after the engine warms up.
30. The General Motors catalytic air injection system is known as either CAIR or CAI.
31. Since 1975, General Motors, Ford, and Chrysler, have used a much simpler type of air injection system that requires no air pump.
32. A typical aspirator system consists of a steel tube, one-way aspirator valve, and a length of hose from this valve to the air cleaner.
33. Some aspirator air injection systems have more than one steel tube and check valve.

Review Questions

The questions listed below will assist you in determining how well you remember the material contained in this chapter. Read each question carefully before choosing the answer. If you cannot answer the question, review the section in the chapter that covers the material.

1. Air injection pumps built before 1968 have _____ vanes.
   a. 2
   b. 3
   c. 4
   d. 5
2. Later-style air pumps clean the incoming air through the use of _____.
   a. a centrifugal filter
   b. separate air filter
   c. air from the air cleaner
   d. both b and c
3. The valve that prevents a backfire during engine deceleration is the _____ valve.
   a. aspirator
   b. check
   c. backfire
   d. diverter
4. The valve that prevents exhaust gas from reaching the air pump is the _____ valve.
   a. aspirator
   b. backfire
   c. check
   d. diverter

5. The valve within the later Chrysler injection systems that controls which entry point the air enters is the _____ valve.
   a. diverter
   b. switching
   c. check
   d. aspirator
6. On later systems, Ford called its diverter assembly a (an) _____ valve.
   a. timed air bypass
   b. backfire
   c. aspirator
   d. suction air
7. The separate valve used to vent the upper chamber of the air bypass valve in the Ford system discussed in this chapter is the _____ valve.
   a. aspirator
   b. diverter
   c. idle vacuum
   d. PVS
8. On a catalytic converter system with a secondary air supply, the device that directs air flow to the converter is the _____ valve.
   a. converter
   b. air bypass
   c. air control
   d. diverter
9. Manufacturers began to use aspirator air injection systems in _____.
   a. 1978
   b. 1977
   c. 1976
   d. 1975
10. The aspirator-type air injection system does not function during _____.
    a. high engine speed
    b. idle
    c. low speed
    d. deceleration

For the answers to these review questions, turn to the Appendix.

Chapter 14

Testing Air Injection Systems

As mentioned in Chapter 13, air injection is basically an add-on system used to provide additional air to the exhaust ports or manifolds. This air contains oxygen that promotes further oxidation of HC and CO compounds within the exhaust gases. Consequently, the system, unlike engine modification, reduces harmful emissions by burning them up outside of the engine combustion chambers.

Since this is the case, any failure within the air injection system causes an increase in HC and CO emissions but can also create a few abnormal engine operating characteristics. For example, if the intake manifold to diverter valve sensing hose is cracked, broken off, or left off, the engine will idle poorly due to a lean misfire. Furthermore, if the diverter valve itself malfunctions, the engine will backfire within the exhaust on deceleration.

It should be obvious then that in order to prevent these unusual conditions and also to reduce HC and CO emissions, the system requires periodic testing and service. The type and frequency of preventive maintenance varies somewhat among the vehicle manufacturers. Therefore, it may be necessary for you to use factory instructions and specifications when checking out a particular system due to possible alterations within the basic system. These changes are necessary in most cases in order for the manufacturer to install the system on different engine-vehicle configurations.

However, to familiarize you with some of the techniques used by the industry, we will present some of the most common testing procedures on the basic system and the additional components necessary to fit it onto a vehicle with a catalytic converter. These include testing the system with an infrared analyzer and checking system components such as the drive belt, air pump, diverter valve, air manifold and hoses, check valve, air injection tubes, switching valve, coolant-controlled engine vacuum switch, bypass valves, ported vacuum switch, idle vacuum valve, secondary air control valve, and aspirator valve.

Testing a Basic Air Injection System with an Infrared Analyzer

You can test the overall effectiveness of a basic air injection system in reducing HC and CO emissions with an infrared analyzer. To perform this check:

1. Start the engine and permit it to reach its normal operating temperature.

2. As outlined earlier, warm up and initially calibrate the infrared analyzer. Then, insert its sampling hose into the tail pipe.

3. Set the idle speed screw on a step of the fast-idle cam that causes the engine to run at about 1,000 rpm.

4. Depress the test mode selector button on the analyzer and then note the HC and CO readings.

5. While maintaining the same engine speed, disconnect the air pump supply hose leading from it to the manifold.

6. Note the HC and CO readings on the analyzer. An increase in HC and CO levels should occur with the hose removed if the system is functioning properly. If the analyzer shows little or no increase in the HC and CO levels, examine all the components of the air injection system as outlined in the following sections.

Checking Air Injection System Components

Drive Belt

1. Inspect the pump drive belt for wear, cracks, or deterioration. Replace the belt if it is defective.

2. Check the tension of the belt (Fig. 14-1) using a strain gauge. Next, adjust the belt's tension as necessary. Note: Belt tension settings vary somewhat between different system manufacturers. However, a typical setting is 55 pounds for a used belt and 75 pounds for a new belt.

Air Pump

To test the air pump for serviceability:

1. Start the engine and permit it to reach normal operating temperature.

2. Set the parking brake and place the transmission into neutral or park.

3. Shut off the engine and remove the outlet hose from the air pump.

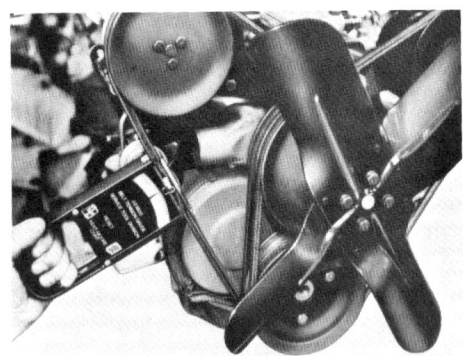

Figure 14-1. Testing belt tension with a strain gauge (Courtesy of Chevrolet Division).

4. Restart the engine and slowly accelerate it to approximately 1,500 rpm while feeling the air flow from the pump's outlet port.

Results and Indications

If the air flow increases as the engine accelerates, the pump is operating satisfactorily. However, if the air flow does not increase, or is not present, or the pump is noisy, proceed as follows:

1. Check the drive belt for proper tension.

2. Check for a leaky pressure relief valve (on pumps so equipped). Air may be heard leaking out of the valve with the pump operating.

3. If belt tension is satisfactory and the relief valve is not leaking, remove the pump for repair or replacement.

4. To determine if excessive noise is the fault of the pump, operate the engine with the pump's drive belt removed. Note: The air pump is not a completely noiseless unit. Under normal conditions, its noise level increases slightly in pitch as the engine accelerates. In addition, before replacing a pump for excessive noise, make sure that it has operated in excess of the 100-mile break-in period. Furthermore, make certain that pump alignment and mounting are correct. Caution: Do not introduce oil into the air pump to reduce its noise. This may quiet down the pump for a little while, but it does not correct the problem permanently and eventually leads to premature pump failure.

5. If pump alignment and mounting are correct and the pump produces excessive noise, remove it for repair or replacement.

Diverter Valve Operation

If the diverter valve malfunctions, the engine usually backfires on deceleration. To check the valve for proper operation:

1. Inspect the condition and routing of all lines to the diverter valve, especially the vacuum signal line or hose. All lines or hoses must be secure, without crimps, and not leaking.

2. Disconnect the vacuum signal line or hose at the diverter valve. Then, start the engine.

3. Check for a vacuum signal at the end of the disconnected line or hose (Fig. 14-2).

4. Reconnect the line or hose to the diverter valve. Next, check for air escaping through the diverter valve silencer, with the engine rpm stabilized at idle.

5. Manually open and then quickly close the carburetor's throttle valves. At

Figure 14-2. Checking for a vacuum signal at the end of the diverter valve vacuum line (Courtesy of United Delco).

this point, a momentary blast of air should discharge through the diverter valve's silencing material for between 1 and 3 seconds (Fig.14-3).

6. If air does not discharge from the silencer, replace the valve because it is not internally serviceable. Note: Although diverter valves are sometimes similar in appearance, their design is such that a particular valve fits the requirements of a given engine. Therefore, always use the correct valve when replacement is necessary.

Air Manifolds and Hoses

Follow this procedure to check for proper operation of air manifolds and hoses:

1. Check all hoses for holes or deterioration.

2. Inspect the injection manifolds for holes or cracks.

3. Check all hose and manifold connections for tightness.

4. Check the routing of all the hoses. Any interference between the hoses and any other object may cause wear and an eventual leak.

5. If you suspect an air leak on the pressure side of the system, check the involved components or connections with a soap and water solution (Fig. 14-4). With the air pump operating, bubbles form at any area where a leak exists. However, be careful to keep the soapy water solution away from the centrifugal filter of the pump.

6. If any hose is found defective, replace it. Note: Manufacturers form air hoses of special high-temperature material. Therefore, if a hose requires replacement, use only the proper type of hose.

Check Valve

You should inspect and test a check valve whenever it is necessary to disconnect the hose from the valve or if you suspect it has failed. A check-valve failure is obvious when an inoperative air pump shows signs of exhaust gases having passed through it.

Figure 14-3. Testing diverter valve operation (Courtesy of United Delco).

Figure 14-4. Testing a hose connection for leaks (Courtesy of United Delco).

To check a typical check valve, orally blow through it toward the air manifold; then, attempt to suck back through the valve. Normal air flow should be in one direction only—toward the air manifold (Fig. 14-5).

Air Injection Tubes

There is no periodic inspection or service for the air injection tubes. However, when it becomes necessary to remove the cylinder head or exhaust manifold from the engine, inspect all the tubes for carbon buildup and for warpage or burned holes.

To service the exposed tubes, remove any carbon buildup with a wire brush. If any of the tubes are warped or burned out, replace them or as necessary the entire injection manifold assembly.

Figure 14-5. Normal direction of air flow through a check valve (Courtesy of Chevrolet Division).

Testing a Chrysler Catalytic Air Injection System

Air-Switching Valve

You can test the overall effectiveness of the Chrysler catalytic air injection system shown in Fig. 14-6 at normal engine temperature with the infrared analyzer using the procedure outlined earlier in this chapter. If the system does not function properly in reducing HC and CO emissions, check all the basic system components as described earlier.

If all the basic system components are operating satisfactorily, check the operation of the switching valve as follows (Fig. 14-7):

1. Disconnect both hoses leading from the check valves to the air-switching assembly.

2. With its coolant at room temperature, start the engine and stabilize its rpm at idle.

3. Check the air flow from the switching valve. Air should now be flowing through the side outlet nipple leading to the exhaust manifold.

4. Permit the engine to warm up until it reaches normal operating temperature.

5. Note the direction of air flow from the switching valve. With the engine at normal operating temperature, the air should now flow from the top outlet nipple leading to the downstream entry point within the exhaust system.

Results and Indications

As the engine warms up, the direction of air flow should switch to the top outlet nipple, with only a small amount of air exiting from the side nipple. If this change does not occur, the switching valve or the coolant-controlled engine vacuum switch is defective.

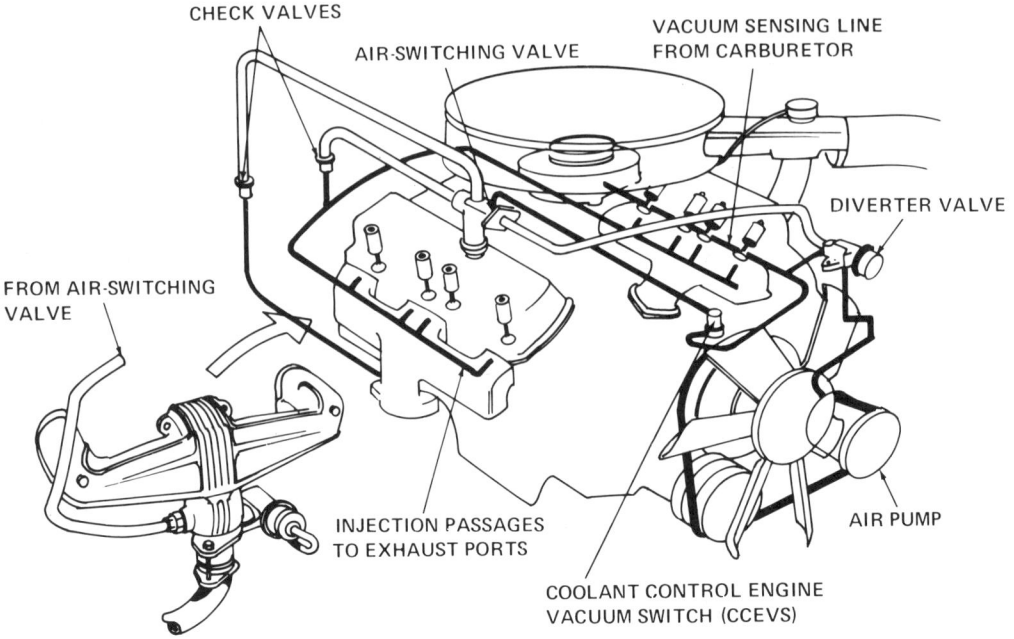

Figure 14-6. You can test the effectiveness of an air injection system with an infrared analyzer.

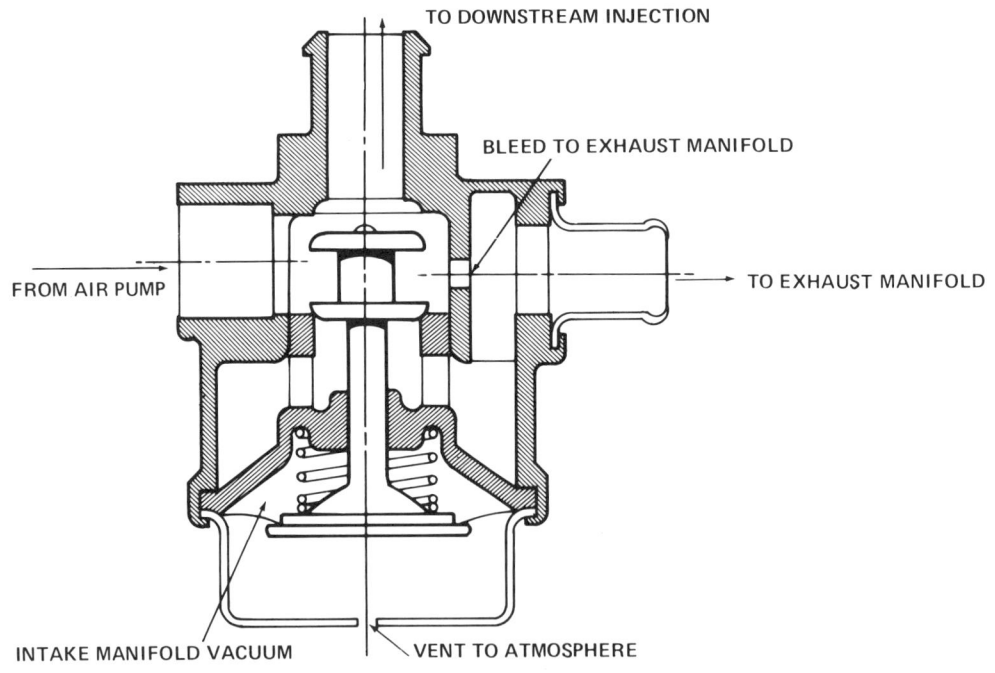

Figure 14-7. Checking the operation of the air-switching valve.

Coolant-Controlled Engine Vacuum Switch

If the switching valve did not perform as specified in the last section, check the operation of the coolant-controlled engine vacuum switch (CCEVS) following these steps:

1. With the engine coolant at room temperature, disconnect both vacuum hoses from the CCEVS. Then, insert the hose from a portable vacuum pump into one of the switch fittings and the hose from a separate vacuum gauge into the other (Fig. 14–8).

2. Carefully remove the radiator pressure cap. Next, insert a thermometer into the upper radiator tank.

3. With the coolant temperatures below 98°F, apply about 15 inches of vacuum to the CCEVS.

4. Note the reading on the separate vacuum gauge. It should now read the same as the gauge on the vacuum pump. If not, the CCEVS is defective and requires replacement.

5. Bleed the vacuum off both the vacuum pump and separate gauge.

6. Start the engine and permit the coolant temperature to reach above 125°F on a 318 V-8 or above 98°F on any other engine. Then, stop the engine and apply 15 inches of vacuum to the CCEVS with the vacuum pump.

7. Note the vacuum reading on the separate gauge. It should now be zero. If not, the CCEVS is defective and requires replacement. Note: If the CCEVS is functioning properly and the switching valve still does not operate, it is defective and requires replacement.

Testing a Ford Catalytic Air Injection System

Air Bypass Valves

You can also test a Ford catalytic air injection system with an infrared analyzer using the procedure outlined earlier. However, just keep in mind that the increase in

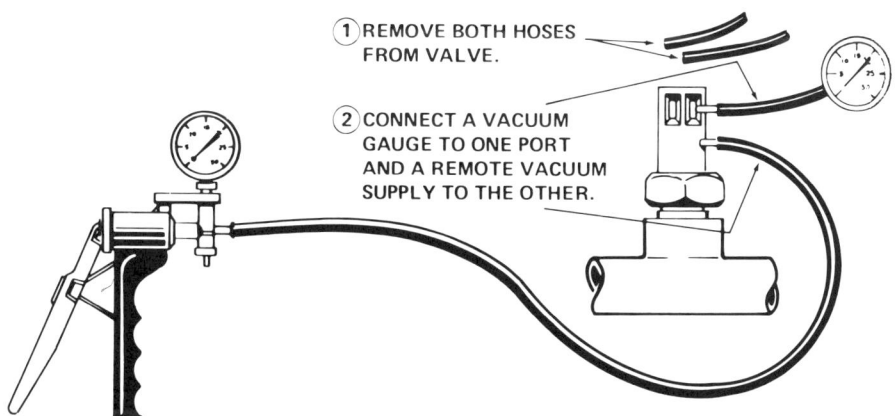

Figure 14–8. Testing a CCEVS assembly with a portable vacuum pump and gauge.

emission levels with the air pump hose disconnected will be considerably lower due to the action of the converter itself.

If the system does not function properly, test all system components except for the air bypass valve, delay valve, and TVS assembly using the instructions set forth earlier in this chapter. If you suspect the air bypass valve is not operating properly, test it for serviceability in this manner:

1. Disconnect the outlet air hose from the bypass valve leading to the check valve and manifold (Fig. 14-9).

2. Remove the vacuum hose from its upper fitting on the bypass valve. Then, in its place, insert the hose from a vacuum pump.

3. With the vacuum pump, apply about 15 inches of vacuum to the bypass valve. Then, start the engine.

4. Note the air flow from the outlet fitting on the bypass valve. Pump air should flow freely from this fitting. If not, the bypass valve is defective and requires replacement.

To test the operation of the later-style bypass valve, use this procedure:

1. Disconnect the outlet hose from the valve leading to the check valve and manifold (Fig. 14-10).

2. Remove the vent line or hose from the upper fitting on the bypass valve.

3. Start the engine and set its idle speed to about 1,000 rpm.

4. Check the air flow from the outlet bypass valve fitting. Pump air should now flow freely from the nipple. If not, and there is an adequate vacuum source to the other small fitting on the bypass valve, it is defective and requires replacement.

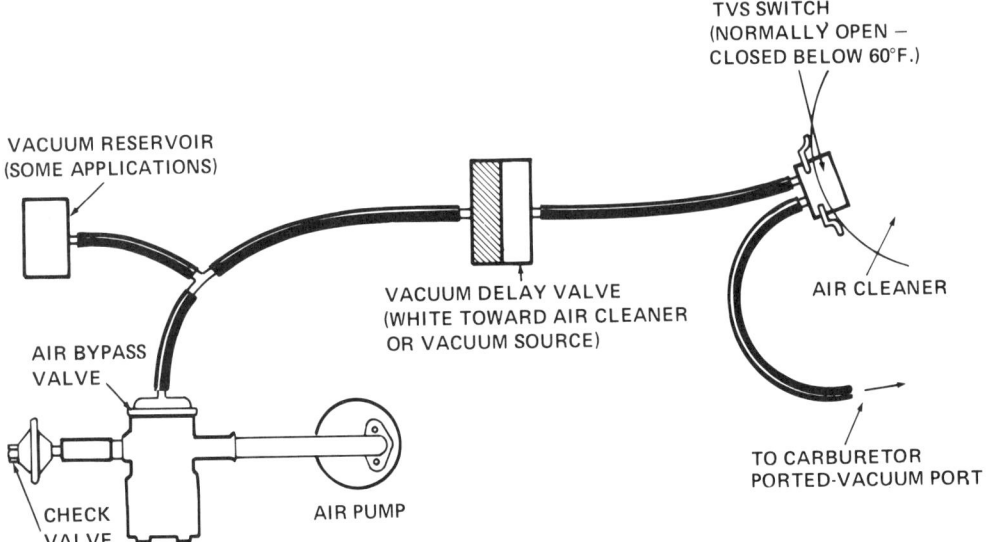

Figure 14-9. Testing a Ford air injection system with the early-style air bypass valve.

Idle Vacuum Valve

If the later bypass valve shown in Fig. 14–10 is still serviceable and the system does not function properly, check the operation of the idle vacuum valve following these steps:

1. Check all the hoses routed to the vacuum valve for proper connection and serviceability.

2. If the hose routing is correct, disconnect both hoses from the idle vacuum valve.

3. Connect the hose from a vacuum pump to the manifold fitting on the valve.

4. Apply about 10 to 15 inches of vacuum to the idle valve with the pump. Then, attempt to blow through the vent fitting on the valve. If air will not pass through the valve or the valve will not hold the applied vacuum, it is defective and requires replacement.

Vacuum-Delay Valve

If the idle vacuum valve functions correctly, check the operation of the delay valve as follows:

1. Remove both vacuum hoses from the delay valve.

2. Connect the hose from a hand-held vacuum pump to the manifold end of the delay valve and a hose from a separate vacuum gauge to the other valve fitting (Fig. 14–11).

3. Apply 15 inches of vacuum with the pump to the delay valve while observing the pointer movement on the second vacuum gauge.

4. Note the number of seconds required

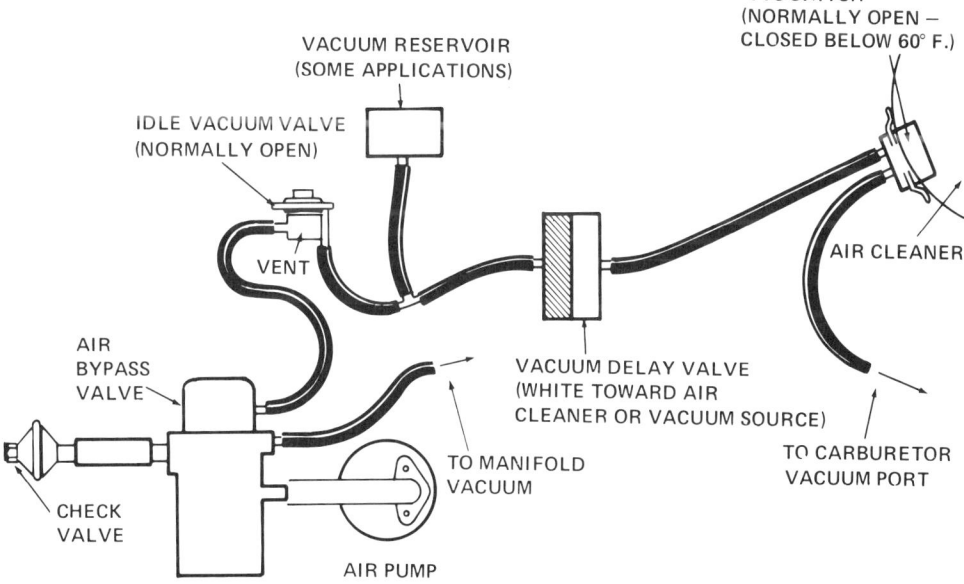

Figure 14–10. Testing a Ford air injection system with the later-style air bypass valve.

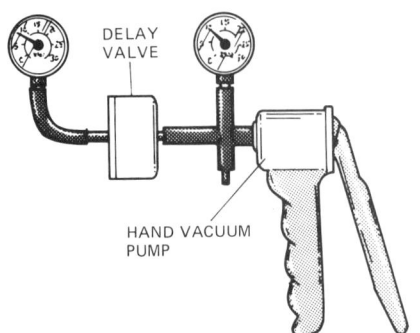

Figure 14-11. Testing a vacuum delay valve.

for the second gauge reading to match that on the gauge of the pump.

5. If the second gauge indicates no reading, or the time period necessary for the second gauge to indicate the pump gauge reading is not to manufacturer's specifications, the delay valve is defective and requires replacement.

TVS Assembly

If the delay valve operates satisfactorily and there is still no vacuum on the idle vacuum valve hose, test the operation of the TVS using these steps:

1. Remove the TVS from the air cleaner.

2. Cool the switch in a refrigerator to 60°F or lower.

3. Connect the hose from a hand-held vacuum pump to the manifold or ported vacuum fitting on the switch.

4. Apply at least 16 inches of vacuum to the switch.

5. With the switch at a temperature below 60°F, it should not leak off any of the applied vacuum. If it does, the switch is defective and requires replacement.

6. Permit the switch to warm up above 60°F. The switch should now open and bleed off the applied vacuum. If it does not, the switch is defective and requires replacement.

Secondary Air Control Valve

The testing of a Thermactor system with a secondary air supply valve is about the same as testing any other late Ford system with one exception, namely, the control valve itself. The secondary control valve is responsible for switching pump air from the exhaust manifolds to the rear half of the catalytic converter after the engine temperature reaches a predetermined value.

To test the secondary control valve (Fig. 14-12):

1. Disconnect both outlet hoses from the control valve leading to the exhaust manifold and catalytic converter check valves.

2. With the engine operating at normal temperature, check at which outlet fitting air is now escaping. Injection air should now pass from the fitting for the converter check valve.

3. If the air is instead flowing from the manifold fitting, check to see if there is vacuum on the control valve sensing line. If there is a vacuum on the line, the PVS assembly is most likely defective and requires replacement.

4. If the PVS is functioning properly and there is no vacuum to the control valve, it is defective and requires replacement. Note: You can also test the control valve with the engine operating as described above by alternately applying and bleeding off vacuum to the control valve

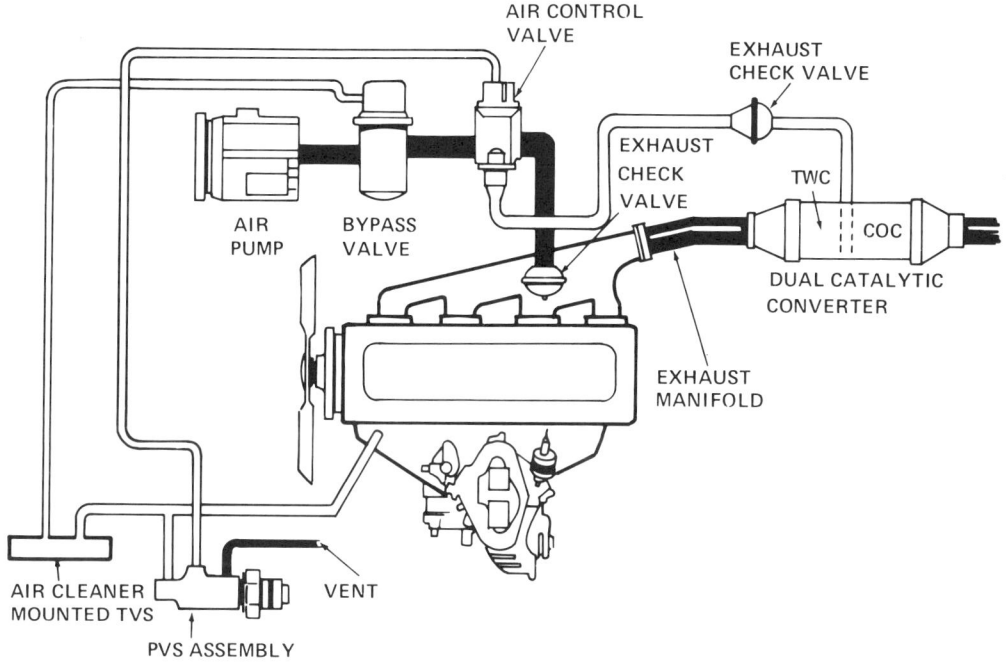

Figure 14-12. Testing a Thermactor injection system with a secondary air control valve.

with a hand-held vacuum pump. With vacuum applied, the control valve should switch the air flow to the manifold fitting. Without applied vacuum, pump air must come out of the fitting for the converter. If the valve does not function in this manner, it is faulty and a replacement is necessary.

Testing a Typical General Motors Catalytic Air Injection System

Test the overall effectiveness of this system in reducing HC and CO emissions with an infrared analyzer as described earlier in this text. Just keep in mind that the analyzer readings, with the air pump outlet hose removed, will be less than those of a basic system used on vehicles without the catalytic converter. The reason for this is, of course, that the converter itself reduces the levels of HC and CO emissions. Also, if the system fails to function properly, check all the individual components as outlined earlier in this chapter.

Checking a Typical Aspirator Air Injection System

To test the aspirator air system shown in Fig. 14-13, follow these steps:

1. Inspect both inlet hose connections for leakage. If either connection leaks, install hose clamps and tighten securely.

TESTING AIR INJECTION SYSTEMS

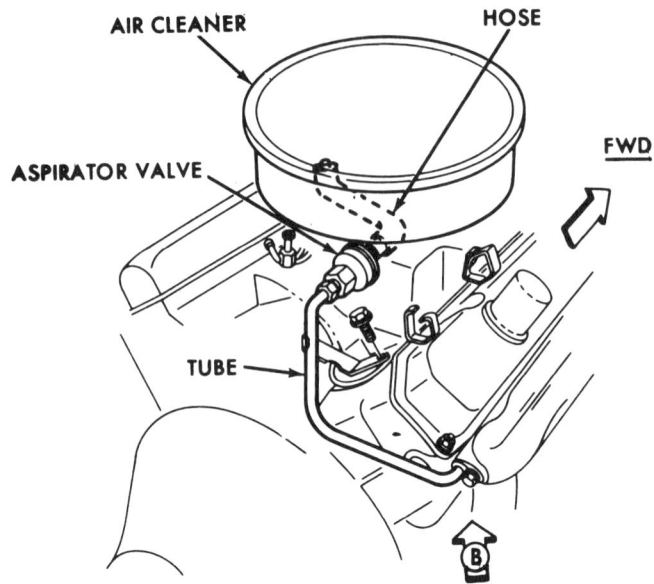

Figure 14-13. Testing a typical aspirator air injection system (Courtesy of Chrysler Corp.).

2. Check the fittings on the ends of the steel tube at the aspirator valve and exhaust manifold for leakage. Tighten as necessary the aspirator-to-tube fitting, and replace the gasket at the manifold connection to stop the leaks.

3. Start the engine and permit its idle speed to stabilize.

4. Disconnect the aspirator inlet hose from the air cleaner. Then, place your thumb over the end of the hose. You should feel the intake air pulses at this time. If instead you feel the presence of exhaust gas, the aspirator valve is defective and requires replacement. Note: There are two other symptoms of aspirator valve failure: (1) an increase in noise from the exhaust system at idle and (2) the inlet air hose will be hard and cracked due to exhaust gas passing through it.

Summary

1. Any failure within the air injection system causes an increase in HC and CO emissions.

2. Certain malfunctions within the air injection system can cause rough engine idle and a backfire in the exhaust system.

3. The type and frequency of preventive maintenance varies somewhat between vehicle manufacturers.

4. You can test the overall effectiveness of an air injection system with an infrared analyzer following the procedures outlined in this chapter.

5. Check the serviceability of the air pump, diverter valve, air manifolds and hoses, check valve, and injection tubes as outlined in this text.

6. If the diverter valve malfunctions, the engine usually backfires during deceleration.
7. There is no periodic inspection or service for the air injection tubes.
8. Check the serviceability of the air switching valve found on a Chrysler catalytic air injection system using manufacturer's instructions or those presented in this chapter.
9. Test the operation of the coolant-controlled engine vacuum switch within the Chrysler catalytic air injection system following the steps presented in this text.
10. When testing a catalytic air injection system with an infrared analyzer, keep in mind that the increase in emission levels, with the air pump hose disconnected, will be lower due to the action of the converter itself.
11. Test the air bypass valve, delay valve, and TVS assembly of a Ford catalytic air injection system using either manufacturer's instructions or those presented in this chapter.
12. Test the Ford secondary control valve following the steps outlined in this text or manufacturer's instructions.
13. You can check an aspirator air injection system using the procedures presented in this chapter or manufacturer's instructions.

Review Questions

The questions listed below will assist you in determining how well you remember the material contained in this chapter. Read each question carefully before choosing the answer. If you cannot answer the question, review the section in the chapter that covers the material.

1. If the diverter valve malfunctions, the engine will _____.
   a. backfire
   b. diesel
   c. stall
   d. accelerate

2. When testing the air injection system with an infrared analyzer, the HC and CO reading should _____ with the air pump output hose disconnected.
   a. zero
   b. stabilize
   c. decrease
   d. increase

3. The tension on a new pump drive belt should be _____ pounds.
   a. 65
   b. 75
   c. 85
   d. 95

4. On engine deceleration, a momentary blast of air should discharge through the diverter valve's silencing material for between _____ seconds.
   a. 1 to 5
   b. 1 to 4
   c. 1 to 3
   d. 1 to 2

5. To test for leakage on the pressure side of an air injection system use _____.
   a. an infrared analyzer
   b. a soapy water solution
   c. cleaning solvent
   d. oil

6. On a Chrysler 318 V-8 engine, the CCEVS should prevent vacuum from reaching the switching valve at _____ °F.
   a. 125
   b. 118
   c. 108
   d. 98
7. Check the serviceability of a Ford idle vacuum and delay valve with a (an) _____.
   a. engine analyzer
   b. test light
   c. infrared analyzer
   d. vacuum pump
8. The device that directs pump air to the rear half of a Ford catalytic converter is the _____.
   a. PVS
   b. TVS
   c. secondary air control valve
   d. idle vacuum valve

For the answers to these review questions, turn to the Appendix.

Chapter 15

Exhaust Gas Recirculation Systems

An exhaust gas recirculation (EGR) system reduces the amounts of nitrogen oxide produced during engine combustion. There are basically two ways in which manufacturers control peak combustion chamber temperatures to hold down the NOx formation within gasoline engines. One way is through the use of a spark control system, which this text discussed in an earlier chapter. The other, and very efficient method, is to dilute the incoming air/fuel mixture with a small amount of an inert gas. An inert gas is one that will not undergo a chemical reaction.

Due to the fact that exhaust gas is relatively inert, manufacturers use it to dilute the air/fuel mixture. This action is done by routing small amounts of this gas (about 6 to 14 percent) from the engine's exhaust system to the intake manifold. This concentration of exhaust gas mixes with the air/fuel mixture, entering the various cylinders, and lowers the mixture's ability to produce heat during combustion.

Exhaust gas performs this function by limiting the quality and quantity of the air/fuel charge that actually enters each of the combustion chambers. For example, because the exhaust gas contains no oxygen, it dilutes the air/fuel charge with a noncombustible gas. In other words, the inert exhaust gas displaces a portion of the oxygen within the highly combustible air/fuel mixture, thus reducing the quality of the total charge reaching each of the cylinders.

Also, since the injected exhaust gases are hot, they expand the air/fuel mixture within the intake manifold. This action reduces the concentration of combustible materials swept into and compressed by the piston in each of the engine's cylinders. As a result, the remaining air/fuel and exhaust gas mixture, which reaches the combustion chambers, is not as powerful when ignited; and it thus creates less heat than an undiluted air/fuel mixture would otherwise produce.

Not all engine operating phases produce excessive NOx emissions. For instance, because the engine creates very small amounts of NOx at idle, exhaust gas recirculation is not necessary or desirable at this time. In addition, the EGR system is inoperative at wide-open throttle operation for more efficient engine operation and vehicle driveability. Therefore, the main operating period that the EGR system should function is at vehicle speeds between about 30 to 70 mph, when NOx formation is the highest.

However, at certain engine temperatures, exhaust gas recirculation may not be necessary even during these periods. For example, when engine temperature is low, the formation of NOx is less. Consequently, operation of the EGR system is not necessary, and the system is therefore made inoperative in most cases. This action improves engine warm-up and vehicle driveability.

Basic EGR System Design

Intake Manifold

Any basic EGR system consists of a number of main components. These include a redesigned intake manifold, EGR valve, a valve-activating system, and some form of thermo valve or switch.

In order to tap a continuous supply of inert exhaust gases from the exhaust system without external pipes or connections, manufacturers have redesigned the intake manifold. The type of modification made to the manifold itself depends largely on whether the system uses a fixed or variable orifice and the location of the latter.

Figure 15-1 illustrates a redesigned intake manifold used with a fixed orifice EGR system, found on some 1972-73 Chrysler-built engines. In this particular V-8 manifold, a pair of stainless steel, fixed orifice jets thread into the floor of the intake manifold under the carburetor.

With this design, exhaust gases entering the crossover passage bleed up into the air/fuel mixture through the floor jet orifices. The amount of exhaust gases entering the charge depends on two factors: (1) the amount of intake manifold vacuum during a given phase of engine operation and (2) the size or diameter of the opening within the floor jet.

Although the installation of the floor jets in the intake manifold represents the simplest of all EGR systems, there is one big disadvantage to its prolonged usage. That is, this type of system permits exhaust gas to enter the intake manifold at all times. This causes rough engine operation at idle and stumble during cold-engine warm-up.

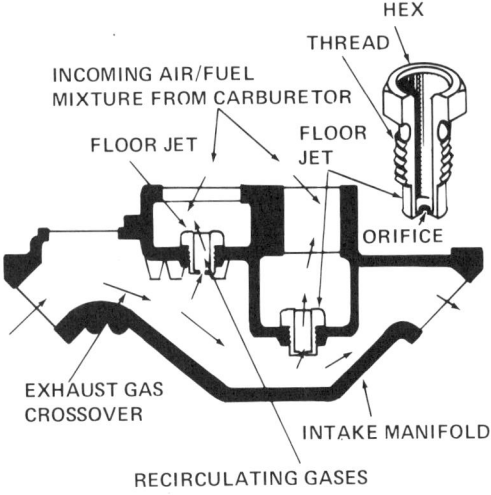

Figure 15-1. Schematic of a Chrysler V-8 engine intake manifold that utilizes floor jets.

BASIC EGR SYSTEM DESIGN   289

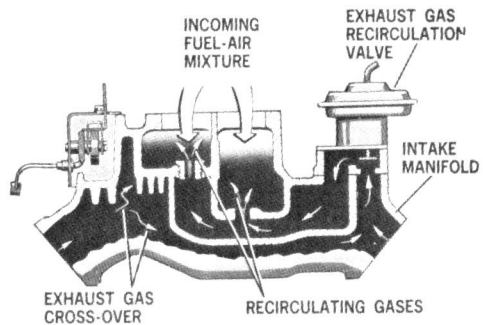

Figure 15-2. Intake manifold with an EGR passage and ports cast into it (Courtesy of Chrysler Corp.).

A modified intake manifold used with a variable-type EGR valve is shown in Fig. 15-2. In this manifold, the manufacturer casts an additional exhaust gas passage into the complex runner system. In other words, instead of only having one exhaust passage to preheat the intake runner floor, this manifold has two—one serves as the preheat exhaust crossover and the second routes the inert gas to the intake runners.

EGR Valve

Separating the above two passages within the manifold is a vacuum-operated, variable orifice metering valve, commonly referred to as the EGR valve. This valve is a spring-loaded, reciprocating poppet-type unit (Fig. 15-3). Within this assembly is a spring-loaded diaphragm that attaches to a tapered EGR valve. The spring side of the diaphragm housing has a port opening that connects into a vacuum source, while the

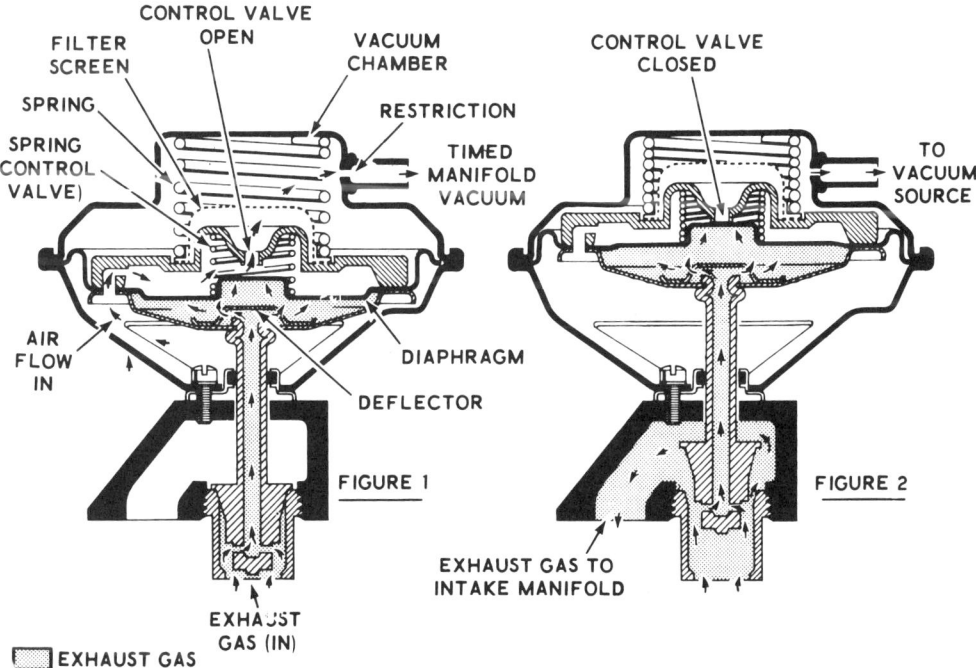

Figure 15-3. Design and operation of a typical EGR valve (Courtesy of General Motors Corp.).

other side has a vent opening into the atmosphere.

In operation, the valve is in the closed position (by the action of the spring) and opens when a given amount of vacuum acts on the spring side of the diaphragm. For example, in Fig. 15-3, the spring is holding the diaphragm and its attached tapered valve in the closed position. As a result, exhaust gases from the crossover passage cannot enter the EGR passage leading to the intake manifold.

When the vacuum acting on the diaphragm reaches about 3 inches Hg, the diaphragm begins to open the valve. This valve reaches its full open position somewhere between 5 to about 8.5 inches of vacuum, depending on system design (Fig. 15-3). As a result, the actual position of the tapered valve meters the flow of exhaust gases into the intake manifold in proportion to the strength of the vacuum signal applied to the diaphragm. Consequently, the EGR valve acts as a variable orifice, which regulates the amount of exhaust gas recirculation in proportion to the strength of the vacuum signal.

However, even with the tapered valve in the wide-open position, the amount of exhaust gas recirculation still has limitations. This is due to the size of the orifice formed between the valve stem and the wall of its seat. This area, as shown in Fig. 15-3, is relatively small.

Vacuum Sources Used to Operate an EGR Valve

Ported Vacuum

Manufacturers can use either one of two different sources of vacuum to operate an EGR valve: ported or venturi. When an EGR valve is activated by a ported vacuum signal (Fig. 15-4), the carburetor has a slot-type port machined into the throttle body, above the throttle valves.

With this arrangement, when the throttle valves open during engine acceleration, they expose the slot to a progressively increased percentage of intake manifold vacuum. As a result, there is no ported vacuum signal to the EGR valve with closed throttle valves at idle. But, as the valves open during engine acceleration, the signal increases, which in turn opens the EGR valve in proportion to the intensity of the vacuum signal applied to its diaphragm. In other words, the amount of valve opening and exhaust gas recirculation depends upon throttle position and intake manifold vacuum.

Due to the important fact that the ported vacuum signal cannot be more than intake manifold vacuum, the EGR valve does not function under wide-open throttle acceleration. In other words, as intake manifold drops extensively during heavy acceleration, so does the ported vacuum signal to the EGR valve.

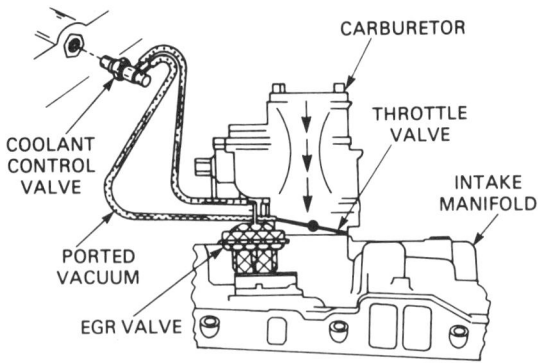

Figure 15-4. EGR valve activated by a ported vacuum signal.

This action prevents any exhaust gas recirculation during extended wide-open throttle operation. Therefore, an EGR valve, which operates by a ported vacuum signal, opens progressively during periods of moderate vehicle acceleration between speeds of about 30 to 70 mph. But the valve closes when there is a weak manifold vacuum produced during extended periods of wide-open throttle operation or at engine idle.

Venturi Vacuum-Operated EGR

An alternate method of providing a vacuum signal to an EGR valve is through the use of a venturi vacuum control system (Fig. 15-5). This system consists of a vacuum tap, amplifier, and dump valve. The tap, like the main fuel nozzle, is nothing more than a projected opening into the throat of the venturi. As such, the tap also has a low pressure (vacuum) applied to its opening whenever a considerable amount of air flows through the carburetor.

In operation, the strength of this venturi vacuum signal depends mainly on the velocity of the air flow through the venturi itself. For instance, during engine idle (closed throttle), the air flow through the venturi is very slow. This results in a zero or very low tap vacuum signal. However, as the throttle valve opens during acceleration and air flow increases, the tap signal intensifies in proportion to the rise in the rate of air flow.

Even when there is a great deal of air passing through the venturi, such as during high engine rpm, the tap signal is too weak to actually operate the EGR valve diaphragm. To overcome this problem, the system requires an amplifier to boost the tap signal sufficiently to open the EGR valve.

The amplifier performs this function by storing intake manifold vacuum within a reservoir inside the unit itself or externally mounted on the firewall. In any case, the reservoir guarantees an adequate vacuum source to activate the EGR valve diaphragm against its spring tension, regardless of variations in manifold vacuum.

Since the reservoir maintains a relatively steady vacuum source to the EGR valve, a dump diaphragm is necessary to prevent the EGR valve from operating at wide-open throttle (WOT). In operation, the dump diaphragm (Fig. 15-6) senses when the throttle is wide open, and it responds by venting the stored reservoir to the atmosphere.

In other words, the dump diaphragm compares venturi tap and manifold vacuum. When tap vacuum is equal to or greater than manifold vacuum, the diaphragm "dumps" reservoir vacuum. This action limits the signal to the EGR valve to manifold vacuum, which during WOT is zero or very low. Thus, the EGR diaphragm spring closes the valve until reservoir vacuum once again increases.

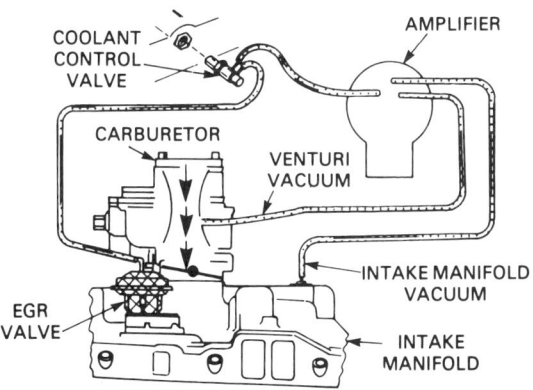

Figure 15-5. EGR valve operated by a venturi vacuum signal.

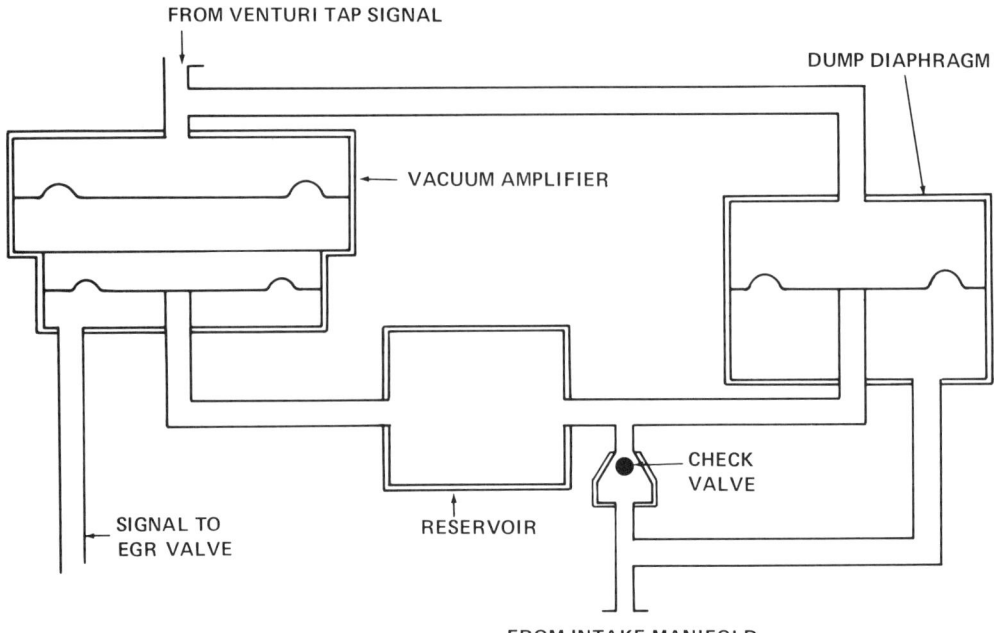

Figure 15-6. Diagram of a venturi vacuum-operated EGR system with a separate reservoir and dump diaphragm.

EGR System Modulating Devices

Automotive engineers have developed a considerable number of modulating or regulating devices to more precisely control the operation of the EGR valve during the various phases of engine performance. These devices are necessary to improve vehicle driveability while still maintaining adequate exhaust gas circulation to control NOx emissions.

Although there have been a large number of modulating devices used so far, this section will only discuss three of these units: a back-pressure transducer, coolant vacuum switch, and dual-diaphragm EGR valve. However, other modulating units will be covered when discussing specific EGR systems.

Back-Pressure Transducer

Some manufacturers use an exhaust back-pressure transducer within the EGR system. This device ensures maximum exhaust gas recirculation during engine acceleration and some vehicle cruise conditions.

The back-pressure transducer illustrated in Fig. 15-7 consists of a housing, an exhaust back-pressure probe, a valve, diaphragm, and spring. Extending from the lower section of the housing is the exhaust back-pressure probe. This probe fits into an exhaust gas passage and carries an ex-

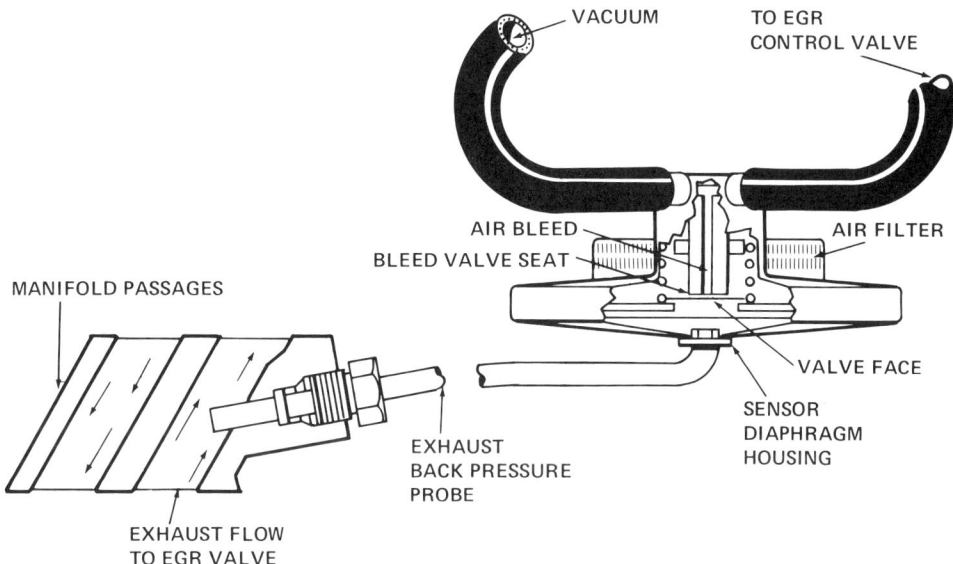

Figure 15-7. Typical EGR back-pressure transducer.

haust gas pressure signal to the lower portion of the diaphragm.

The back-pressure valve controls the strength of the vacuum signal that operates the EGR valve. The valve accomplishes this task by permitting, at times, some of the vacuum signal to leak off to the atmosphere via the air bleed.

The valve shown in Fig. 15-7 has a seat and face. The valve seat has a T shape and is hollow. The upper portion of the seat extends out of the housing to form the fittings for two vacuum hoses—one routed to the vacuum source and the other to the EGR valve.

Controlling the opening in the base of the valve seat is the valve face. The valve face mounts onto the center of the diaphragm. Consequently, any movement of the diaphragm causes the valve face to either open or close the opening within the valve seat. When the face is away from the seat, the valve is open and acts as an air bleed for the vacuum contained in the hose to the EGR valve.

The diaphragm itself responds to exhaust gas back pressure and the force of the calibrated spring to either open or close the valve. For example, when exhaust gas back pressure is high, the diaphragm moves up, and the valve face is against the seat. However, when back pressure is low, the tension of the calibrated spring pushes the diaphragm down, causing the valve face to move away from its seat. In this case, the valve is open and now acts as an air bleed for the vacuum signal to the EGR valve.

Transducer Operation

When exhaust gas back pressure is high, during engine acceleration and under some cruise conditions, the exhaust gas pressure traveling up the metallic probe overcomes the valve diaphragm's spring tension. As a result, the diaphragm moves upward and its attached valve face closes off the air bleed opening within the seat. This permits a full vacuum signal to pass through the hose to the EGR valve, resulting in maximum exhaust gas recirculation.

However, when exhaust gas pressure is low, during idle or engine deceleration, the diaphragm spring moves it and the valve face downward. With the valve face off its seat, the valve forms an air bleed for the vacuum circuit to the EGR valve. It should be obvious then that the strength of the signal to the EGR valve is variable, depending upon the position of the valve within the back-pressure transducer.

Coolant-Controlled Vacuum Valves

Both the EGR systems using either the ported or venturi vacuum signals require a coolant-controlled vacuum valve (see Figs. 15-4 and 15-5). The valves in both of these systems operate in much the same manner as the coolant-controlled valves described earlier. However, when used in conjunction with an EGR system, the control valve prevents vacuum from reaching the EGR valve if engine temperature is below a preset value. This action prevents exhaust gas recirculation when coolant temperatures are low to improve engine performance during its warm-up.

Dual-Diaphragm EGR Valves

Another method of modulating exhaust gas recirculation is through the use of a dual-diaphragm EGR valve (Fig. 15-8). This valve design provides increased EGR rates with larger engine loads. The signal used to designate changes in engine load is intake manifold vacuum.

The valve itself is similar to the single-diaphragm unit except that a second diaphragm has been added to the assembly. This diaphragm connects to the upper one with a spacer. Consequently, both diaphragms move together, along with the attached EGR valve.

The intake manifold signal acts on the air space between the two diaphragms. The upper diaphragm has a larger effective piston area than the lower one. Therefore, the effect of the intake manifold vacuum on this larger area adds to the force already applied by the calibrated spring. This tends to keep the attached EGR valve in the closed position, regardless of the vacuum from the manifold acting on the lower diaphragm, which normally would attempt to open the EGR valve against spring tension.

With this arrangement, full EGR valve movement depends on the amount of intake manifold vacuum applied between the diaphragms and the strength of the EGR valve signal to the upper chamber. For example, during engine acceleration, when the load on the engine increases, intake manifold vacuum decreases. As a result, the combined load of the calibrated spring and the vacuum acting on the base of the upper diaphragm is less. This permits the EGR valve to open further with a given vacuum signal acting on the upper chamber.

Consequently, during cruising speeds, when intake manifold vacuum is highest, the valve opening is less than what it would be while the engine is accelerating. This provides the valve with the capability of providing more exhaust gas recirculation on engine aceleration, where loads are

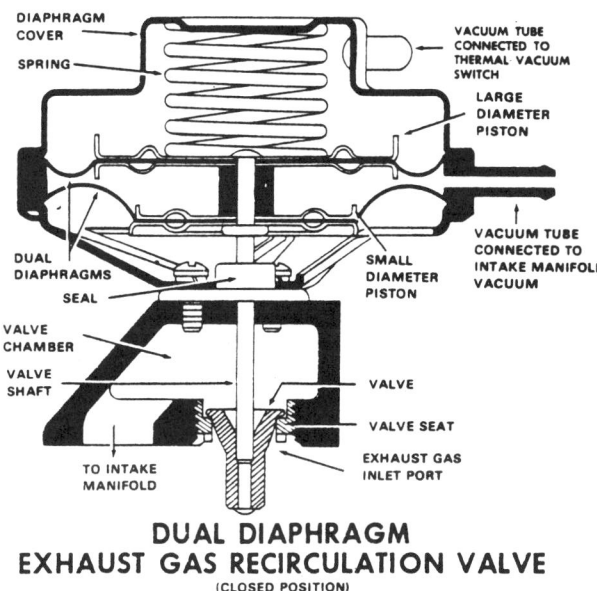

Figure 15-8. Dual-diaphragm EGR valve modulates exhaust gas recirculation to changes in engine vacuum (Courtesy of Chevrolet Division).

higher and the tendency to produce NOx is greatest.

Specific EGR Systems

Chrysler Corporation

As in the case of most other exhaust emission control systems, automotive manufacturers have produced many different variations of the basic EGR system, using many of the components already described in this chapter. Therefore, it is impossible in the space remaining to cover all the EGR systems used by all the automobile manufacturers in detail. However, this section covers recent versions and special variations in the EGR systems used by Chrysler, Ford, and General Motors.

Although Chrysler did at first use a floor jet EGR system that was very simple, it did create some problems with rough engine idle and stumble during the warm-up period. This system has already been mentioned (see Fig. 15-1).

To overcome these shortcomings, Chrysler developed the system shown in Fig. 15-9. This system has a single diaphragm-operated EGR valve, which activates by means of a venturi vacuum control system. This system consists of a vacuum tap, amplifier, and dump valve, already described in an earlier section (see Figs. 15-5 and 15-6).

Along with these components, the Chrysler system has a coolant-controlled exhaust gas recirculation (CCEGR) valve and switch in addition to an EGR delay system. The CCEGR valve threads into the radiator's top tank and tees into the vac-

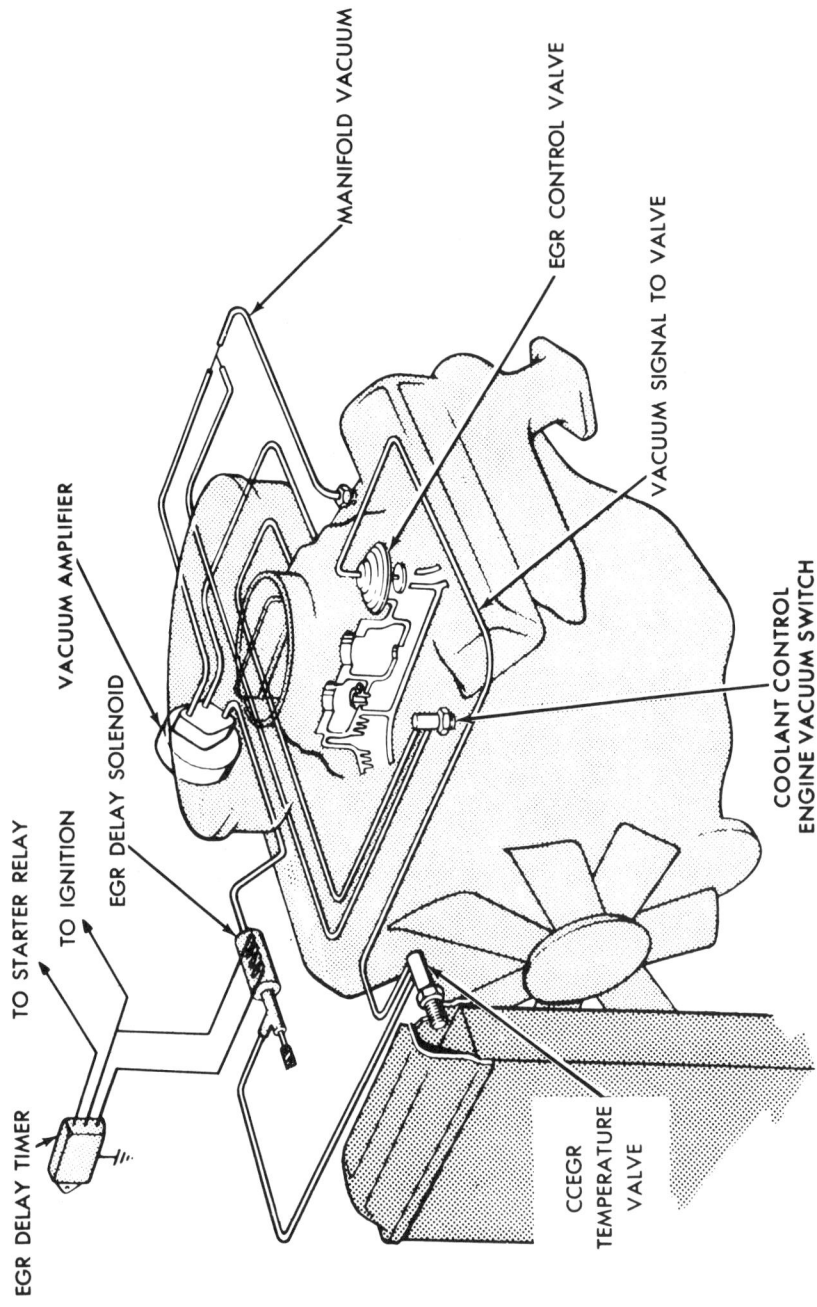

Figure 15-9. Components of a Chrysler EGR system (Courtesy of Chrysler Corp.).

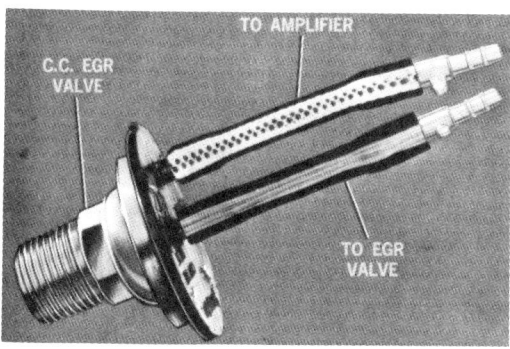

Figure 15-10. Design of a CCEGR valve (Courtesy of Chrysler Corp.).

uum line between the amplifier and the EGR valve (Figs. 15-9 and 15-10).

The function of the CCEGR valve is to prevent exhaust gas recirculation at low temperatures. For example, until coolant temperature reaches about 65°F, the valve remains closed. This prevents the amplifier signal from reaching the EGR valve. Thus, the EGR valve does not react to any amplifier signal until coolant temperature is above 65 °F.

In some models, a similar device fits into the thermostat housing (see Fig. 15-9). This coolant-controlled engine vacuum switch (CCEVS) operates in a manner similar to the valve mentioned above. However, the switch does not activate until coolant temperature within the engine reaches about 90 °F. When the switch does open at this temperature, it permits the application of full intake manifold vacuum to the amplifier.

Time-Delay Systems

The Chrysler EGR system shown in Fig. 15-9 also has a time-delay system. The purpose of this system is to prevent exhaust gas recirculation for about 35 seconds after the driver turns on the ignition switch. This action improves engine starting and initial vehicle driveability.

This basic system consists of a delay timer and solenoid. The delay timer mounts onto the firewall inside the engine compartment and electrically activates the solenoid for about 35 seconds once the ignition switch is on.

The solenoid has a plunger that controls two vacuum ports, which tee between the vacuum amplifier and the CCEGR valve. Therefore, when the relay energizes the solenoid, it cuts off the vacuum signal to the CCEGR and EGR valves. However, after 35 seconds, the relay deenergizes the solenoid, and its plunger permits the amplifier signal to reach the EGR valve. This permits normal exhaust gas recirculation.

Figure 15-11 illustrates a late Chrysler EGR system with a redesigned delay system. This delay system is similar to the one just described except that the newer one has an additional component; namely, a charge temperature switch (CTS). This switch fits into the number-six branch runner of an intake manifold for a six-cylinder engine or the number-eight branch runner of an eight-cylinder engine.

The function of the CTS is to provide a signal to the delay timer relative to the temperature of the air/fuel charge within the manifold. For example, when the intake charge temperature is below 60 °F, the CTS closes. As a result, the timer does not function to open the vacuum solenoid. Consequently, the amplifier signal is not strong enough to operate the EGR valve for normal exhaust gas recirculation.

However, the CTS does open as the temperature of the air/fuel charge reaches above 60 °F. Thus, the timer eventually opens the solenoid, which permits an adequate signal to reach and open the EGR valve.

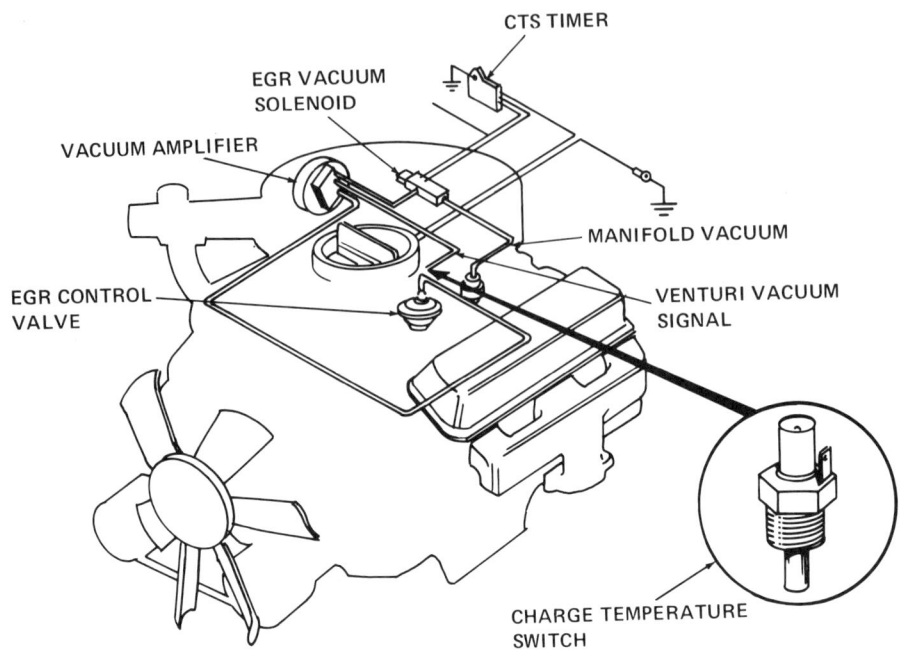

Figure 15-11. Schematic of a Chrysler EGR system with a CTS delay system.

Chrysler Maintenance Reminder System

In 1975, Chrysler installed an EGR maintenance reminder system on its vehicles; however, it was not used on 1976 and later automobiles. The purpose of the system was to remind the driver that the EGR system requires service. This was accomplished by the activation of a dash light at 15,000-mile intervals, which reminded the driver to "CHECK EGR."

This dashboard light is controlled through a mileage-counting switch that fits in the speedometer cable about halfway between the transmission and the speedometer head in the dash (Fig. 15-12). The

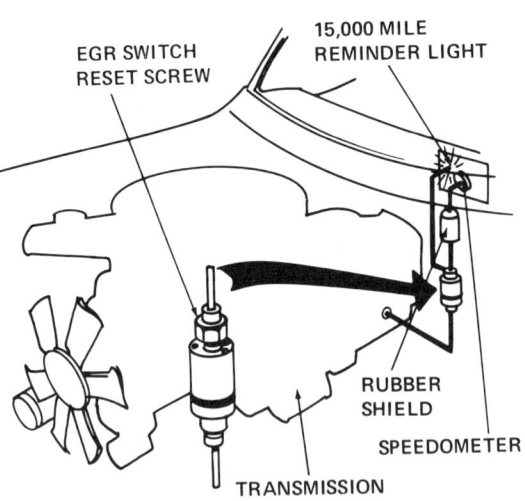

Figure 15-12. Diagram of a Chrysler maintenance reminder system.

switch functions like an odometer; however, when the first 15,000 miles have gone by, the switch closes the circuit to light the indicator. The light then stays on until the switch is reset by a mechanic.

The indicator light reminds the driver to have a technician check and service the EGR system as soon as possible, but it does not affect vehicle safety in any way. The system will again activate the light after another 15,000 miles have passed, once the switch has been reset.

Ford EGR Systems

There have been many modifications to Ford's basic EGR system, first introduced in 1973. Figure 15-13 is a schematic of a basic Ford system consisting of an EGR valve, a specially designed intake manifold or spacer plate, and a temperature-controlled vacuum valve.

Ford EGR Valves

With its basic system, Ford has utilized three somewhat different styles of EGR valves: the poppet, modulating, and tapered stem. A poppet-type valve (Fig. 15-14) consists of a spring-loaded diaphragm along with a valve stem and valve, all operating inside an enclosed valve body. The flexible diaphragm reacts to vacuum applied to its spring side. When the signal is strong enough, the diaphragm will move upward, compressing the calibrated spring.

Attached to the center of the diaphragm is the valve stem. Fitted to the opposite end of the stem from the diaphragm is the valve itself. With this arrangement, any upward movement of the diaphragm causes the valve to move away from the valve seat.

At approximately 3 inches of vacuum, the diaphragm begins to move upward. This action unseats the valve and permits

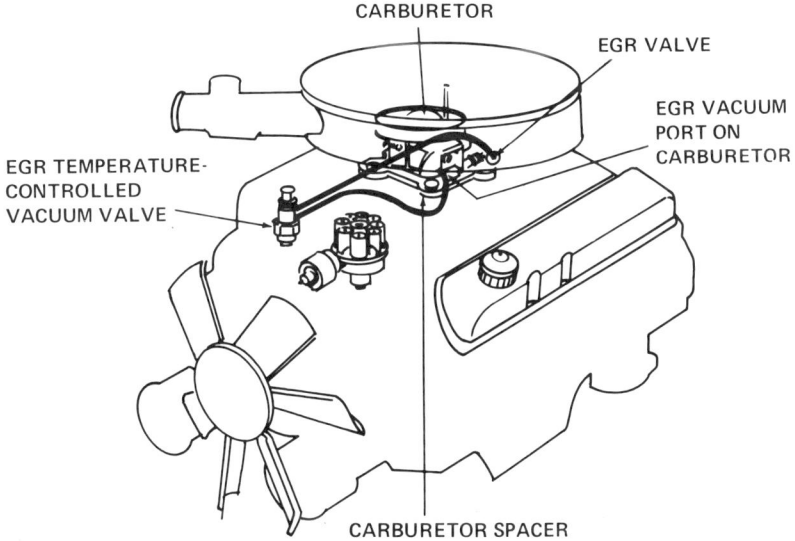

Figure 15-13. Basic Ford EGR system.

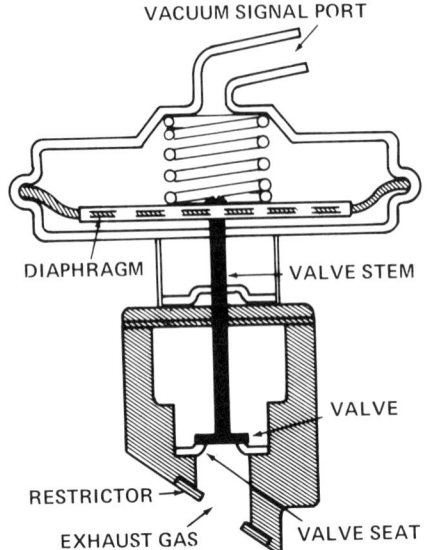

Figure 15-14. Ford poppet-type EGR valve.

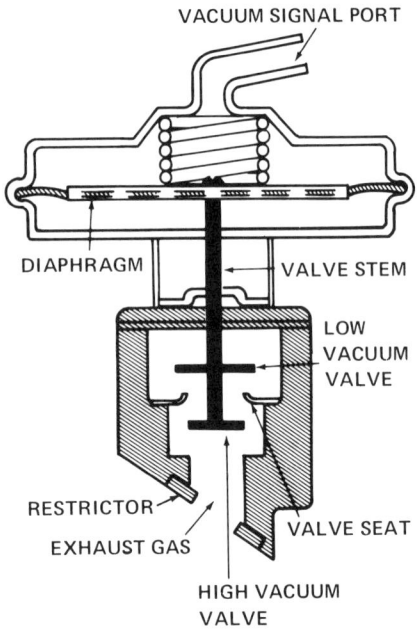

Figure 15-15. Ford modulating-type EGR valve.

exhaust gases to flow into the air/fuel mixture within the intake manifold. Once the diaphragm unseats the valve, the only means of limiting exhaust gas flow is by the size of the restrictor placed into the inlet port within the valve body.

Figure 15-15 illustrates the modulating-type Ford EGR valve. This valve is similar to the poppet-type unit with one exception. The modulating unit has an additional disc added to the stem below the main valve disc. The purpose of the additional disc is to modulate (regulate) exhaust gas recirculation during certain phases of engine operation to improve the driveability of vehicles with certain engine models.

When the vacuum signal to the valve is between 3 to 10 inches Hg, the modulating valve operates in exactly the same way as the poppet type. However, as the signal reaches about 10.5 inches, the lower disc (high vacuum flow restrictor) approaches the shoulder of the seat. This restricts the flow of exhaust gases. Consequently, this valve permits an increased exhaust gas flow into the intake manifold during engine acceleration, but it tends to restrict this flow during engine cruising speeds when the EGR vacuum signal is high.

The Ford tapered-stem EGR valve (Fig. 15-16) also utilizes a single spring-loaded diaphragm connected through a stem to a valve. The main difference between this valve and the other two just described is the valving mechanism. The other two Ford valves used either a single- or double-disc arangement to regulate exhaust gas recirculation. However, the valve shown in Fig. 15-16 utilizes a tapered valve, which gradually unseats as the EGR diaphragm moves upward to permit an increasing exhaust gas flow.

Because all three of these valves are slightly different in construction, there is

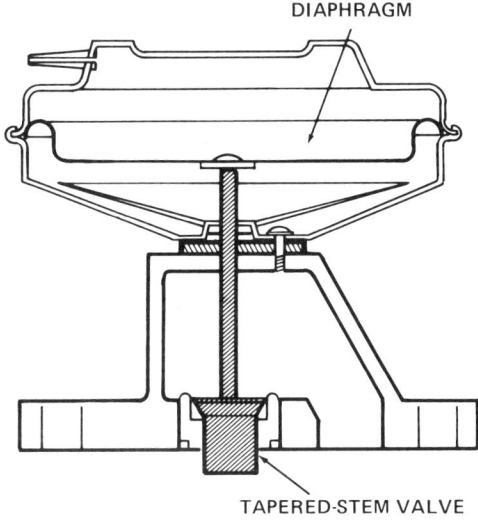

Figure 15-16. Ford tapered-stem valve.

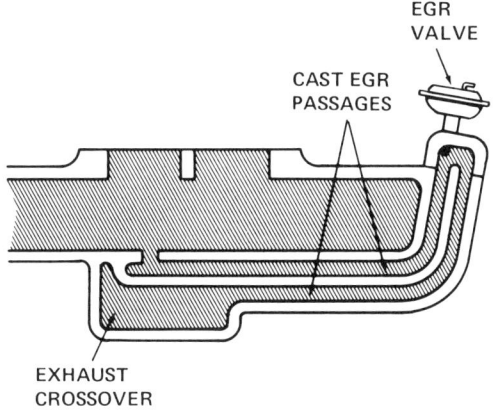

Figure 15-17. Ford EGR system with ports within the intake manifold floor to recycle the exhaust gases.

one important fact to remember when a replacement becomes necessary. That is, the three valve types are not interchangeable among engine types or model years.

Methods Used To Supply Exhaust Gas To the EGR Value

Ford uses various methods to deliver exhaust gas to the EGR valve. For instance, on some 1974 engines, the manufacturer installs a floor entry system to the EGR valve (Fig. 15-17). In this system the intake manifold has two passages cast into the floor beneath the intake runners. One of the passages connects into the exhaust crossover and transports the exhaust gases to an opening in the manifold below the EGR valve.

The other passage interconnects the EGR valve with two holes in the manifold floor directly beneath the carburetor's primary throttle bores. This design permits the incoming exhaust gas to mix with the air/fuel mixture before it enters the various combustion chambers.

Many four- and six-cylinder Ford engines have EGR valves that receive exhaust gases from an external source. In this case, the exhaust gases pass from the manifold through a stainless steel tube to a special spacer installed below the carburetor.

Another and more common method Ford utilizes on its V-8 engines to recirculate the exhaust gas is through the use of a special intake manifold and carburetor spacer (Fig. 15-18). The manifold itself has a special passage that indexes with the exhaust crossover and carries the exhaust gases to an opening below the carburetor spacer anytime the engine is operating.

The spacer plate contains two separate passages for the EGR valve: an inlet and outlet. The inlet passage opening indexes over the exhaust gas port from the crossover passage in the intake manifold. The spacer's outlet passage opens into both primary throttle bores. The two gaskets shown in Fig. 15-18 are necessary to prevent leakage between the spacer, intake manifold, and carburetor.

The EGR valve mounts on the back of the spacer. The valve itself meters the amount of incoming exhaust gas to the primary throttle bores.

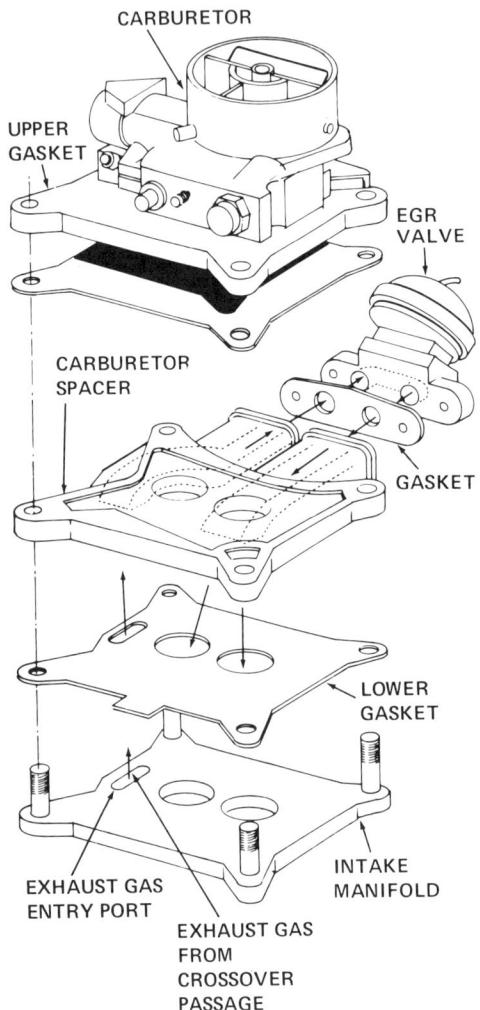

Figure 15-18. Ford EGR system using a spacer below the caburetor to recycle the exhaust gas.

Temperature-Controlled Vacuum Valve

Ford also uses a temperature-controlled vacuum valve within its EGR system (see Fig. 15-13). This valve fits into the engine's cooling system either within a heater hose or in the intake manifold's water jacket.

In either case, this valve, sometimes called a ported vacuum switch (PVS), performs the same function and operates in a similar manner as the Chrysler CCEGR valve. That is, the PVS cuts off vacuum to the EGR valve when engine temperature is below a preset value.

Ford EGR System Modifications

Since its initial introduction, Ford has modified its basic EGR system with the addition of different types of subsystems. These subsystems include a high-speed modulating, venturi vacuum control, back-pressure transducer, and temperature control. Some early Ford vehicles with V-8 engines used a high-speed EGR modulating subsystem (Fig. 15-19). This subsystem improved vehicle driveability at road speeds above 64 mph. This device, widely used in 1973, was only used on police interceptor models in 1974 and was deleted from production after that.

This system consists of a speed sensor, an electronic amplifier, and a vacuum solenoid. The speed sensor, in the speedometer cable behind the instrument panel, directs a signal to the amplifier, which is above the glove box in most models.

When the signal reaching the amplifier indicates a road speed in excess of 64 mph, it closes the normally open vacuum solenoid valve. The solenoid valve con-

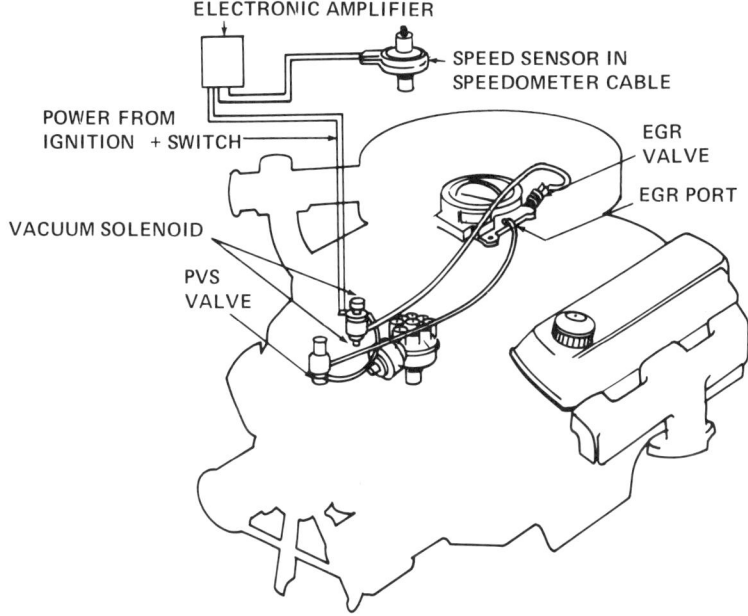

Figure 15-19. Schematic of a Ford high-speed modulating system.

tains a plunger that controls the passage of vacuum between the assembly's two port openings. One of the port openings connects via a hose to the EGR valve while the other connects to the temperature-controlled vacuum valve.

With this arrangement, when the plunger is in the open position and the engine at normal operating temperature, a vacuum signal can pass through the temperature-controlled vacuum valve, solenoid, and to the EGR valve. However, when the sensor signals the amplifier to close the solenoid valve, the plunger seals off the EGR vacuum port and vents any trapped vacuum in the circuit to the atmosphere. This action prevents the EGR valve from operating.

But as vehicle road speed drops below 64 mph, the signal from the speed sensor causes the amplifier to deenergize the solenoid valve. With the solenoid valve open, the plunger closes the atmosphere vent and opens the EGR vacuum circuit. This action permits the EGR valve to function in a normal fashion.

Venturi Vacuum Subsystem

Some 1974 and later Ford engines used a venturi vacuum subsystem rather than a ported vacuum signal to operate the EGR valve. This subsystem consists of an amplifier and a reservoir (Fig. 15-20). The amplifier shown in this illustration has four outlet ports connected by hoses to various system components. For instance, the lower hose connects to the venturi tap in the carburetor, which provides the basic signal to activate the amplifier diaphragm.

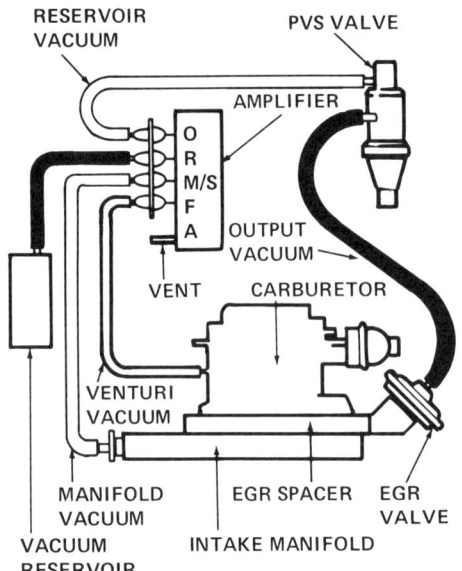

Figure 15-20. Diagram of a Ford venturi vacuum control subsystem.

The second fitting from the bottom is connected to the intake manifold by a hose. This hose supplies intake manifold vacuum to the amplifier and reservoir, which this subsystem will use to operate the EGR valve.

The third fitting from the bottom accommodates a hose that transfers the intake manifold vacuum from the amplifier to the reservoir. The reservoir in this subsystem stores intake manifold vacuum. This assures that there will be sufficient vacuum to operate the EGR valve even when intake manifold vacuum drops during engine acceleration. The uppermost fitting on the amplifier has a hose routed to the PVS valve. From this valve, another hose connects to the EGR valve.

In operation, the EGR valve remains closed at idle since there is no venturi vacuum signal at this time. However, once air flow through the venturi is sufficient, a venturi tap signal moves the diaphragm in the amplifier. This permits intake manifold vacuum to begin opening the EGR valve in proportion to the strength of the venturi vacuum signal.

When the venturi vacuum signal is equal to or greater than intake manifold vacuum, the EGR valve will not function. This is due to the action of the vacuum amplifier, which dumps the output vacuum to the EGR valve. This action causes the valve to close.

Back-Pressure Transducer Subsystem

Late 1975 and newer Ford engines use some form of back-pressure transducer subsystem. One form of transducer is shown in Fig. 15-21. This separate device modulates (regulates) exhaust gas recirculation by varying the strength of the vacuum signal to the EGR valve, according to the amount of exhaust gas back pressure. This ensures maximum exhaust gas recirculation during certain periods of engine acceleration and under some vehicle cruise conditions.

The transducer illustrated in Fig. 15-21 consists of a housing, an exhaust pressure probe, air bleed valve, diaphragm, and spring. Extending from the lower section of the housing is the pressure probe. This probe fits into an exhaust gas passage and carries a back-pressure signal to the lower portion of the diaphragm.

The air bleed valve, which is normally open, controls the strength of the vacuum signal that operates the EGR valve. The valve accomplishes this task by permitting some of the vacuum signal for the valve to leak off to the atmosphere through the air filter.

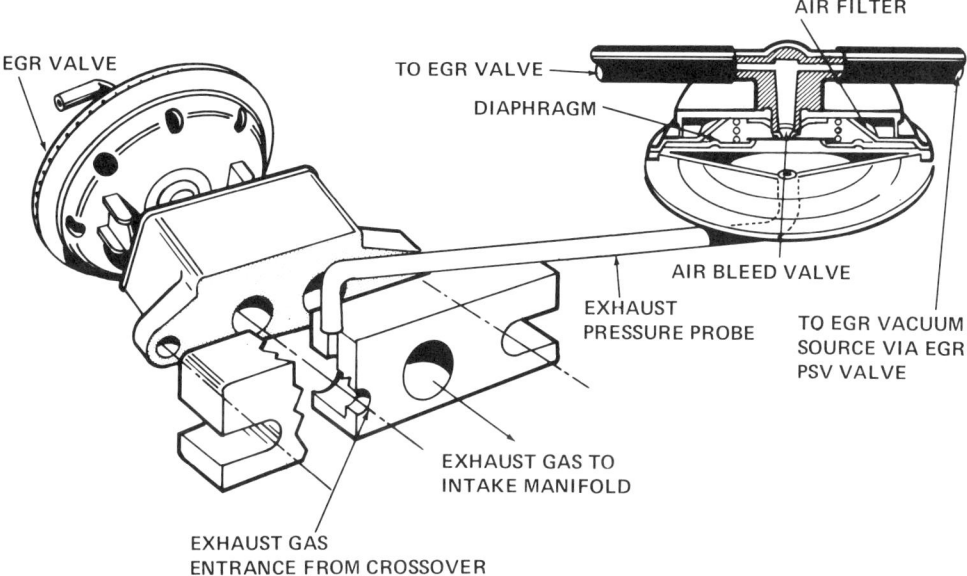

Figure 15-21. Ford back-pressure transducer.

The diaphragm itself responds to the exhaust gas back pressure and the force of the calibrated spring to either permit the air bleed to open or close. For example, when exhaust gas back pressure is high, the diaphragm moves up and blocks the opening in the air bleed valve. However, when back pressure is low, the tension of the spring pushes the diaphragm down, thus uncovering the opening in the air bleed. Now, the valve acts as an air bleed for the vacuum signal to the EGR valve.

On some 1977 models, Ford built the back-pressure transducer into the EGR valve assembly (Fig. 15-22). The upper portion of this assembly contains the EGR valve diaphragm and calibrated spring. This diaphragm attaches to the EGR valve through a hollow stem and moves in response to a vacuum signal entering the housing through the nipple. In other words, when there is vacuum applied to the upper chamber, the diaphragm moves upward, compresses the calibrated spring, and opens the EGR valve.

Attached to but below the valve diaphragm is the transducer diaphragm. The diaphragm moves either up or down, independent of the valve diaphragm, inside the exhaust gas chamber. However, both diaphragms do move up and down together because they are attached.

Attached to the lower side of the main diaphragm is the transducer valve, which is normally open and has a bleed opening into the upper chamber. With this design, the only time this valve seals the upper chamber from the atmospheric vent is when the transducer diaphragm is in the up position. This occurs as exhaust back pressure builds up.

Attached to the valve and transducer diaphragm assembly is a hollow stem. This hollow stem forms the passage for the exhaust back pressure that will operate the transducer diaphragm.

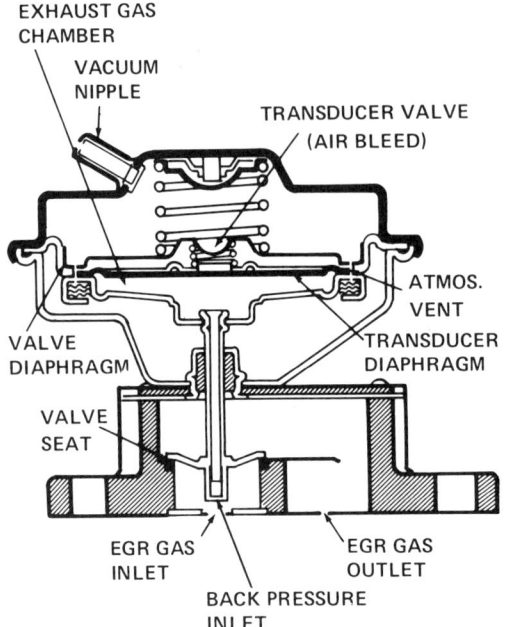

Figure 15-22. Ford EGR valve with a built-in exhaust back-pressure transducer.

A Ford engine using this form of EGR valve will have no exhaust gas recirculation at engine idle. The reason for this is twofold. First, at idle, there will be no vacuum signal to the upper chamber to operate the EGR valve diaphragm. Therefore, the calibrated spring keeps the EGR valve seated. Second, since there is very little exhaust back pressure at idle, the transducer valve spring pushes the transducer diaphragm downward. This action opens the air bleed valve to the atmosphere, thus venting any vacuum trapped in the upper chamber.

As the engine accelerates and the vacuum signal to the upper chamber increases, the EGR valve still does not begin to open because the applied vacuum leaks off through the open bleed valve. However, at a given engine speed, exhaust gas back pressure is high enough to move the transducer diaphragm up and close the bleed valve. Now, the vacuum in the chamber will cause the valve diaphragm to move upward.

As the EGR valve diaphragm moves up, its attached hollow stem advances the valve away from its seat. The amount of valve movement will be in proportion to the strength of the vacuum signal above the valve diaphragm and the intensity of the exhaust gas back pressure against the base of the transducer diaphragm.

At high-speed wide-open throttle operation, exhaust gas recirculation reduces. The reason for this is that the applied vacuum signal on the upper chamber drops off in response to throttle valve position. Consequently, the calibrated spring moves the EGR diaphragm and valve downward, thus reducing recirculation.

Ford's Electronically Controlled EGR System

In 1978, Ford started to use an Electronic Engine Control (EEC) System on some of its vehicles. This system controls engine timing, injection pump air flow, and exhaust gas recirculation.

The EGR portion of this system consists of an EGR valve, valve position sensor, and a pair of electrically operated solenoid valves. Unlike any other EGR valve described up to this point, this Ford valve has a diaphragm that operates by air pressure (Fig. 15-23). The diaphragm receives this pressure via a hose that connects to the air bypass valve within the air injection system.

When the air enters the actuator pressure port on the valve, it moves the diaphragm up against spring tension. As the diaphragm moves up, in proportion to the

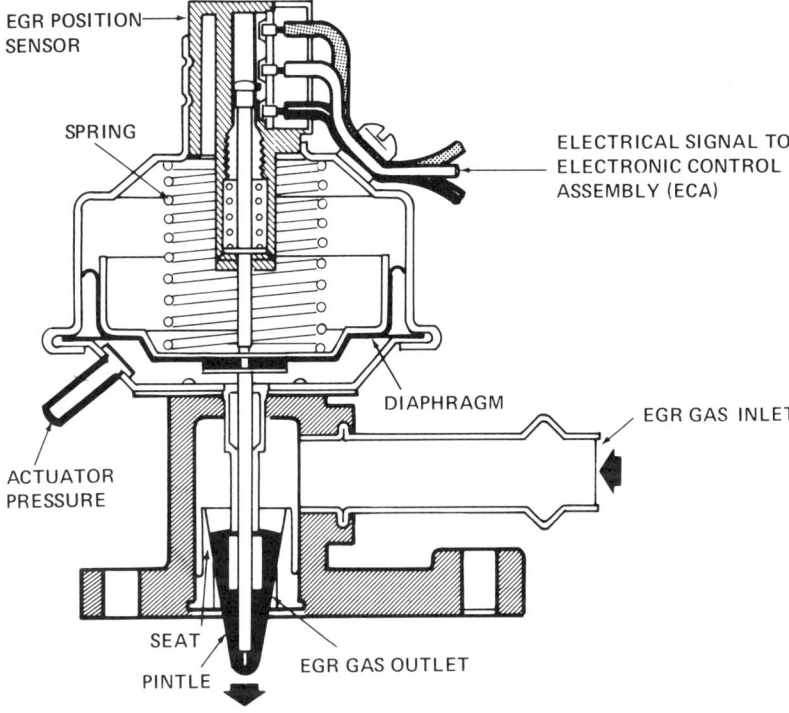

Figure 15-23. Ford air-operated EGR valve used in the EEC system.

amount of applied air pressure, it moves the EGR valve off its seat, permitting the flow of exhaust gases to begin recirculating.

Mounted above, but part of, the EGR valve assembly is the position sensor. This sensor has a number of wires that connect to the electronic control assembly. These wires carry information, in the form of electrical signals, from the sensor to the electronic control assembly indicating the position of the EGR valve metering rod.

This information informs the control assembly just how much exhaust gas is recirculating at any given time. The control assembly then compares this information to that received from other sensors within the EEC system and inceases or deceases the EGR rate.

The solenoid valves increase or decrease this rate of flow. The two solenoid valves are mounted together on a bracket next to the left rocker-arm cover (Fig. 15-24).

The solenoids themselves connect by wiring to the electronic control assembly. This assembly continuously analyzes the information it receives from all the system sensors and electrically changes the position of the solenoid valves to maintain the flow of exhaust gases at the level the engine requires to reduce NOx emissions.

The rear solenoid is the vent valve for the system (Fig. 15-25a). When the plunger valve within the solenoid is open (down position), air pressure from the injection system to the EGR valve vents into the atmosphere. This causes the EGR valve to

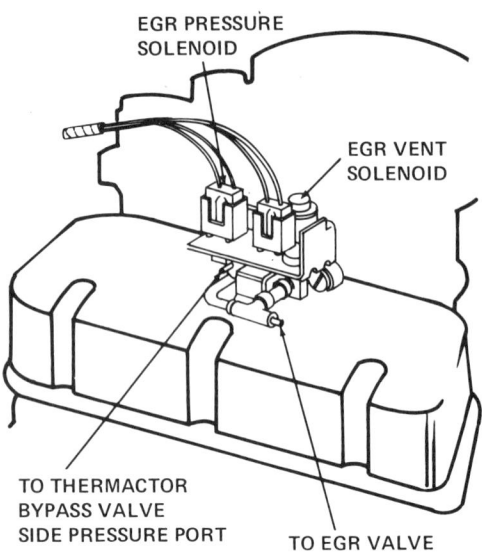

Figure 15-24. Operation of the vent and pressure solenoids.

close. However, when the electronic control assembly sends power to the rear solenoid, the plunger moves upward and closes off the vent port. Now, air pressure can operate the EGR valve.

The front solenoid forms the pressure control valve for the system (Fig. 15-25b). Without electric power, this solenoid is closed; consequently, air from the pump will not reach the EGR valve diaphragm. However, the plunger position does hold and trap any air pressure already within the hose connected to the EGR valve. This air maintains the EGR valve in the position it was when the pressure solenoid closed.

When the control assembly directs power to the pressure solenoid, its plunger moves upward. This action permits air pressure from the pump's bypass valve to pass into the upper port on the solenoid body, move through the open valve, and out the lower port opening to the EGR valve. The air pressure then opens the EGR valve.

General Motors EGR Systems

General Motors first introduced its EGR system on some vehicles in 1972; but all 1973 and later engines used an EGR system to reduce NOx emissions.

The basic General Motors system consists of a ported vacuum-operated EGR valve and some form of temperature control valve. The design and function of these components are similar to those found on other vehicles and already discussed in this chapter.

In a manner similar to the other automobile manufacturers, General Motors also made some modifications to the basic system in order for it to function more efficiently on some engine designs. These changes include the use of a dual-diaphragm EGR valve and an exhaust back-pressure transducer subsystem.

The dual-diaphragm EGR valve (Fig. 15-26) provides increased EGR rates with larger engine loads. The design and operation of this particular unit was already covered in this chapter.

The back-pressure transducer subsystem ensures the correct amount of exhaust gas recirculation during engine acceleration and vehicle cruise conditions. Like Ford, General Motors used both the separate transducer system and the one incorporated into the EGR valve.

Summary

1. An EGR system reduces the amounts of NOx emissions produced during engine combustion.

2. There are two ways in which manufacturers control peak combustion chamber temperatures to hold down the formation of NOx.

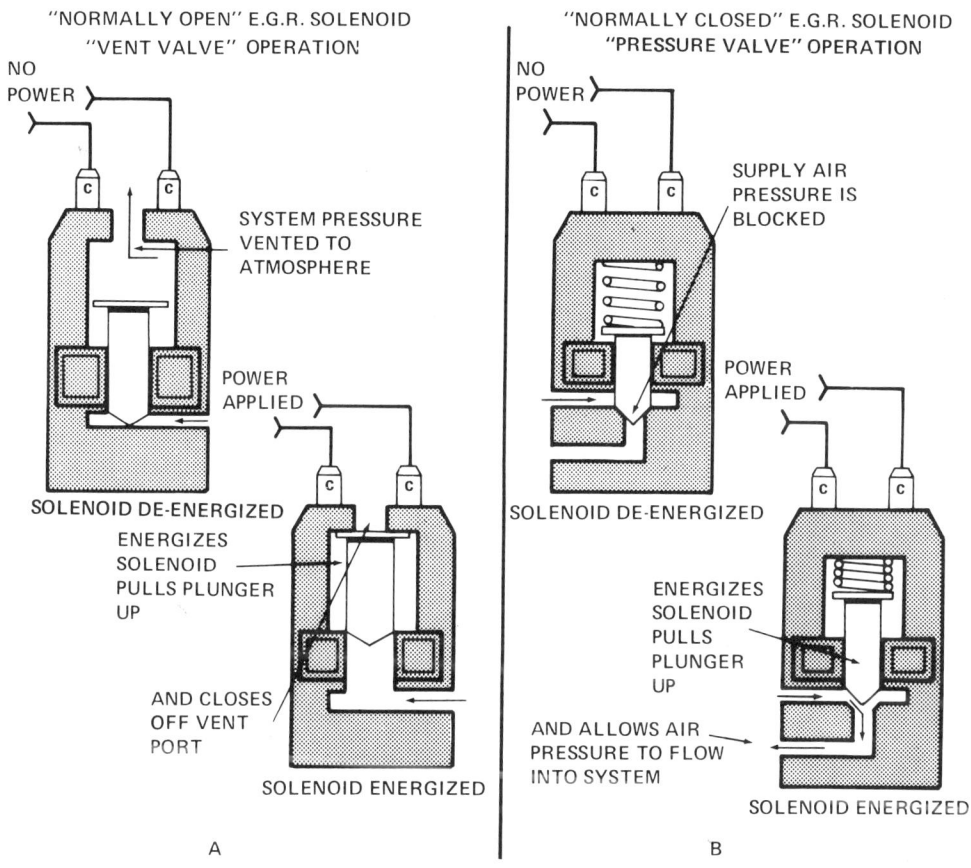

Figure 15-25. Two solenoids raise or lower the rate of exhaust gas recirculation by regulating the air flow to the EGR valve.

3. Exhaust gas is relatively inert and is thus used to dilute the air/fuel mixture.
4. Recirculating the exhaust gas into the intake manifold limits the quality and quantity of the air/fuel charge that enters the combustion chambers.
5. Not all operating phases of an engine produce excessive amounts of NOx emissions.
6. A basic EGR system consists of a redesigned intake manifold, EGR valve, valve-activating system, and some form of thermo valve or switch.
7. The design of any EGR manifold depends on whether the system uses a fixed or variable orifice EGR valve and the location of the latter.
8. An EGR valve is usually a vacuum-operated variable orifice metering device.
9. Manufacturers can use either a ported or venturi vacuum signal to operate the EGR valve.

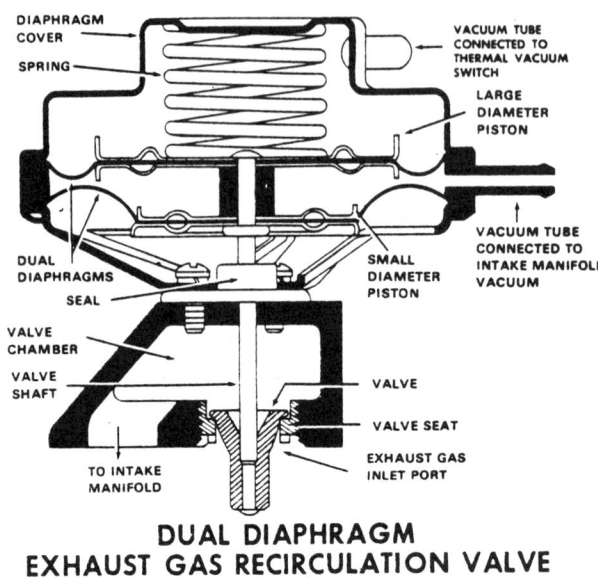

Figure 15-26. Dual-diaphragm EGR valve modulates exhaust gas recirculation to changes in engine vacuum (Courtesy of Chevrolet Division).

10. The venturi vacuum control system consists of a vacuum tap, amplifier, and dump valve.
11. Automotive engineers have developed a considerable number of modulating devices to more precisely control the operation of the EGR valve during the various phases of engine performance.
12. The back-pressure transducer ensures maximum exhaust gas recirculation during engine acceleration and under vehicle cruise conditions.
13. The temperature control valve or switch within an EGR system prevents exhaust gas recirculation when the engine is cold.
14. The dual-diaphragm EGR valve increases EGR rates with larger engine loads.
15. The Chrysler EGR system has a single diaphragm-operated EGR valve that activates by means of venturi vacuum along with a CCEGR valve and a delay system.
16. A Chrysler time-delay system prevents exhaust gas recirculation for about 35 seconds after the driver turns on the ignition switch.
17. In 1975, Chrysler used a maintenance reminder system.
18. A basic Ford EGR system consists of an EGR valve, a specially designed intake manifold or spacer, and a temperature-controlled vacuum valve.
19. With its basic system, Ford has utilized three somewhat different styles of EGR valves: the poppet, modulating, and tapered stem.

20. Ford uses various methods to deliver exhaust gas to the EGR valve.
21. Ford also utilizes a temperature-controlled vacuum valve within its basic EGR system.
22. Some 1974 and later Ford engines used a venturi vacuum subsystem to operate the EGR valve.
23. The Ford back-pressure transducer subsystem regulates exhaust gas recirculation by varying the strength of the vacuum signal to the EGR valve, according to the amount of pressure within the exhaust system.
24. On some 1977 models, Ford built the back-pressure transducer into the EGR valve assembly.
25. In 1978, Ford began to use an Electronic Engine Control System on some of its vehicles.
26. The Ford Electronic Engine Control System uses an air-operated EGR valve, a valve position sensor, and a pair of solenoid valves.
27. General Motors EGR systems have been in service since 1972.
28. The basic General Motors EGR system consists of a ported vacuum-operated EGR valve and some form of temperature-activated vacuum control valve.
29. Modifications to the basic General Motors EGR system include the use of a dual-diaphragm EGR valve and an exhaust back-pressure transducer.

Review Questions

The questions listed below will assist you in determining how well you remember the material contained in this chapter. Read each question carefully before choosing the answer. If you cannot answer the question, review the section in the chapter that covers the material.

1. An EGR system should operate between _____ mph.
   a. 20 to 60
   b. 30 to 70
   c. 40 to 80
   d. 50 to 90

2. The _____ regulates the amount of exhaust gas entering the air/fuel mixture.
   a. EGR valve
   b. orifice valve
   c. intake manifold
   d. carburetor spacer plate

3. A venturi vacuum control system requires a _____ to maintain an adequate supply of vacuum to operate the EGR valve.
   a. valve
   b. tank
   c. tap
   d. reservoir

4. The _____ modulates the EGR valve operation due to the changing pressure of the exhaust gas.
   a. oxygen sensor
   b. back-pressure probe
   c. back-pressure transducer
   d. transducer valve

5. The _____ stops EGR valve operation when the engine is cold.
   a. dump valve
   b. coolant-controlled vacuum valve
   c. vacuum tap
   d. vacuum reservoir

6. The Chrysler time-delay system prevents EGR valve operation for _____ seconds after the driver turns on the ignition switch.
   a. 35
   b. 25
   c. 15
   d. 10
7. With its basic EGR system, Ford used _____ type(s) of EGR valves.
   a. one
   b. two
   c. three
   d. four
8. Most Ford four- and six-cylinder engines provide exhaust gas to the EGR valve through _____.
   a. the spacer plate
   b. the intake manifold
   c. floor jets
   d. a steel tube
9. The Ford high-speed modulating system prevents EGR valve operation above _____ mph.
   a. 60
   b. 64
   c. 68
   d. 72
10. Ford began to use back-pressure transducers in its EGR systems beginning in _____.
    a. 1973
    b. 1974
    c. 1975
    d. 1976
11. Ford vehicles with the Electronic Engine Control System have EGR valves that are _____ operated.
    a. air
    b. vacuum
    c. gas
    d. oil
12. The General Motors EGR valve that provides increased EGR rates with large engine loads has _____ diaphragm(s).
    a. four
    b. three
    c. one
    d. two

For the answers to these review questions, turn to the Appendix.

Chapter 16

Servicing EGR Systems

As pointed out in Chapter 15, the automotive industry has already produced many different types of EGR systems used on automobiles and light trucks to reduce NOx emissions. All of these systems do require preventive maintenance at given intervals. A typical manufacturer, for example, requires a complete inspection and testing of its system on the average of every 12,000 miles or 12 months in order to assure proper functioning of the device.

On many vehicles, it is up to the vehicle's owner to keep track of the interval between EGR system service and take the vehicle into a shop for the necessary service. However, some vehicles do come from the factory with an EGR maintenance reminder system, which flashes a warning light onto the dashboard when the system is overdue for preventive maintenance.

In either case, when the service interval comes due, a skilled technician must service the system or the engine will produce excessive NOx emissions and perform very poorly under certain conditions. For instance, if due to neglect there is excessive carbon buildup within the EGR valve assembly, the diaphragm-operated stem and attached valve can stick in the open or partially open position. This causes the engine to operate poorly at idle and stumble repeatedly on acceleration during the warm-up period.

General EGR System Service

System Inspection and Parts Replacement

Before actually discussing the inspection, testing, and service of specific EGR systems, let us first examine some service tips which apply to them all. These deal mainly with general system inspection, parts replacement, and EGR valve cleaning.

When inspecting an EGR system, always check the physical condition and proper routing of all the vacuum hoses. The best way to ascertain if all vacuum hoses connect to the proper component is to check their routing against an EGR vacuum diagram (Fig. 16-1). This diagram shows not only the correct routing of vacuum hoses to components within the EGR system but also the remaining exhaust emission control devices found on the vehicle. You will find this type of diagram either within the service manual for the vehicle or, in many cases, on a decal located within the engine compartment.

The removal and subsequent reinstallation of most EGR system components is not a very difficult task. However, when replacing parts, remember the following points:

1. Make certain that the replacement component is an exact duplicate of the one removed.

2. Use a thin coating of nonhardening sealer around the threads of a coolant-operated temperature switch before its installation.

3. When replacing a part that has several vacuum hose connections, such as a vacuum amplifier, it is a good idea to disconnect one hose at a time and immediately reconnect it to the same port on the new component.

4. Always use a new gasket when replacing an EGR valve or a back-pressure transducer.

5. Tighten all mounting screws of an EGR valve or back-pressure transducer to 10 foot-pounds or factory specifications to prevent a vacuum or exhaust leak.

Cleaning an EGR Valve

There are two acceptable methods of cleaning an EGR valve. The method you use will depend upon the particular valve

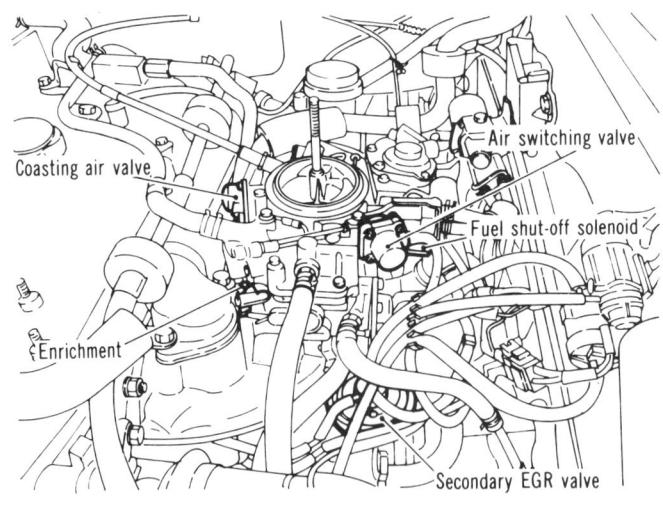

Figure 16-1. Typical EGR vacuum diagram showing the routing of the vacuum hoses (Courtesy of Chrysler Corp.).

design and the equipment that is available. However, before attempting to clean an EGR valve, remember these tips:

1. Always remove an EGR valve from the engine before attempting to clean it.

2. Be careful not to push on the valve diaphragm during the cleaning process because this may tear or damage it.

3. When using a manifold heat control valve solvent to clean the unit, be careful not to get any on the diaphragm.

To clean an EGR valve with an abrasive-type cleaning machine, follow this procedure:

1. Remove all the exhaust deposits from the base of the valve, along with its mounting surface, with a wire brush or power-driven wire wheel.

2. Clean the valve seat and pintle area by inserting it into an abrasive-type spark plug cleaning machine. Then, apply an abrasive blast to the area for about 30 seconds.

3. Connect the hose from a hand-held vacuum pump to the fitting on the EGR valve. Then, apply 15 inches of vacuum to the diaphragm.

4. Reinsert the valve end of the EGR unit into the cleaning machine. Then, apply an abrasive blast for another 30 seconds.

5. Inspect the valve. If all the exhaust deposits have not been removed by the process so far, repeat steps 2, 3, and 4.

6. Clean the valve seat and pintle area with compressed air.

7. Install the EGR valve assembly onto the engine with a new gasket. Torque its mounting screws to 10 foot-pounds or factory specification. Then, reconnect the vacuum hose to the valve.

To clean the EGR valve using a manifold heat control valve solvent, follow these steps:

1. Apply a liberal amount of manifold heat control valve solvent to the poppet valve and seat area. Note: Exercise extreme care when applying this solvent so as not to spill any of it on the diaphragm as this may cause diaphragm failure.

2. Allow the solvent about 30 minutes to soften up the deposits.

3. Connect a hand-held vacuum pump hose to the diaphragm fitting and apply sufficient vacuum to open the poppet valve fully. Caution: Do not push on the diaphragm to open the valve; use a vacuum source only.

4. Use a sharp-edged tool to carefully scrape the loosened deposits from the poppet valve and seat. Note: If after cleaning the valve you notice excessive wear on the stem, valve, or seat, replace the assembly.

5. Using a new gasket, replace the EGR valve assembly onto the engine. Then, torque the mounting screws to 10 foot-pounds or the amount specified by the manufacturer.

6. Connect the vacuum hose to the valve.

Inspecting, Testing, and Servicing Specific EGR Systems

Chrysler Corporation

Through the years Chrysler Corporation has used three different styles of EGR systems. Chrysler used a floor jet system in 1972 and 1973. This was a very simple system, which used fixed orifices threaded into the intake manifold floor directly under each primary carburetor barrel (Fig. 16-2).

From 1973 to 1976, Chrysler utilized a ported vacuum EGR system on some of their vehicles (Fig. 16-3). This system used a variable orifice EGR valve, which operated by means of a ported vacuum signal from the carburetor. In addition, this system used an air temperature control valve to cut off the EGR valve operation if air temperature dropped below 68 °F.

The third Chrysler EGR system also began its service in 1973. However, this system is still in current use today.

The third system uses a venturi vacuum signal and an amplifier to operate the EGR valve (Fig. 16-4). The system also includes a delay system and one or more temperature control valves to modulate system operation during the various phases of engine operation.

Inspection and Servicing of a Floor Jet EGR System

The floor jets (see Fig. 16-2) within this system require inspection at 12,000-mile intervals for the buildup of carbon deposits. To perform this inspection follow this procedure:

1. Remove the air cleaner assembly.

2. With a screwdriver block open the choke and throttle valve.

3. Shine a flashlight down into the throat of the carburetor and visually check the openings within the floor jet. If the floor jets are open, no further service is necessary.

4. If carbon has restricted the jets, remove the carburetor assembly to provide easy access to the orifices.

5. With a socket, extension, and ratchet remove the jets from the manifold floor. Caution: Since the jets are nonmagnetic, do not allow them to drop into the manifold runners during the removal or installation process.

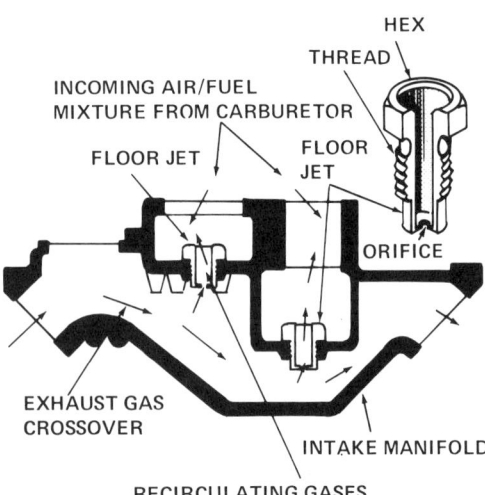

Figure 16-2. Chrysler EGR system, which uses fixed orifice floor jets.

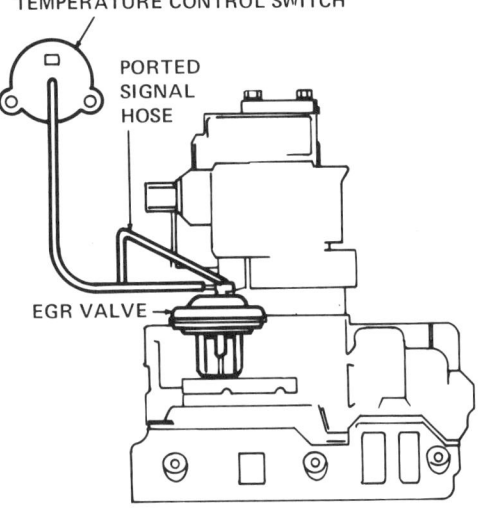

Figure 16-3. Chrysler EGR system in which the valve activates by a ported vacuum signal.

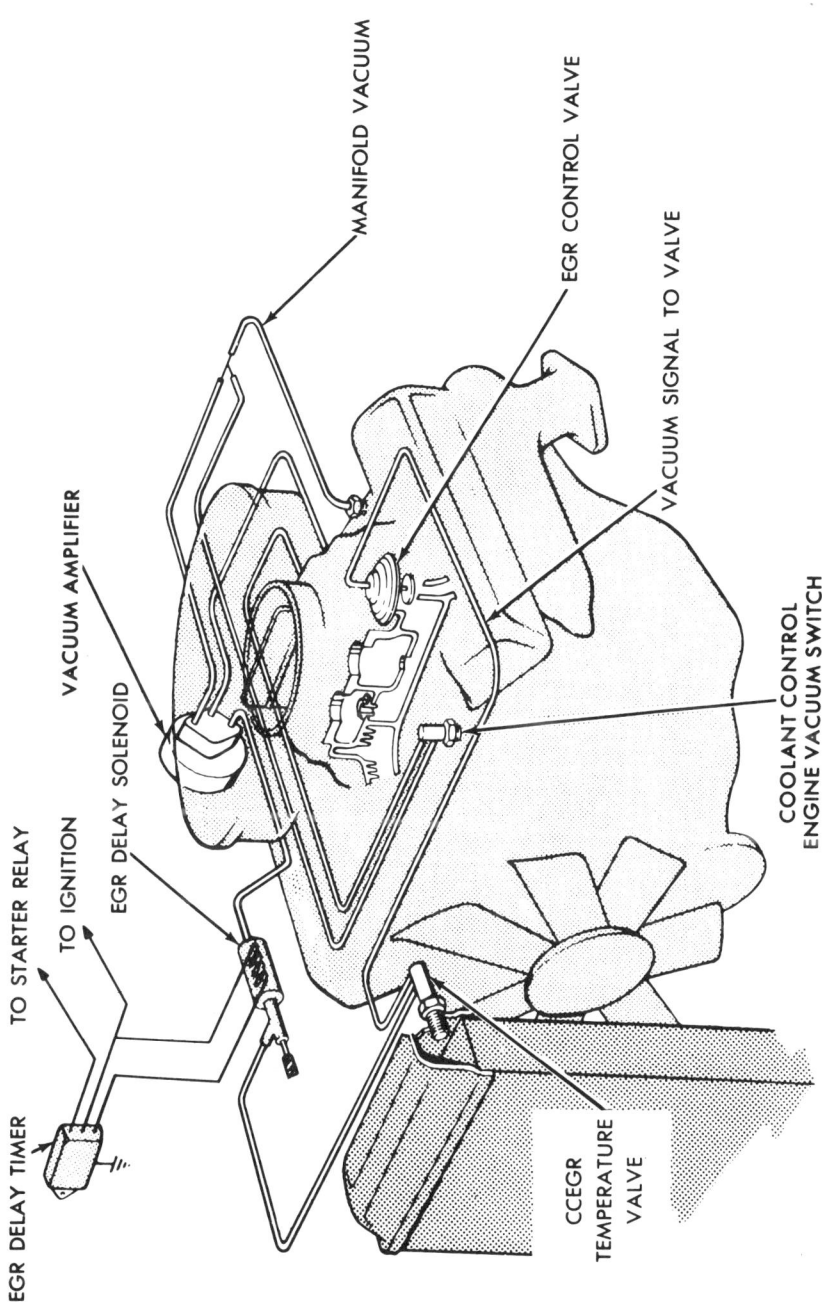

Figure 16–4. Chrysler EGR system in which the valve operates by a venturi vacuum signal (Courtesy of Chrysler Corp.).

6. Soak the jets in manifold heat control valve solvent till their openings clear. Caution: Never use a drill or piece of wire to clean out the jet openings because this action could enlarge the orifices and affect idle quality.

7. Reinstall the jets and torque them to 25 foot-pounds.

8. Using a new gasket, replace the carburetor and torque its attaching bolts or nuts to specifications.

Testing a Ported EGR System— The EGR Valve and Control System

To operationally check the EGR valve and its vacuum source for serviceability, follow this procedure:

1. Connect a tachometer to the engine.

2. Start the engine and permit it to reach its normal operating temperature.

3. With the engine idling in neutral, gradually accelerate it to approximately 2,000 rpm but not over 3,000 rpm.

4. Observe the EGR valve stem (Fig. 16-5). Visible movement of the EGR valve stem should occur during this procedure, which you can determine by the change in the relative position of the groove on the EGR valve stem.

5. Gradually permit the engine to return to idle. The valve stem should now move downward, toward the closed position.

6. Repeat the procedure several times to confirm stem movement.

If the valve stem did not move properly during the above test, check its vacuum source as follows:

1. Disconnect the vacuum line to the EGR valve.

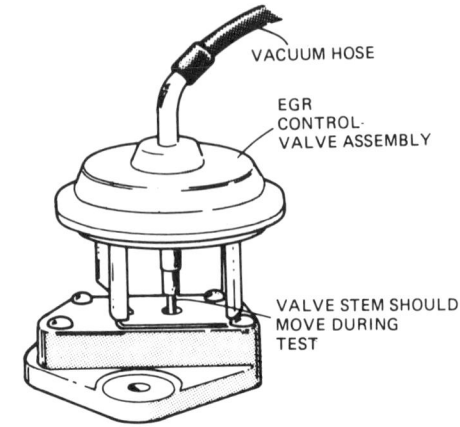

Figure 16-5. Checking the operation of a typical EGR valve.

2. Connect a vacuum gauge to the open end of the hose.

3. Start the engine and slowly accelerate it to about 2,000 rpm.

4. Note the reading on the vacuum gauge. It should be about zero at engine idle, but the needle should rise with increases in engine rpm. If not, either the vacuum hose or the ported vacuum slot within the carburetor is plugged.

Testing the EGR Valve and Passages for Restrictions

If the ported vacuum control system functions properly, you should test the EGR valve and its passages for exhaust gas flow. To perform this check:

1. Connect a tachometer to the engine.

2. Start the engine and allow it to reach normal operating temperature.

3. Disconnect the vacuum hose to the EGR valve and insert the hose from a hand-operated vacuum pump over its fitting (Fig. 16-6).

4. With the engine idling in neutral, ap-

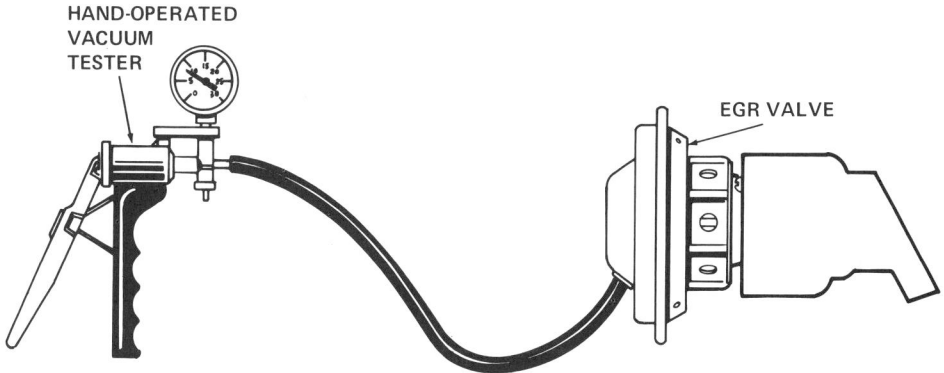

Figure 16–6. Using a vacuum pump to activate the EGR valve to check its passages for exhaust gas flow.

ply a vacuum signal of 10 inches to the EGR valve.

5. Note the rpm on the tachometer.

Results and Indications

1. Idle speed should drop 150 rpm or more with the vacuum signal applied. This rpm drop confirms that exhaust gas recirculation is taking place.

2. If the speed change does not occur, or is less than the specified minimum, there are exhaust deposits within the EGR valve or intake manifold passages. In this situation, it is necessary to remove the EGR valve, inspect and clean its passages, along with those within the intake manifold.

Testing the Temperature Control Valve

The Chrysler EGR system illustrated in Figure 16–3 uses a temperature control valve. The first temperature control valve for this type of system was mounted on the firewall. This type of unit opened below an air temperature of 60 °F.

However, after March, 1973, Chrysler moved the temperature valve to the radiator, where engine coolant temperature opened the unit below 62 °F.

In either case, when the valve opens, the ported vacuum EGR valve signal was bled off to the atmosphere. This prevented the EGR system from functioning at low temperatures to improve the driveability of the vehicle.

To test the temperature control valve:

1. Disconnect the vacuum hose from the valve.

2. Connect the hose from a vacuum pump to the open fitting on the valve.

3. At temperatures about either 62 or 68°F, depending on the valve type, it should accept and hold several inches of vacuum. If the valve does not perform in this fashion, it is defective and requires replacement.

Testing the Venturi Vacuum EGR System

You can test the EGR valve, its control system, and actual exhaust gas recirculation of this system (see Fig. 16–4) using the same procedures outlined for a ported vacuum device. However, if the EGR valve will not open without the application of vacuum from a hand-held pump, it will be necessary to perform a special test on the control and amplifier system.

Testing the Venturi Vacuum Control and Amplifier System

To perform this test:

1. Disconnect the vacuum hose from the venturi tap on the carburetor. In its place install the hose from a vacuum gauge.

2. Start the engine and slowly accelerate it to about 2,000 rpm.

3. Note the reading on the vacuum gauge. It should be about 3 to 4 inches. If not, the vacuum tap within the carburetor is restricted and requires cleaning.

4. If the vacuum tap signal is satisfactory, disconnect the vacuum gauge hose from the tap. However, do not reconnect the EGR vacuum hose to the tap fitting at this time.

5. Disconnect the EGR valve hose. Then, connect the hose from the vacuum gauge to the valve fitting.

6. Connect the venturi vacuum hose to a hand-held vacuum pump.

7. Start the engine and permit it to reach normal operating temperature.

8. Note the reading on the vacuum gauge. At this point it should not be enough to open the valve.

9. With the vacuum pump, apply 3 to 4 inches of vacuum to the amplifier via the tap hose.

10. Note the reading on the vacuum gauge. It should now immediately register a full manifold vacuum reading. If it does not, check all the hose connections. If they are all satisfactory, the vacuum amplifier is defective and requires replacement.

Testing a CCEGR Valve

If the venturi vacuum control system and amplifier are functioning properly and the EGR valve still does not function, check the operation of the coolant-controlled EGR (CCEGR) temperature valve following these steps:

1. Start the engine and permit it to reach its normal operating temperature.

2. With a piece of vacuum hose, bypass the CCEGR valve so that the amplifier connects directly to the EGR valve.

3. Check the EGR valve for normal operation. If the valve now operates correctly, the CCEGR valve is defective and requires replacement.

Testing the EGR Time-Delay System

If the EGR valve still does not function with the CCEGR valve bypassed, test the serviceability of the EGR delay system as follows:

1. Disconnect the electrical connection to the delay solenoid (Fig. 16–7).

2. Start and accelerate the engine to about 2,000 rpm.

3. Check the action of the EGR valve stem. If the valve stem now moves, reconnect the solenoid and disconnect the timer.

4. Repeat the system test and observe the valve stem's action. If there is now no movement, replace the solenoid. If there is movement, replace the timer.

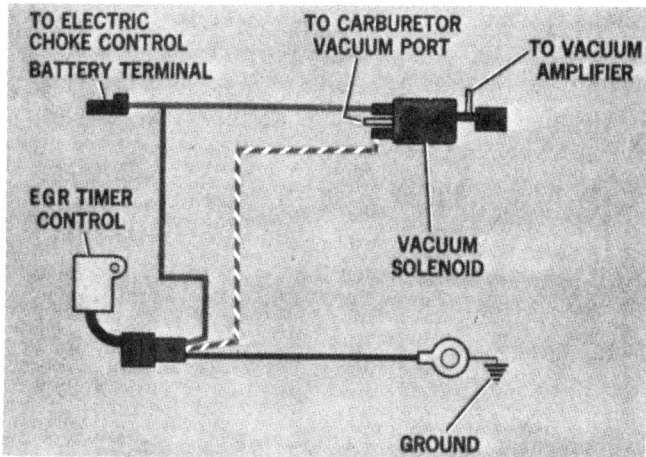

Figure 16-7. Testing the EGR time-delay relay and solenoid (Courtesy of Chrysler Corp.).

Resetting a Chrysler EGR Maintenance Reminder System

Chrysler installed a maintenance reminder system on its 1975 automobiles to be sure that this system received service at 15,000-mile intervals. To turn off the reminder light after servicing the EGR system, follow these directions:

1. Slide the rubber cover on the switch away from the unit to expose the set screw (Fig. 16-8). Note: It may become necessary to disconnect one of the attached speedometer cables for easy access to the reset screw.

2. Turn the reset screw approximately one and a half turns and release it. This action zeroes the unit's odometer gear and, at the same time, opens the electrical circuit.

3. As necessary, reconnect the speedometer cable to the unit.

Check the maintenance reminder light; it should now be off.

Testing Ford EGR Systems

As mentioned in the last chapter, over the years, Ford has developed and used several different types of EGR systems. Because of these variations in design, it

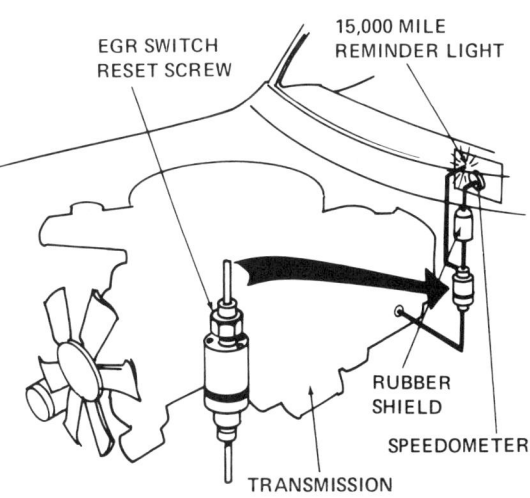

Figure 16-8. Resetting a Chrysler EGR maintenance reminder system.

will be necessary for you, in most cases, to look up certain specifications in the shop manual before testing a given system. These specifications deal mainly with the amount of vacuum necessary to operate a specific EGR valve and the opening and closing temperature of the temperature control unit utilized on the system. With these facts in mind, let us examine the most common tests used by the industry to determine if a Ford EGR system and its components are functioning properly.

Ported EGR System

To test a ported-type Ford EGR system, shown in Fig. 16-9, follow these directions:

1. Tee a vacuum gauge into the vacuum hose at the EGR valve (Fig. 16-10).

2. Start the engine up and permit it to operate until its coolant temperature is higher than about 115 °F.

3. While observing both the vacuum gauge and EGR valve stem, quickly open and then release the throttle. The EGR valve should start to move and be wide open at a given amount of vacuum, as shown on the gauge.

4. Compare the results of this test to specifications. A typical specification may call for EGR valve stem movement to begin at about three-fourths of an inch of vacuum and complete its movement by 8 inches of vacuum. If the valve stem does not move in this manner, replace the EGR valve and repeat the test.

5. If there was no vacuum reading on the gauge during this test, disconnect the vacuum gauge from the T fitting and connect its hose directly into the carburetor EGR port. Then, start and accelerate the engine quickly.

6. Note the reading on the vacuum gauge. If there is still no reading on the gauge or it does not increase, the carbure-

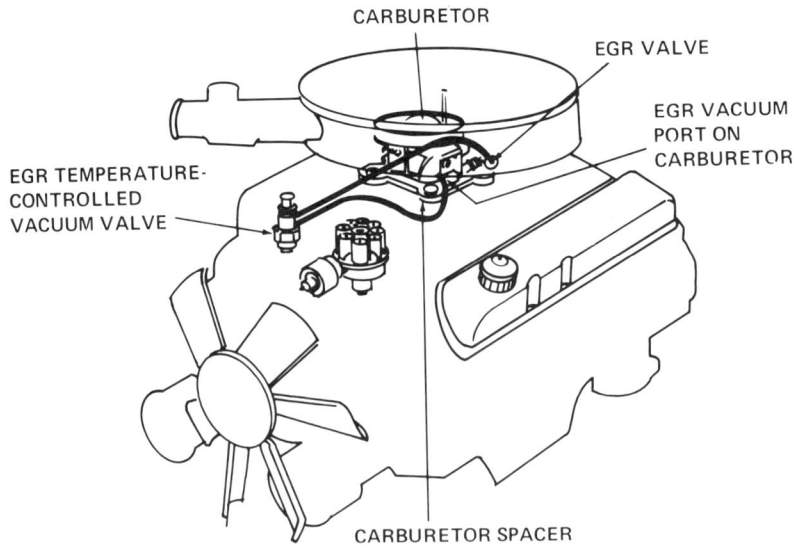

Figure 16-9. Testing a ported-type Ford EGR system.

TESTING FORD EGR SYSTEMS

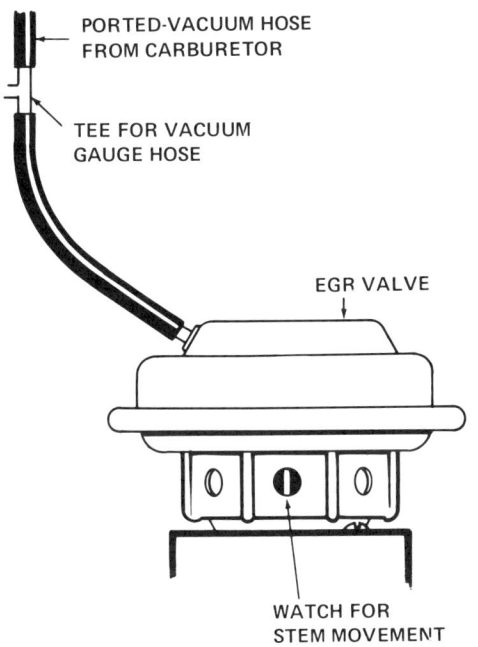

Figure 16-10. Checking the EGR valve and ported vacuum control system.

tor vacuum passage may be leaking or plugged.

7. If the ported signal is satisfactory, disconnect the vacuum gauge hose from the carburetor fitting and reconnect the EGR signal hose. Then, tee the vacuum gauge into the EGR valve port of the temperature control valve.

8. While observing the vacuum gauge, open the throttle valve quickly. If the gauge reading does not increase as the engine accelerates, replace the temperature valve.

Checking the EGR Valve and Its Passages For Restrictions

To perform this particular test:

1. Disconnect the vacuum hose from the EGR valve (Fig. 16-11). Next, connect the hose from a hand-held vacuum pump to the EGR valve fitting.

2. Connect a tachometer to the engine.

3. Start the engine and permit it to reach normal operating temperature.

4. With the hand pump, apply the specified amount of vacuum (usually about 8 inches) to the EGR valve.

5. Note the reading on the tachometer. As the EGR valve begins to open, the engine idle speed should decrease and it should idle roughly and perhaps stall. If this does not occur, remove and clean the EGR valve.

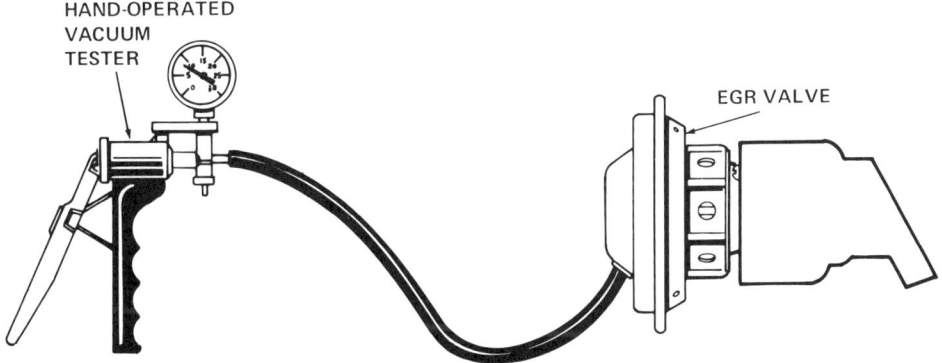

Figure 16-11. Testing the EGR valve and its passages for restrictions.

6. After cleaning the valve, repeat the above test. If the engine still does not respond as it should, replace the EGR valve.

Testing a Ford Venturi Vacuum EGR System—Amplifier Test

You can test a venturi vacuum-operated EGR system using the same procedures as outlined above. If during this testing procedure there is no vacuum indication on the gauge teed into the vacuum hose to the EGR valve, test the amplifier as follows:

1. On a single-plug amplifier (Fig. 16-12), disconnect the amplifier-to-EGR valve hose at the amplifier "O" port. Then, connect the hose from a vacuum gauge to this open fitting.

2. Start the engine and permit it to reach its normal operating temperature. Next, note the reading on the vacuum gauge. With the engine at idle, the vacuum gauge should indicate within ± 0.3 inch of the specified reading.

3. With the engine operating, open and close the throttle valve while noting the reading on the vacuum gauge. If the vacuum reading changes more than 1 inch during acceleration, replace the amplifier. Also, as the engine returns to idle, the vacuum reading should be the same as in step 2. If not, replace the amplifier.

4. While accelerating the engine, disconnect the external reservoir hose at the amplifier "R" port while noting the vacuum gauge reading. If the reading does not increase to about 4 inches or more, replace the amplifier.

To test the amplifier with a dual-vacuum plug (Fig. 16-13):

1. Disconnect the EGR valve hose at the amplifier "O" fitting. Then, attach the hose from a vacuum gauge to this connection.

2. Disconnect the vacuum hose from the carburetor venturi vacuum port.

3. Start the engine and permit it to run at idle at normal operating temperature.

4. Note the reading on the vacuum gauge. If the reading is not within ± 0.3 inch of the specified reading, replace the amplifier assembly.

5. With the engine operating, open and close the throttle valve, while noting the reading on the vacuum gauge. The reading should be the same as in step 3. If the reading is not the same or it increases more than 1 inch of vacuum during acceleration, replace the amplifier.

6. Reconnect the carburetor's venturi vacuum hose.

7. Check the vacuum reading on the gauge with the engine at normal idle speed.

O – OUTPUT TO EGR VALVE
R – FROM RESERVOIR
M OR S – MANIFOLD VACUUM
V – VENTURI VACUUM
A – ATMOSPHERE VENT

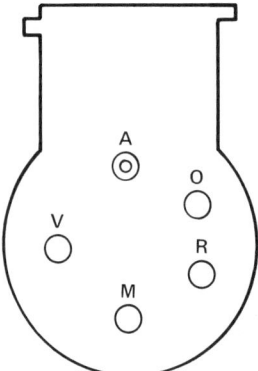

Figure 16-12. Vacuum fittings on a single-plug Ford amplifier.

TESTING FORD EGR SYSTEMS

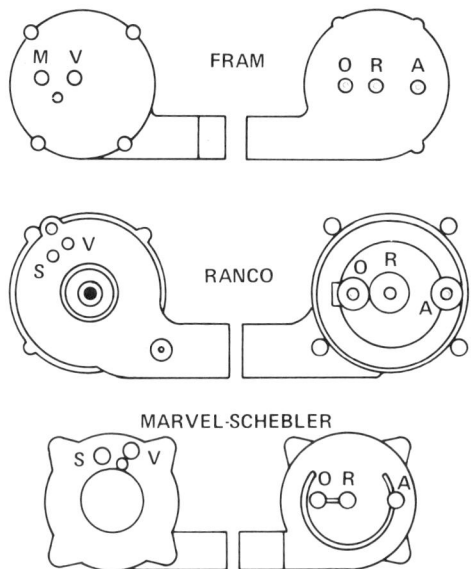

AMPLIFIER CODED FOR PORT CONNECTIONS:
  O – OUTPUT TO EGR VALVE
  R – FROM RESERVOIR
  M OR S – VACUUM SOURCE
           (SPARK OR EGR PORT)
  V – VENTURI VACUUM
  A – ATMOSPHERE (VENT)

Figure 16-13. Vacuum fittings located on the three types of dual-plug Ford amplifiers.

If the reading is more than one-half inch above the indication taken during step 3, check and reset the engine idle speed.

Testing a Ford Thermostatic Vacuum Switch

A thermostatic vacuum switch (TVS) is a type of temperature control vacuum valve. When used in conjunction with an EGR system, this device controls the routing of a vacuum signal to a given EGR system component.

Because this device performs varied functions at different operating temperatures within the many complex Ford EGR system configurations, it will be necessary for you to look up the specifications for the operating temperature and vacuum routing through the switch before testing it. The specifications are available in various service manuals.

To test a typical TVS assembly off the vehicle, follow these steps:

1. Remove the TVS assembly.

2. Cool the TVS in a refrigerator until its temperature is 50 °F or less.

3. Connect the hose from a hand-held vacuum pump to the lower vacuum fitting on the switch (Fig. 16-14). Next, connect a hose from a vacuum gauge to the upper switch fitting.

4. With the vacuum pump, apply about 10 inches of vacuum to the switch while noting the reading on the vacuum gauge. With the switch at about 50 °F there should be no vacuum reading on the second vacuum gauge. If there is, the switch is defective and requires replacement.

5. Bleed the vacuum off the switch.

6. Permit the TVS assembly to warm up to room temperature (above at least 60 °F). Next, with the vacuum pump reapply a vacuum signal of at least 10 inches to the switch.

7. Note the reading on the second vacuum gauge. It should now indicate the same reading as the gauge on the hand pump. If not, the TVS assembly is defective and requires replacement.

Testing a Ford High-Speed EGR Modulating System

To test this particular system (Fig. 16-15):

1. Tee a vacuum gauge into the valve signal hose.

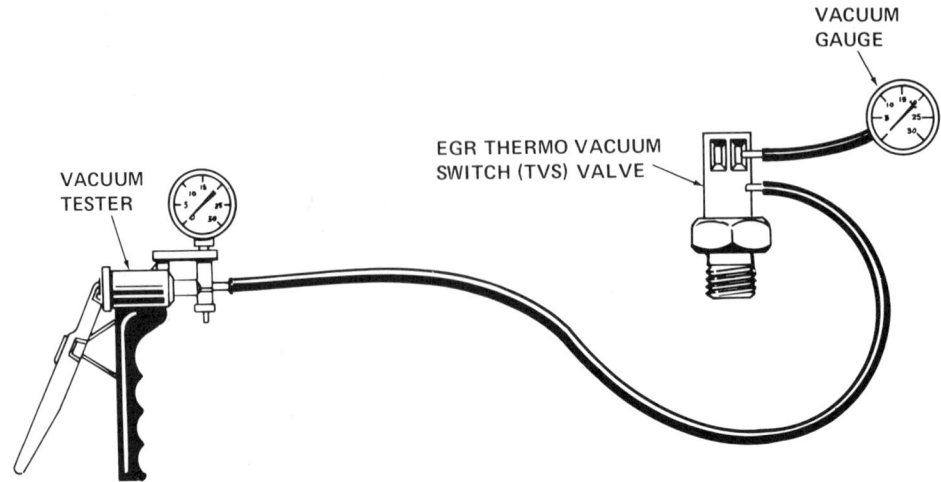

Figure 16–14. Testing a typical Ford TVS.

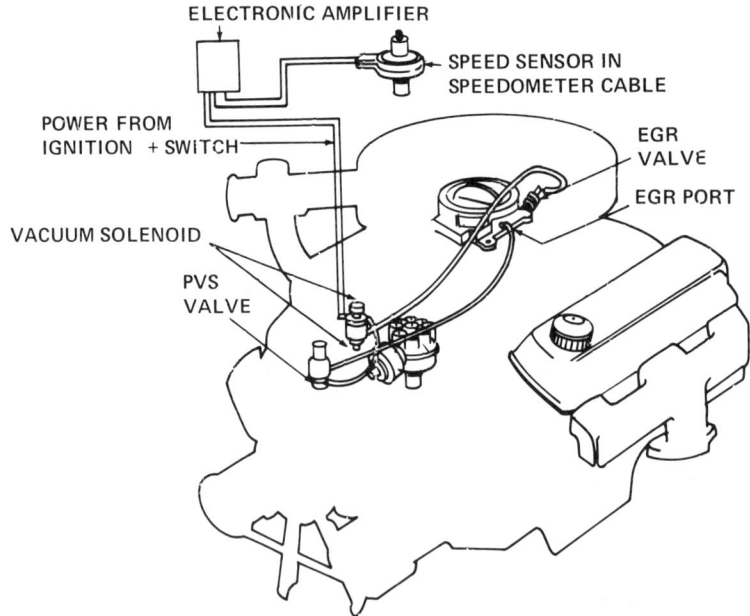

Figure 16–15. Testing a Ford high-speed modulating system.

2. Raise the rear wheels and block them or install the vehicle on a dynamometer. While noting the vacuum gauge reading, operate the engine at idle until its coolant temperature exceeds about 125 °F. The vacuum gauge should indicate no reading.

3. Place the transmission in gear and accelerate the engine while noting the vacuum reading on the gauge. The reading should increase as the engine accelerates.

4. Increase engine speed until the speedometer reaches over 67 mph. At this point, the vacuum gauge should drop to zero.

If the vacuum gauge does not drop to zero, test the speed sensor as follows:

1. Remove the sensor from the speedometer cable and disconnect the multiple connector from the sensor assembly.

2. Calibrate the ohmmeter as necessary, and connect the one ohmmeter lead to the sensor case and the other to the black wire within the center of the sensor (Fig. 16-16).

3. Note the reading on the ohmmeter. If the reading is infinite (showing an open circuit), the sensor is in good condition. However, if the ohmmeter indicates any continuity (less than infinite resistance), replace the sensor assembly.

4. Connect the leads of the ohmmeter to the terminals of the multiple connector (Fig. 16-17).

5. Note the ohmmeter reading. If the reading is within specifications, usually between 40 and 60 ohms, the sensor coil is in good condition. However, if the reading is not within specifications, replace the sensor assembly.

If the sensor is still serviceable, check the solenoid vacuum valve using this procedure:

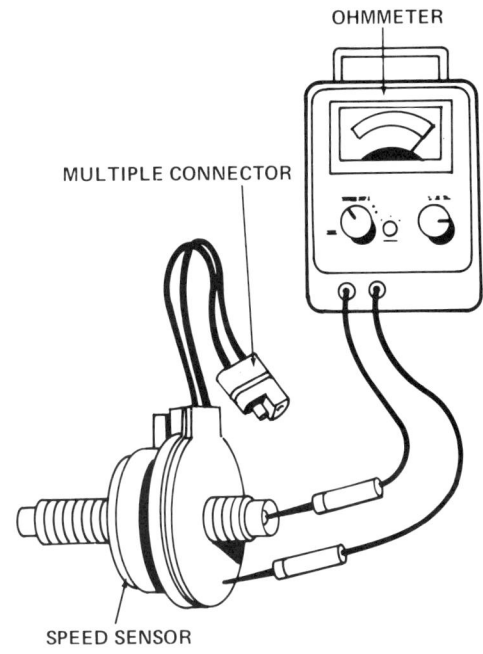

Figure 16-16. Testing the speed sensor for grounds.

1. Disconnect the electrical connections to the solenoid and the vacuum hoses.

2. Connect the hose from a vacuum pump to one of the solenoid fittings and the hose from a vacuum gauge to the other.

3. Apply about 10 inches of vacuum to the solenoid with the pump while noting the vacuum reading on the second gauge. The second gauge should indicate the same reading as the vacuum pump. If not, the solenoid is defective and requires replacement.

4. Connect one jumper lead to either solenoid terminal and ground. Then connect a second jumper lead from a 12-volt source to the second solenoid terminal. The solenoid should now engage.

5. Bleed the vacuum off both the hand

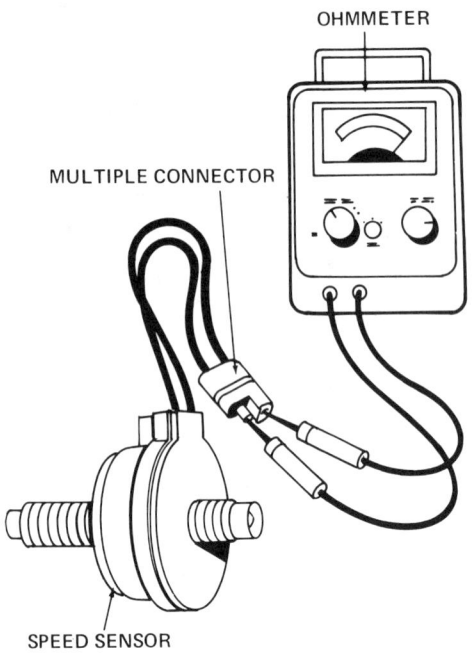

Figure 16-17. Testing the resistance of the sensor coil.

pump and the gauge. Next, reapply about 10 inches of vacuum to the solenoid while noting the action of the needle on the second gauge. The second gauge should indicate a zero reading. If not, the solenoid is defective and requires replacement.

If the speed sensor and solenoid both check out to specifications, the electronic amplifier is the cause of the system malfunction. Replace the amplifier and recheck system operation.

Testing the Separate EGR Back-Pressure Transducer

To check this particular component on the vehicle (Fig. 16-18) follow this procedure:

1. Start the engine and permit it to reach normal operating temperature. Then, run the engine at a fast idle by placing the throttle stop on the fast step of the fast-idle cam.

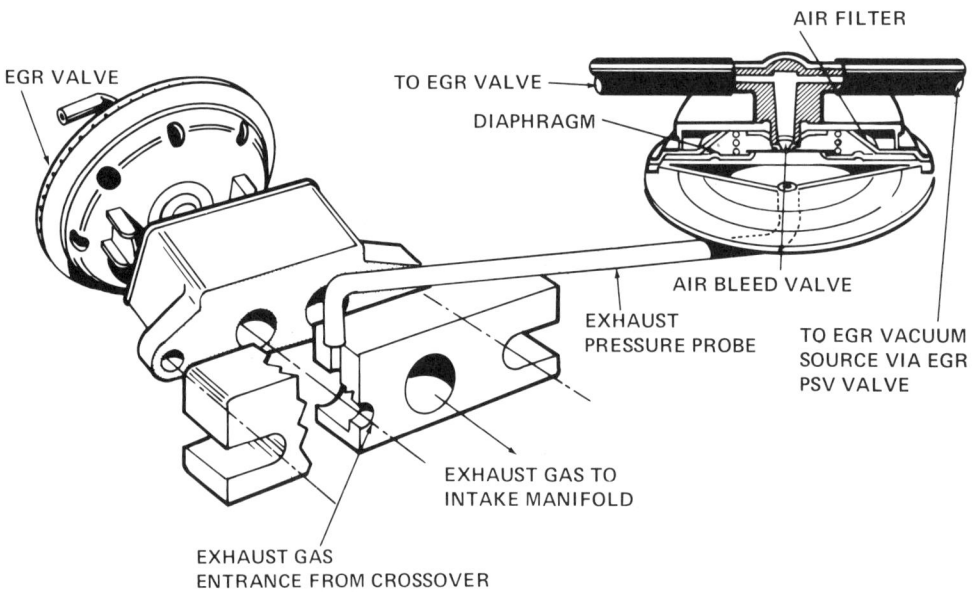

Figure 16-18. Testing a Ford separate back-pressure transducer.

2. Remove the hose from the vacuum side of the transducer. Then, connect the hose from a vacuum gauge to this fitting.

3. Check for a vacuum reading on the gauge. Next, disconnect the gauge hose from the transducer and install its vacuum hose.

4. Disconnect the vacuum hose from the EGR valve and connect the fitting of the vacuum gauge to the open hose.

5. Check the reading on the vacuum gauge. It should be the same as the one taken on the vacuum side of the transducer, or within 2 inches Hg. If the reading is not to specifications, there may be a clogged exhaust passage, an incorrect hose connection, or a leaking transducer.

To check a transducer off the vehicle:

1. Install the hose from a hand-operated vacuum pump to one transducer fitting. Next, plug the other hose connection with your finger.

2. Attempt to apply a vacuum to the transducer with the hand pump. In this situation, the transducer bleed valve should be open, and you should not be able to pump up any vacuum. If you can, the bleed valve is clogged, and the transducer requires replacement.

3. Plug one of the exhaust tube openings with your hand. Next, with your mouth, apply air pressure to the other tube opening.

4. With the hand pump, attempt to apply a vacuum to the transducer. Now, you should be able to build up some vacuum on the unit, but this vacuum should drop to zero as you discontinue applying oral air pressure. Note: Applying air pressure to the exhaust tube opening does not close the transducer bleed completely. Therefore, there is still a little leakage. However, the leakage is insufficient to stop you from pumping up a vacuum within the unit if you operate the hand pump rapidly. Caution: Never use a shop air pressure to test the transducer in this manner. It may rupture the diaphragm.

5. If, during this test, the transducer bleed does not close, the transducer is defective due to clogging and requires a replacement.

Testing a Ford Combination Back-Pressure and EGR Valve

To test this device:

1. Connect a tachometer to the engine.

2. Start the engine and permit it to reach normal operating temperature.

3. Slowly open and then close the throttle valve, while watching the EGR valve stem (Fig. 16-19). If there is no movement

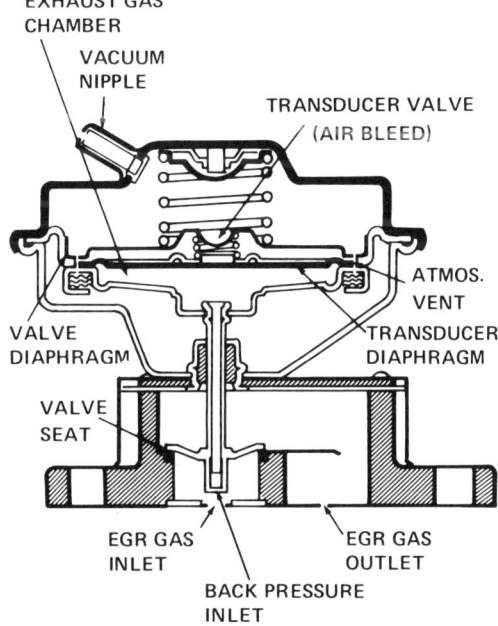

Figure 16-19. Testing a Ford combination back-pressure and EGR valve.

of the stem noted, increase engine speed to 2,000 rpm and hold it steady. Next, remove or pinch off the EGR vacuum hose. Engine speed should now increase at least 50 rpm.

4. If engine rpm does not increase, return the throttle valve to the normal idle position. Next, remove and plug the EGR vacuum line.

5. Connect the hose from a hand-operated vacuum pump to the EGR valve.

6. While noting the action of the tachometer, apply about 15 inches of vacuum to the EGR valve. If idle speed decreases or the engine idles roughly, the EGR valve is defective and requires replacement.

Testing a Ford Electronic EGR System—EGR Valve Position Sensor

The EGR position sensor is part of the EGR valve assembly, which mounts on the carburetor spacer. To test this sensor (Fig. 16-20):

1. Disconnect the electrical harness plug from the sensor.

2. As necessary, calibrate the ohmmeter.

3. While reading the ohmmeter, connect its leads between the orange-white and black-white wires. The reading should be between 2,800 and 5,300 ohms, with the engine off.

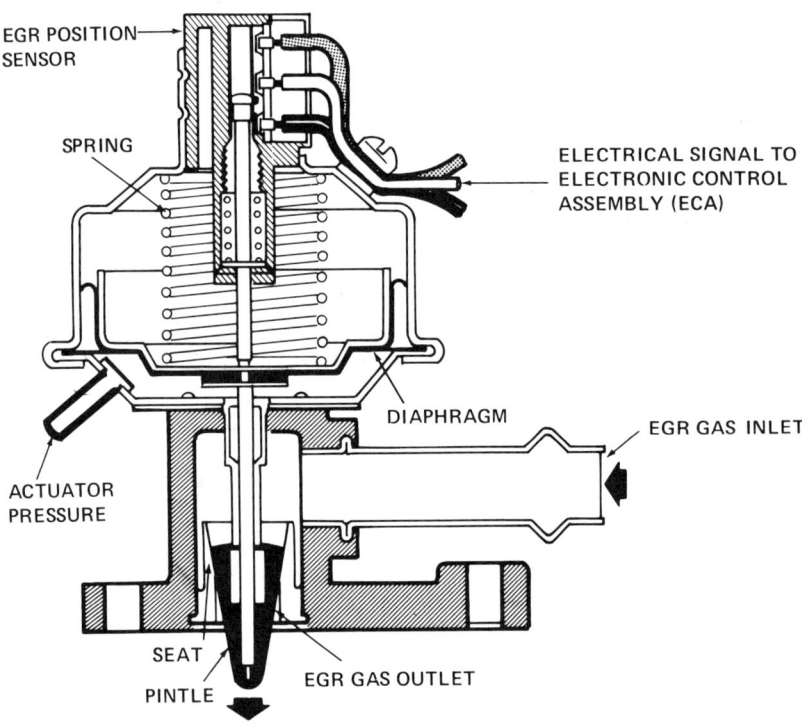

Figure 16-20. Testing a Ford EGR valve position sensor.

4. While observing the ohmmeter reading, connect its leads between the orange-white and the brown-light green wires. The reading should be between 350 to 940 ohms, with the engine off.

5. If the readings are outside of these limits, the sensor is defective, and the entire EGR valve requires replacement.

Testing the EGR Solenoids

There are two solenoids necessary to operate the EGR valve in this system. These solenoids are mounted together on a bracket next to the left rocker-arm cover (Fig. 16–21). To test the rear vent solenoid:

1. Disconnect the electrical plug from the solenoid.

2. With one jumper lead, apply electrical voltage to one solenoid terminal and ground the second solenoid terminal with another lead.

3. With a hand-operated air-pressure pump, apply a few pounds of pressure to the solenoid while noting the gauge indica-

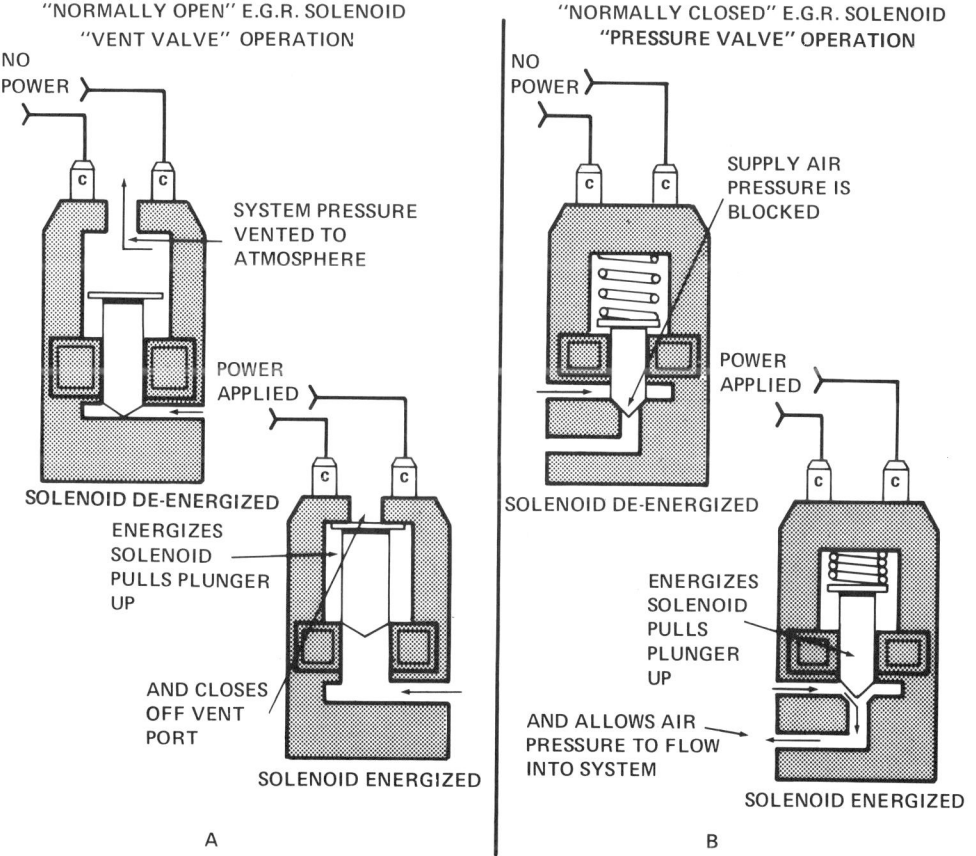

Figure 16–21. Testing a Ford EGR control solenoid.

tion. The valve within this solenoid does not make a perfect seal. However, as long as it does not leak off any more than a half pound of pressure in more than 5 seconds, the solenoid is in satisfactory condition.

To test the front pressure solenoid:

1. Disconnect the wiring harness from the solenoid.

2. Disconnect the upper hose from the solenoid. Next, in its place, install a hose from a hand-operated air-pressure pump.

3. With the pump, apply a few pounds of pressure to the solenoid. The valve should not leak off any more than a half pound in 5 seconds.

4. While observing the gauge reading on the hand pump, apply battery voltage to one of the solenoid terminals with a jumper lead while grounding the second with another jumper lead. The solenoid should immediately open, and the gauge pressure should decrease immediately to zero. If either solenoid does not perform as outlined, it is defective and requires replacement.

Testing a General Motors EGR System—Ported Vacuum Control System

To test the ported EGR control system (Fig. 16-22):

1. Remove the air cleaner and plug the manifold-to-air cleaner vacuum hose.

2. Remove the EGR signal hose from the ported vacuum fitting on the carburetor. In its place, insert a hose from a vacuum gauge.

3. Connect a tachometer to the engine.

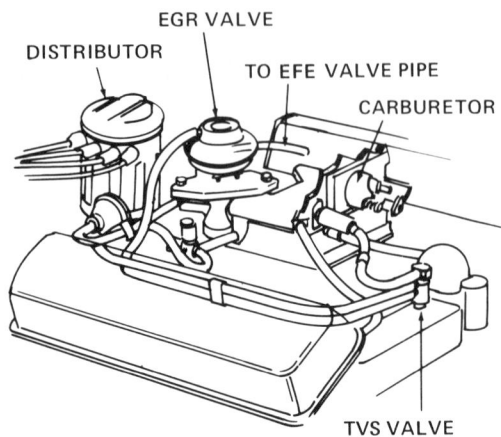

Figure 16-22. Testing a General Motors EGR system.

4. Start the engine and permit it to reach normal operating temperature.

5. Open the throttle valve until the engine reaches about 2,000 rpm, while noting the action of the vacuum gauge. The vacuum gauge reading should rise as the engine accelerates. If not, the ported vacuum fitting or passage in the carburetor is plugged.

EGR System Functional Test

To perform this check:

1. Disconnect the vacuum gauge hose from the ported vacuum fitting on the carburetor. Next, reinstall the EGR vacuum signal hose onto this fitting.

2. With the engine operating, open the throttle valve part way and then release it while observing the EGR valve stem. The valve stem should move upward as the engine accelerates and then move down as the engine returns to idle.

3. Remove and unplug the EGR valve

hose. Then, connect a hose from a hand-operated vacuum pump to this valve fitting.

4. Increase engine speed from idle to about 1,500 rpm.

5. While observing the tachometer, apply at least 11 inches of vacuum with the pump to the EGR valve. The tachometer should indicate at least a drop of 200 rpm on a vehicle with an automatic transmission or 150 rpm on one with a manual transmission. If these gauge readings do not occur, the valve is not opening and requires cleaning or replacement.

Testing the EGR Valve Diaphragm

To perform this check:

1. Remove the vacuum signal hose from the EGR valve.

2. Insert a hose from a hand-operated vacuum pump onto the EGR valve fitting.

3. While noting the gauge, apply at least 10 inches of vacuum to the EGR valve. The valve should open fully and remain open with no leakage of vacuum from the assembly. If the gauge indicates a loss of vacuum and the valve does not remain open, it is defective and requires replacement.

Testing a Combination Back-Pressure and EGR Valve

To perform this check (Fig. 16-23):

1. Remove the back-pressure EGR valve from the intake manifold.

2. Connect the hose from a hand-operated vacuum pump to the EGR valve fitting.

3. Then, attempt to apply at least 10 inches of vacuum to the valve. If the valve accepts the vacuum and opens, the back-pressure control valve is struck closed. In this case, clean the valve and repeat the test. If the valve still opens, the assembly is defective and requires replacement.

4. If the valve did not accept vacuum and remained closed, apply air pressure by mouth to the exhaust gas intake. At the same time, again attempt to apply at least 10 inches of vacuum to the valve.

5. The valve should open completely. If it does not, either the control valve is stuck open or the exhaust passages are plugged. Clean the valve and its passages and repeat the test. If the valve still does not open, the assembly is defective and requires replacement.

Testing the Thermo Vacuum Switch (TVS)

To perform this procedure:

1. Tee a vacuum gauge in the TVS-to-carburetor EGR port (top fitting on the switch).

2. Start the engine and permit it to warm up to normal operating temperature.

3. Observe and note the reading on the vacuum gauge.

4. Remove the vacuum gauge, and tee it into the TVS-to-EGR valve port (bottom fitting on the switch).

5. Observe the reading on the vacuum gauge, and compare it with the reading taken in step 3. If the two readings are not

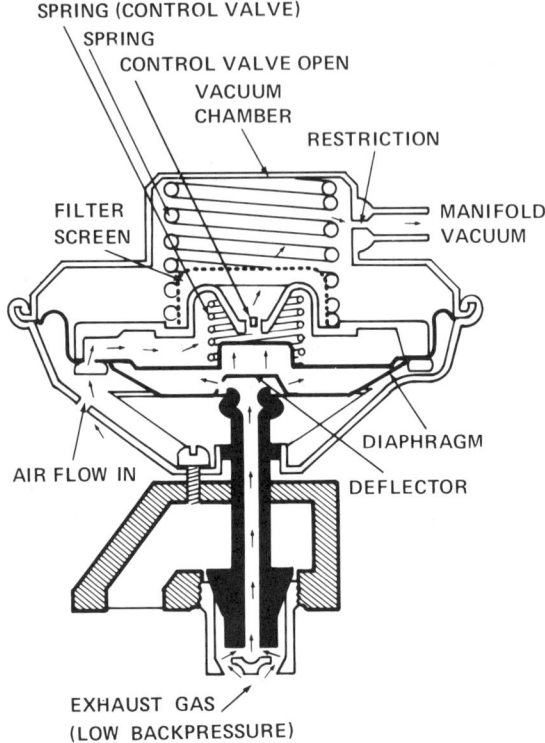

Figure 16-23. Testing a GM combination back-pressure and EGR valve.

within one-half inch of each other, replace the TVS assembly.

Summary

1. All EGR systems require preventive maintenance at given intervals.
2. Usually, it is up to the vehicle's owner to keep track of intervals between EGR service.
3. If the EGR valve becomes clogged with carbon, the engine can operate poorly at idle, especially if the valve remains in the open position.
4. When inspecting an EGR system, always check the physical condition and proper routing of all vacuum hoses.
5. The removal and subsequent reinstallation of most EGR system components is not very difficult if you remember a few general rules.
6. There are two acceptable methods of cleaning an EGR valve.
7. Through the years, Chrysler has used three different styles of EGR systems.
8. The Chrysler EGR system in current use utilizes a venturi vacuum signal to operate the EGR valve.
9. Test a Chrysler ported EGR system following the steps outlined in this chapter.
10. Check the EGR valve and its passages for exhaust gas flow using the procedures presented in this text.

11. Test the serviceability of a Chrysler temperature control valve following manufacturer's instructions or those outlined in this chapter.
12. Check the Chrysler venturi vacuum control and amplifier system following the procedure outlined in this text.
13. Test the serviceability of a Chrysler CCEGR valve using manufacturer's instructions or those outlined in this chapter.
14. Check a Chrysler EGR time-delay system following the steps presented in this text or manufacturer's instructions.
15. Reset a Chrysler EGR maintenance reminder system using the procedure outlined in this chapter.
16. To test a ported-type Ford EGR system, follow the directions outlined in this text or manufacturer's instructions.
17. Check a Ford venturi vacuum EGR system following the manufacturer's directions or those presented in this chapter.
18. Test a Ford high-speed EGR modulating system using the steps outlined in this text or manufacturer's directions.
19. Check a Ford back-pressure transducer subsystem following the procedures specified in this text or in the vehicle's service manual.
20. You can test the serviceability of a Ford electronic EGR system using the steps outlined in this chapter or manufacturer's instructions.
21. Check a General Motors EGR system utilizing either manufacturer's instructions or those outlined in this text.

Review Questions

The questions listed below will assist you in determining how well you remember the material contained in this chapter. Read each question carefully before choosing the answer. If you cannot answer the question, review the section in the chapter that covers the material.

1. If an EGR valve sticks open, the engine will _____.
   a. start hard
   b. idle poorly
   c. burn more fuel
   d. misfire at all times

2. To open an EGR valve for cleaning, use a (an) _____.
   a. vacuum pump
   b. vacuum gauge
   c. screwdriver
   d. abrasive blaster

3. The floor jets of the early Chrysler EGR system require inspection at _____-mile intervals.
   a. 10,000
   b. 12,000
   c. 14,000
   d. 16,000

4. When testing an EGR valve on a Chrysler engine with a vacuum pump, idle speed should drop _____ rpm or more.
   a. 75
   b. 100
   c. 125
   d. 150

5. With its ported vacuum EGR system, Chrysler used _____ types of temperature control vacuum valves.

a. four
b. three
c. two
d. five

6. A Chrysler EGR maintenance reminder light will come on at _____ -mile intervals.
   a. 13,000
   b. 15,000
   c. 17,000
   d. 19,000

7. An EGR valve on a Ford ported vacuum system should start to open at _____ inch(es) of vacuum.
   a. $\frac{3}{4}$
   b. 1
   c. $1\frac{1}{4}$
   d. $1\frac{1}{2}$

8. When testing a single-plug Ford amplifier with a vacuum gauge, remove the hose from the _____ port.
   a. S
   b. R
   c. O
   d. P

9. When testing a typical Ford TVS, cool it until its temperature is _____ °F or less.
   a. 80
   b. 70
   c. 60
   d. 50

10. The resistance within a Ford high-speed modulating sensor should be between _____ ohms.
    a. 20 to 40
    b. 40 to 60
    c. 60 to 80
    d. 80 to 100

11. To check a Ford back-pressure transducer, use a (an) _____.
    a. vacuum pump
    b. infrared analyzer
    c. tachometer
    d. engine analyzer

12. To test the valve position sensor in a Ford EEC system, use a (an) _____.
    a. vacuum pump
    b. infrared analyzer
    c. ohmmeter
    d. voltmeter

13. After applying 11 inches of vacuum to a General Motors EGR valve, engine speed should drop _____ rpm on a vehicle with a standard transmission.
    a. 125
    b. 150
    c. 175
    d. 200

14. Use a (an) _____ to test the serviceability of a GM combination back-pressure and EGR valve.
    a. infrared analyzer
    b. vacuum pump
    c. volt-ammeter
    d. vacuum gauge

For the answers to these review questions, turn to the Appendix.

Chapter 17

Catalytic Converter Systems

One of the more successful methods of reducing harmful emissions treats the exhaust gas after it leaves the combusion chambers. As pointed out in Chapter 13, manufacturers have used air injection systems for many years for just this purpose. This system, as you recall, injected air into the exhaust port or manifolds to further oxidize HC and CO compounds within the exhaust system, thereby helping to reduce them into harmless gases. However, air injection is only successful up to a given point.

Beginning in 1975, with the advent of even stricter exhaust emission standards, manufacturers began to use a new device to further treat the exhaust gas. This device was the catalytic converter.

A catalytic converter has two main functions. First, the device further reduces the amount of hydrocarbons, carbon monoxide, and now even the amount of nitrogen oxide in the vehicle's exhaust. The converter accomplishes this function by providing an additional area for the oxidation or reduction of these pollutants to occur and a catalyst to promote these changes. Second, the catalytic converter also permits, in most cases, the manufacturer to retune the engine for more power, performance, and better fuel economy.

Converter Location

The catalytic converter fits into the engine's exhaust system between the exhaust manifold and muffler (Fig. 17–1). On earlier installations on vehicles with one exhaust pipe, one catalytic converter was used. If the vehicle had dual exhausts, two converters were necessary, one for each side. However, some single-exhaust installations do use two converters of the same type: a mini and a main (Fig. 17–1).

In recent years, the installation of two

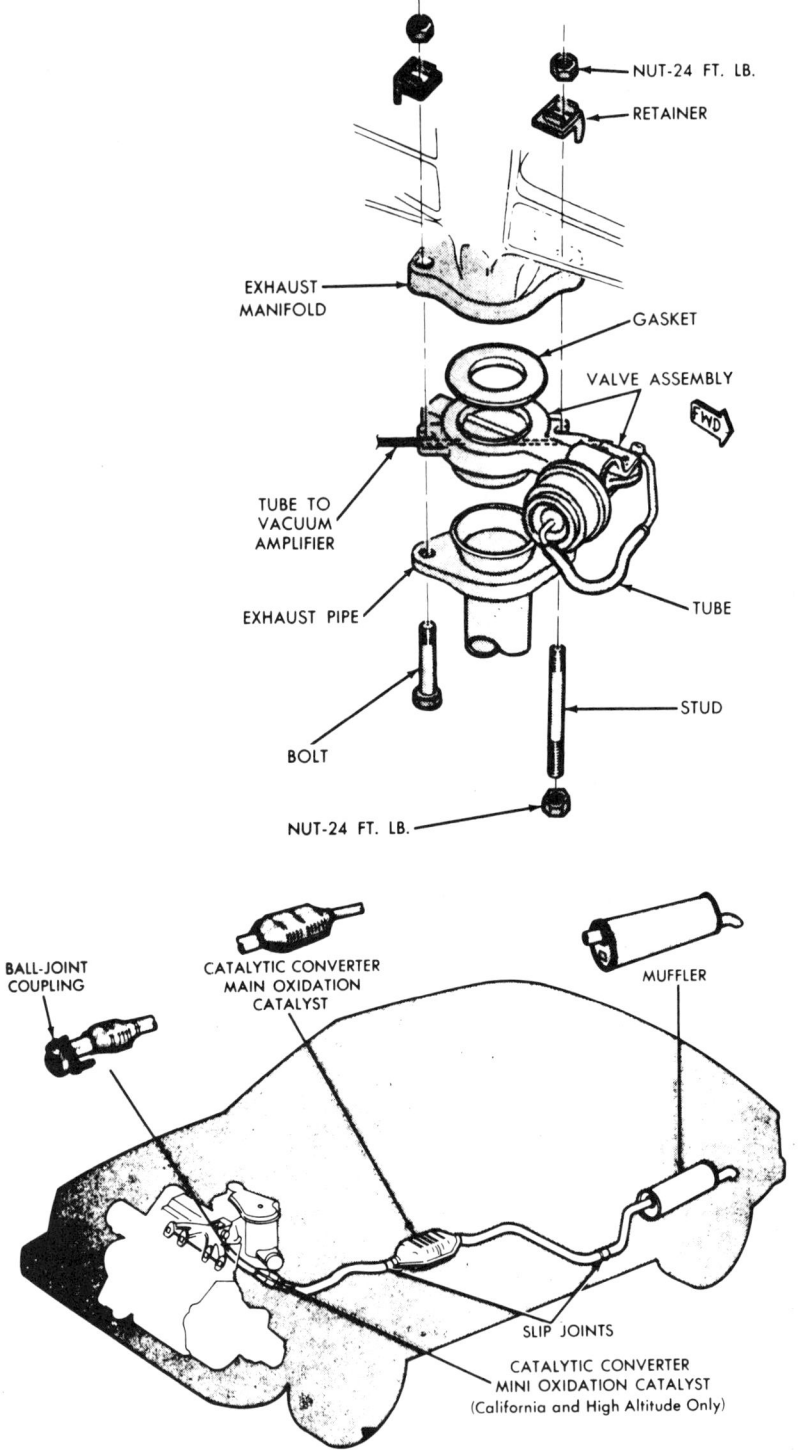

Figure 17-1. Typical catalytic converter installation (Courtesy of Chrysler Corp.).

converters into a single exhaust has become the rule rather than the exception. But these converters (a reduction and an oxidation) perform different functions as we shall explain. Finally, some manufacturers have now combined both of these devices into a single unit, commonly referred to as either a three-way or hybrid converter.

Although the catalytic converter looks much like another muffler, there are some obvious differences in design. One difference, for example, that distinguishes it from an ordinary muffler is the metal used for its outer skin. The outer skin of the converter is made of stainless steel, which is more durable and corrosion resistant than the metal used in ordinary muffler construction.

Another notable difference in installations between the catalytic converter and the plain muffler is the use of heat shields and interior insulating pads around the converter on some vehicles (Fig. 17-2). These devices are necessary due to the high operating temperature of the converter shell, which can reach approximately 600–800 °F (312–422 °C). The upper shield in this installation protects the passenger compartment from damage due to excessive heat, whereas the lower shield decreases the possibility of a fire if some combustible object happens to be caught on the converter while the vehicle is in motion. Finally, the special insulation beneath the carpeting in the vehicle minimizes heat buildup inside the passenger compartment.

Function and Types of Catalyst Materials

Before discussing the several different designs of catalytic converters, it is a good time to explain the function and different kinds of catalysts used in these units. A catalyst is a substance that causes a chemical reaction without being changed by the reaction. In other words, since the catalyst only encourages rather than takes part in the reaction, the process does not use the material up.

There are three kinds of catalysts used in converters: platinum, palladium, and rhodium. For example, in the oxidation converter, platinum and palladium are the noble metals used as the catalyst. A noble metal is one that resists oxidation. The catalyst in the oxidation converter is approximately 70 percent platinum and 30 percent palladium. Platinum is the better of the two as a catalyst, but it is more expensive. On the other hand, palladium is not as efficient as platinum, but manufacturers use it to reduce the overall cost of the unit.

The reduction converter also uses two materials for the catalyst. In this converter type, the manufacturer forms the catalyst using platinum and rhodium.

Types of Catalytic Converters

Oxidation Converters

Manufacturers use two basic types of converters: oxidation and reduction. The function of the oxidation converter is to oxidize or add oxygen to certain harmful elements and compounds within the exhaust gases. The resulting chemical process then changes these harmful elements and compounds into harmless materials.

In automotive applications, the chemical compounds which require oxidizing are hydrocarbons and carbon monoxide.

During the oxidation process within the catalytic converter, the hydrocarbons are

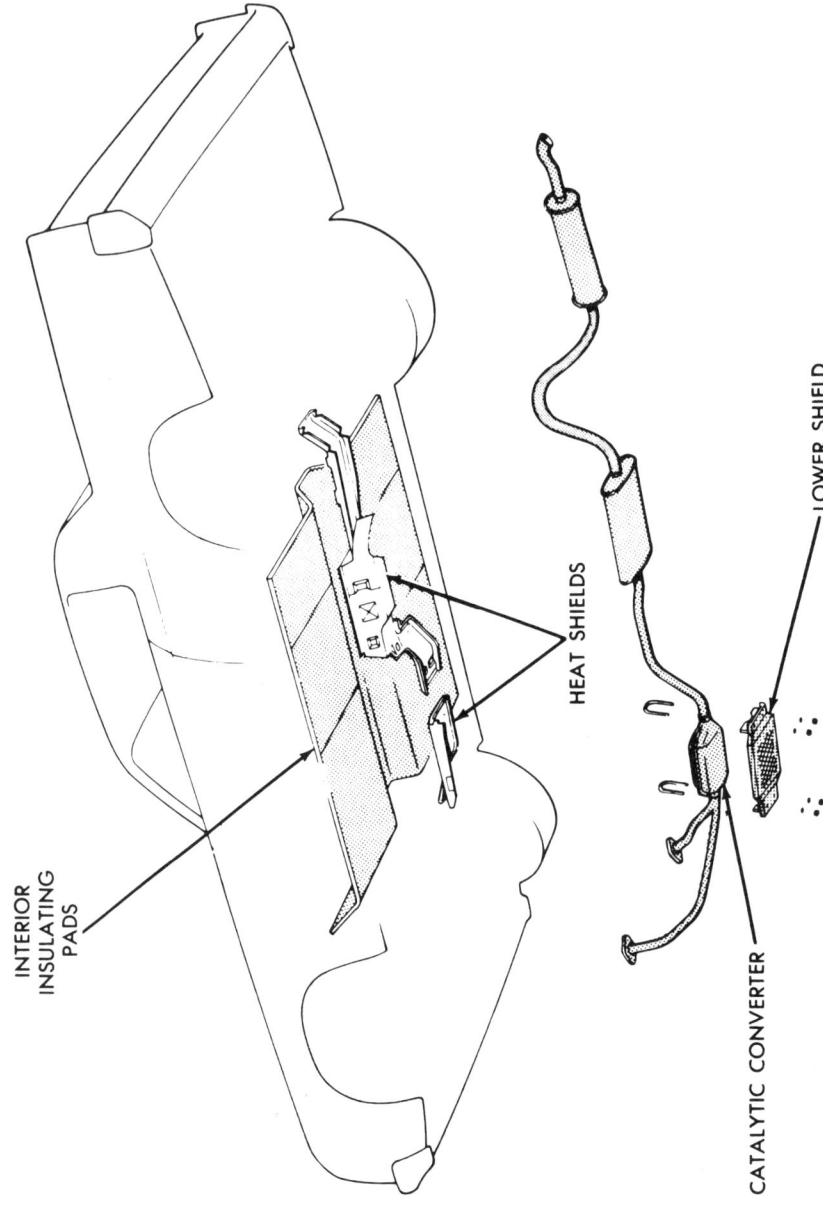

Figure 17-2. To prevent damage due to high converter temperature, some vehicles have heat shields and interior insulating pads (Courtesy of Chrysler Corp.).

broken apart into hydrogen and carbon atoms (Fig. 17-3). The hydrogen then becomes oxidized and converts into water vapor. At the same time, the carbon also oxidizes and converts to carbon dioxide. Any carbon monoxide resulting from incomplete combustion oxidizes to form carbon dioxide. Consequently, the entire process forms water vapors and carbon dioxide, which are harmless gases that do not pollute the atmosphere.

Since this converter oxidizes the HC and CO pollutants, the process generates considerable heat. For instance, the temperature during oxidation within the converter is about 200 °F (92.4 °C) higher than the temperature of exhaust gases entering the converter. If exhaust gas is 1,200 °F (648 °C) when it enters the converter, the oxidation process will raise the temperature to approximately 1,400 °F (752 °C).

However, the actual operating temperature within the converter does vary during different phases of engine operation. The oxidation converter itself does not begin to function till exhaust gas temperatures reach about 500 °F (257.4 °C). Then, as the vehicle operates under changing conditions, the operating temperatures inside the converter vary from 1,200 °F (640 °C) to about 1,600 °F (862 °C). As a result, the outside stainless steel shell temperatures also vary from about 600-800 °F (312-422 °C).

In order for the catalyst to oxidize the harmful compounds and produce these high temperatures, additional air is necessary. There are two methods used to supply the catalytic converter with additional air. The first method uses an air injection system, which pumps the air into the exhaust manifold and in some cases, directly into the converter itself. The catalytic converter can also receive air, which was not consumed during the combustion process, from the exhaust manifold.

Reduction Converters

An oxidation-type converter has little effect on NOx levels because the control of this pollutant requires a separate reaction called reduction. Reduction is a chemical process in which oxygen is taken away from a compound. Therefore, this process is the opposite of an oxidation reaction.

In operation, the reduction converter provides a reaction which changes NOx compounds to harmless nitrogen and carbon dioxide. The elements rhodium and platinum in the catalyst are responsible for chemically promoting the shift of oxygen from the NOx to the CO compound. As a result, the reduction converter changes the incoming NOx and CO compounds, by chemically promoting the shift of oxygen, into nitrogen and carbon dioxide (Fig. 17-4).

Oxidation Converter Designs

Monolith

There are two basic internal designs for oxidation-type converters: a monolith design and a pellet design. The monolith converter consists of a shell, honeycomb

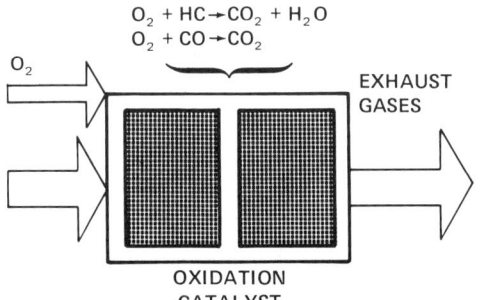

Figure 17-3. The oxidation converter changes HC and CO into H_2O and CO_2.

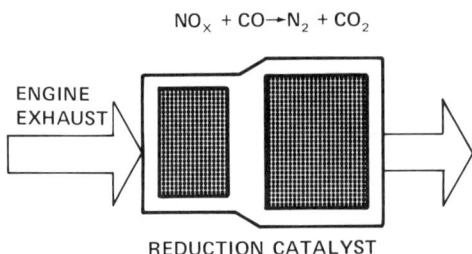

Figure 17-4. A reduction converter promotes the shift of oxygen from the NOx into the CO compounds to form N_2 and CO_2.

monolith, flow diffuser, and stainless steel mesh (Fig. 17-5). Manufacturers make the converter housing or shell of two stamped metal pieces welded together to form a round or oval assembly. The outer housing or shell is made of aluminized or stainless steel because of the high temperatures associated with the oxidation process, which the metal must withstand.

Each side of the shell supports a portion of the catalytic element. The element in this converter consists of the honeycomb monolith upon which the manufacturer deposits the platinum catalyst.

The monolith element has a honeycomb-type design. The element has hundreds of cellular passages for the exhaust gases to flow through (Fig. 17-5b). The manufacturer forms the monolith from a ceramic material, which can withstand the high temperatures generated during the oxidation process.

Covering this cellular ceramic element is a very thin coating of platinum or palladium. Either or both of these metals provide the chemical activity, operating temperature, and durability required by an effective catalyst within the oxidation converter.

Between the converter inlet and the catalytic element is a flow diffuser. This device is necessary with the monolithic-type element to allow a uniform flow of exhaust gas over its entire area. If the unit did not have a diffuser, the gases would tend to flow only through the center portion of the element.

The stainless steel mesh has three functions. First, it protects the monolithic material from damage due to shock or severe jolts by acting as a cushion. Second, the mesh also protects the catalytic element from thermoshock caused by temperature extremes. Finally, the mesh keeps the monolithic element in proper position during the final assembly of the converter shell.

Pellet-Type Catalytic Converters

With the exception of a drain plug, the outside of a pellet-style oxidation converter looks much the same as the monolithic unit. However, the internal structure is quite different.

Internally, the pellet-style converter has a pellet substrate, upper and lower baffles, and insulation (Fig. 17-6). Rather than a ceramic monolithic element, this style of converter uses small aluminum oxide pellets approximately one-eighth to three-sixteenths of an inch in diameter. These small aluminum oxide pellets are coated with a very thin layer of platinum and/or palladium to form the catalyst of the converter.

The baffles within this converter serve several functions. For example, the baffles direct the flow of exhaust gases through the converter and catalyst. As the exhaust gases enter the converter, they are directed upward by one baffle (Fig. 17-7). The only way for the exhaust gases to leave the converter is to flow downward through the baffle, pass through the bed of pellets, and out through the lower baffle to the outlet side of the converter. Finally, while acting

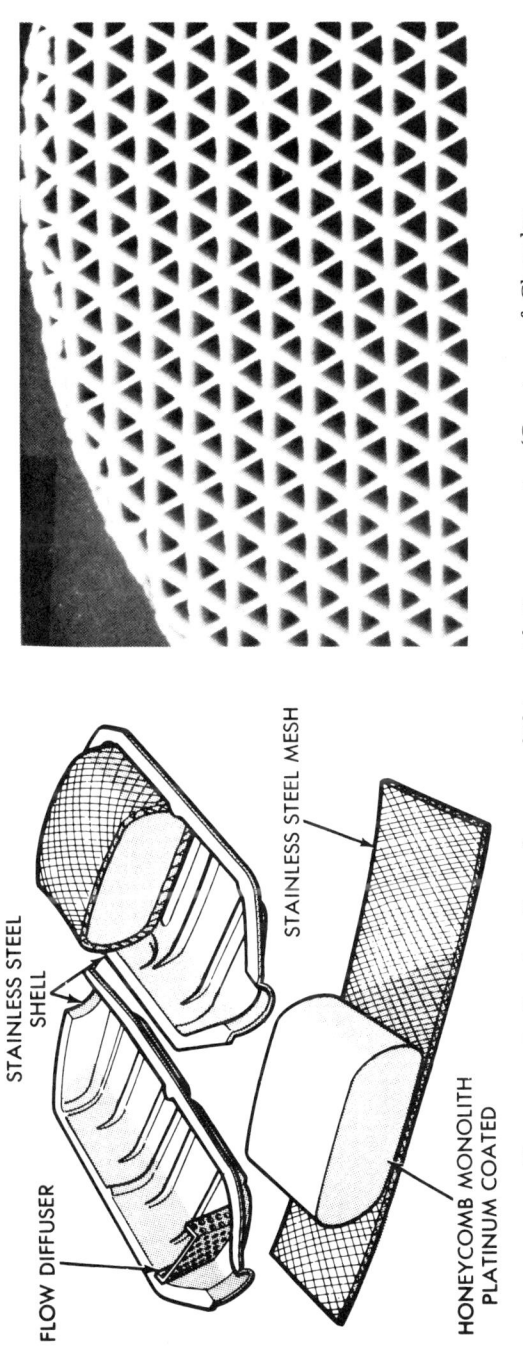

Figure 17-5. Construction of a monolithic oxidation converter (Courtesy of Chrysler Corp.).

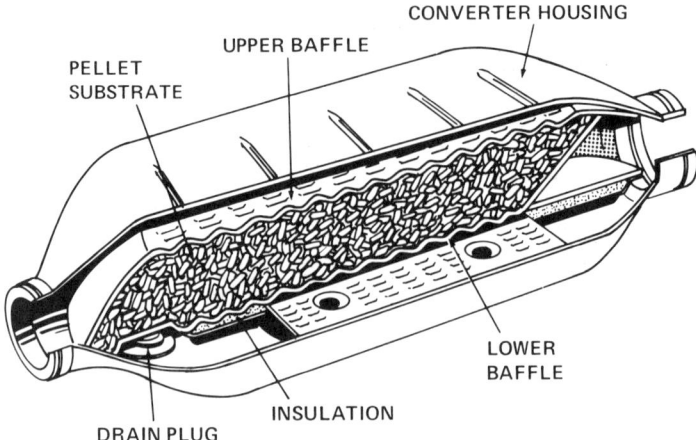

Figure 17-6. Construction of a pellet-type oxidation converter.

as a guide for the exhaust gases, the baffles also support and contain the bed of pellets.

Between the inner and outer shell of this converter is a layer of insulation. The insulation retards heat transfer from the catalyst to the outer shell. Consequently, the converter housing remains at about the same temperature as the outside of the muffler. Thus, very few vehicles with this type of converter require a heat shield.

The drain plug on the outside of the converter permits the removal of the catalyst pellets with special equipment. This permits the replacement of the old catalyst charge after 50,000 miles of service or if the catalyst material becomes damaged.

Advantages and Disadvantages of Monolithic- and Pellet-Type Converters

Both of these types of oxidation converters perform efficiently, but both have advantages and disadvantages. For instance, the monolithic converter is more resistant to damage from vibrations and is smaller overall. Therefore, it heats up

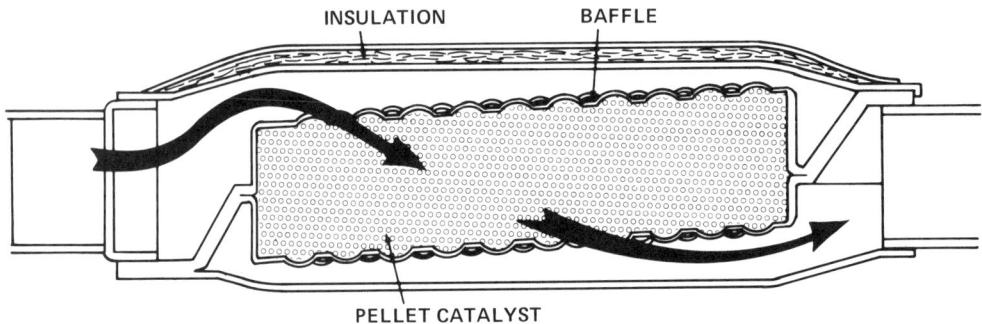

Figure 17-7. Flow of exhaust gases through a pellet-type oxidation converter.

faster permitting it to perform its function sooner. In addition, the monolithic converter offers less resistance to exhaust flow and therefore has less back pressure.

There are two disadvantages to the use of the monolithic-type converter. First, the device is not repairable; consequently, when defective, it requires replacement. Second, the monolithic converter requires more platinum in the catalyst. Therefore, it is more expensive to build.

The pellet-style converter on the other hand, is repairable because its pellets are replaceable. Moreover, the unit uses less platinum and therefore is less expensive to build initially.

There are three basic disadvantages of the pellet converter design. First, because of its larger size, the unit warms up more slowly than the monolithic design. Second, the pellet converter offers more resistance to exhaust gas flow; therefore, the unit creates more back pressure in the exhaust system. Lastly, the pellet converter is not quite as durable as the monolith design.

Three-Way Catalytic Converters

Beginning in 1978, two domestic automobile manufacturers, Ford and General Motors, began using a two-stage three-way converter. This unit controlled not only HC and CO but NOx emissions also (Fig. 17–8).

The front half (first stage) of this converter, close to the unit's inlet, has a reduction-oxidation catalyst that controls NOx, HC, and CO emissions. This portion of the converter has a monolithic substrate with a rhodium and platinum catalyst.

The rear portion of the converter (second stage) has only a platinum catalyst. This catalyst further oxidizes any remaining HC and CO emissions.

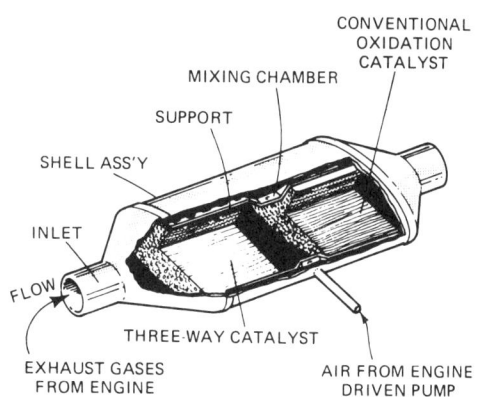

Figure 17–8. Three-way converter controls HC, CO, and NOx emissions.

Specific Catalytic Converter Systems—Chrysler

Chrysler Corporation installs the monolithic-type converter on their vehicles (see Fig. 17–5). Since this unit is not repairable, the entire unit requires replacement after about 50,000 miles or in case of damage.

Since this type of converter transmits a considerable amount of heat through its housing, heat shields are necessary to protect the passenger compartment, automatic transmission and its cooler lines, the torsion bars, the rear shock absorbers, and other chassis components (Fig. 17–9). The size, shape, and placement of the shields themselves do vary according to the vehicle model.

Vehicles with both a converter and an air injection pump system require additional shielding, since the injection of air causes the converter to operate at even higher temperatures. Consequently, vehicles equipped with air injection systems use a protective grillwork under the converter. This grill prevents direct shell material contact with burnable materials. In ad-

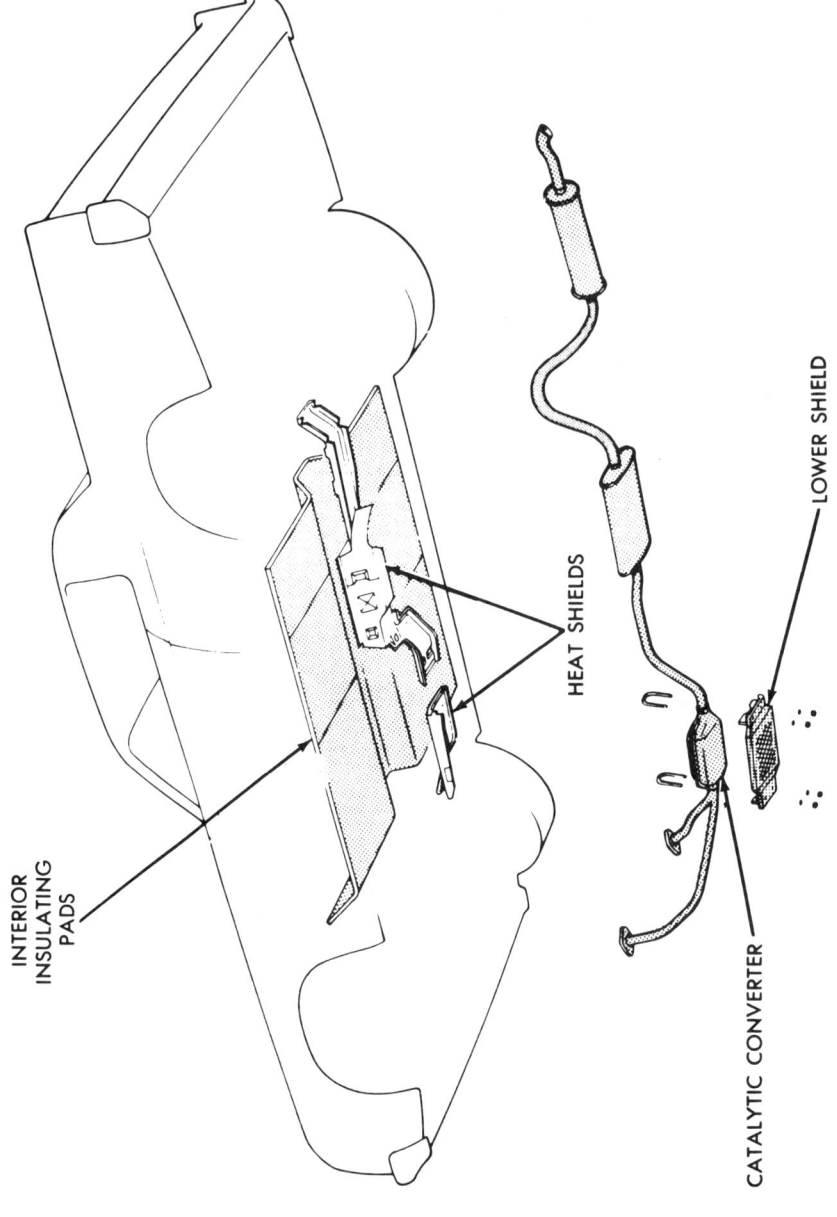

Figure 17-9. Chrysler catalytic converter heat shield installation (Courtesy of Chrysler Corp.).

dition, the guard helps cool the converter housing by transferring heat to the air flow beneath the vehicle.

Mini Converter Installation

On some 1977 and later vehicles, Chrysler installed a small oxidation-type converter welded into the engine exhaust pipe about 6 to 12 inches from the exhaust manifold (Fig. 17-10). This device, sometimes known as the underhood converter, begins the oxidation of HC and CO compounds before they reach the main underfloor unit. This results in a more complete oxidation of the harmful emissions in the exhaust gas.

As shown in Fig. 17-11, the oxidation reaction begins in the mini converter. As shown here, oxygen along with HC and CO compounds pass into the mini converter, where the oxidation process begins. Then, the partially to completely oxidized compounds pass into the main converter along with additional oxygen. The main converter completes the oxidation of HC and CO compounds and converts them into harmless carbon dioxide and water.

Catalytic Converter Protection Subsystem

Some of the 1975 Chrysler models had a catalytic protection system (CPS). This system prevented excessive hydrocarbons from reaching the converter during a closed-throttle deceleration after high-speed vehicle operation. The excessive hydrocarbons are due to an overrich mixture brought about by high intake manifold vacuum during deceleration. This vacuum pulls so much fuel out of the carburetor bowl through the idle circuit that the engine will operate too rich with closed throttle valves. To prevent this, the system uses a solenoid and electronic speed switch (Fig. 17-12). The solenoid mounts onto the side of the carburetor and is identical to an antidieseling solenoid.

The electronic speed switch reacts to engine rpm. When, for example, engine speed reaches about 2,000 rpm, the switch energizes the solenoid and plunger extents.

When the driver releases the accelerator peddle with engine speed above 2,000 rpm, a solenoid plunger holds the throttle valve open at the fast-idle position (about 1,500 rpm). This keeps the throttle valve slightly open thus providing additional air to balance the air/fuel mixture, which speeds up combustion to protect the converter from overheating.

Once engine rpm drops below 2,000, the speed switch deenergizes the solenoid. As a result, the throttle valve returns to the normal slow-idle position as the solenoid plunger retracts.

Chrysler's Catalytic Air Supply System

Before 1977, Chrysler used its basic air injection system to supply the additional air to the catalytic converter. However, in 1977, Chrysler modified this system by the addition of an air-switching valve and coolant control engine vacuum switch. These units (Fig. 17-13) control the air flow from the injection pump during and after engine warm-up. This assists in oxidizing the HC and CO emissions within the catalytic converter without interfering with the control of NOx by the EGR system. (Refer back to Chapter 13 for a detailed description of this particular system.)

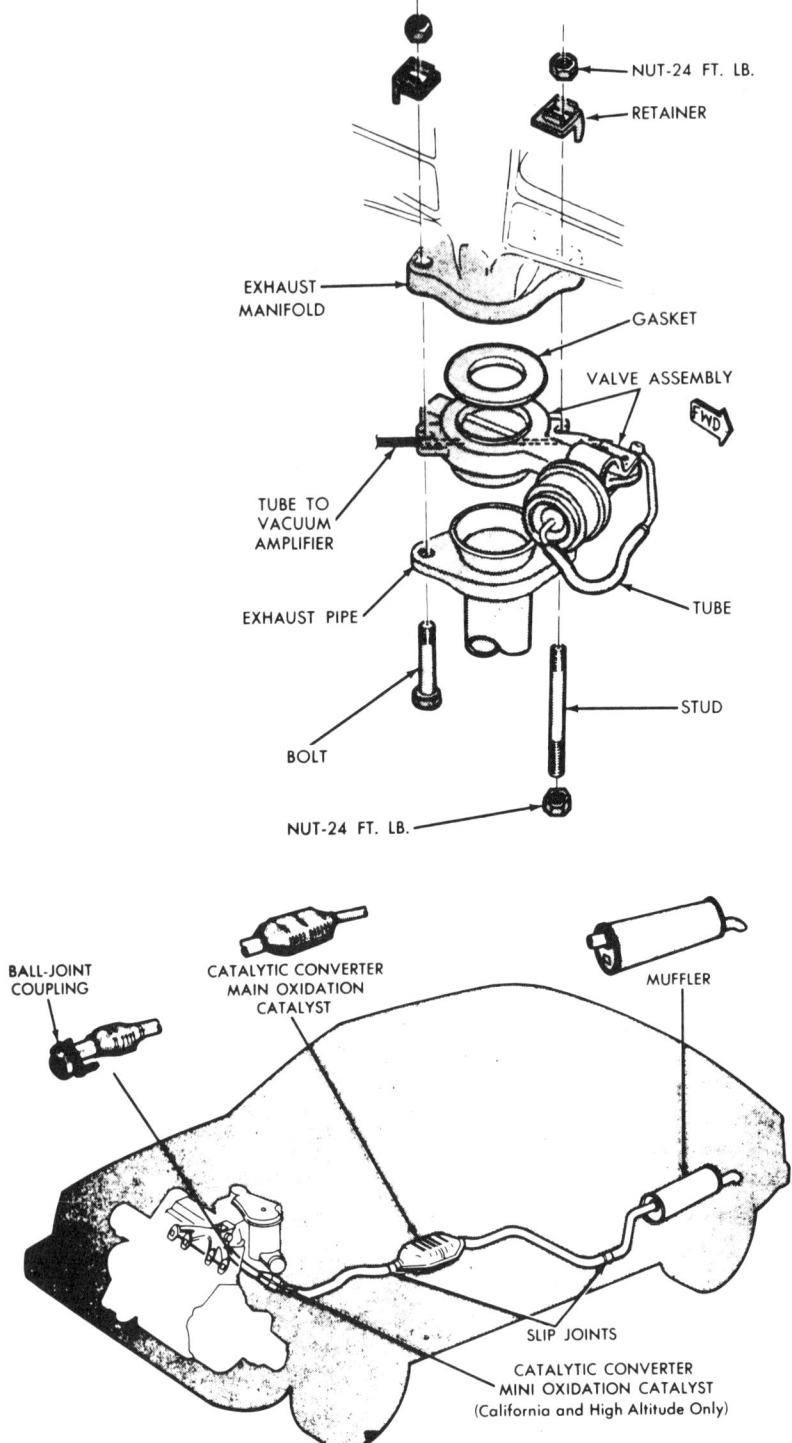

Figure 17-10. Mini converter installation (Courtesy of Chrysler Corp.).

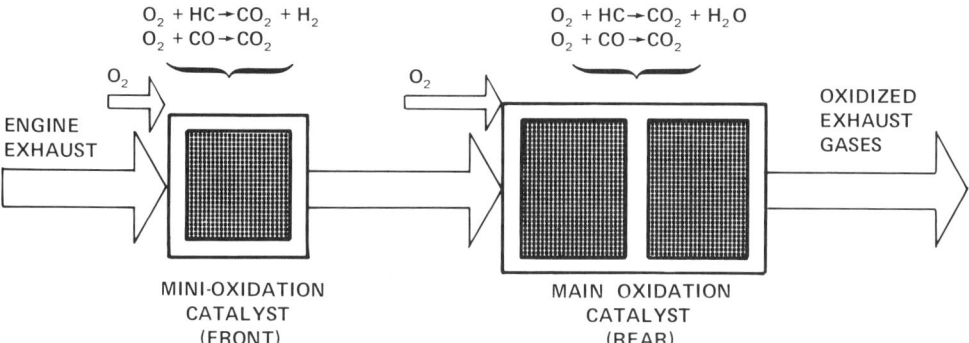

Figure 17-11. Oxidation reaction within a mini and main converter installation.

Three-Way Converter Installation

Beginning in 1980, some Chrysler vehicles came from the factory with three-way catalytic converters. This converter worked in conjunction with an air injection system to reduce NOx as well as HC and CO emissions.

This three-way converter fits into the exhaust system in front of the main oxidation converter. The front portion of this converter is responsible for changing NOx into

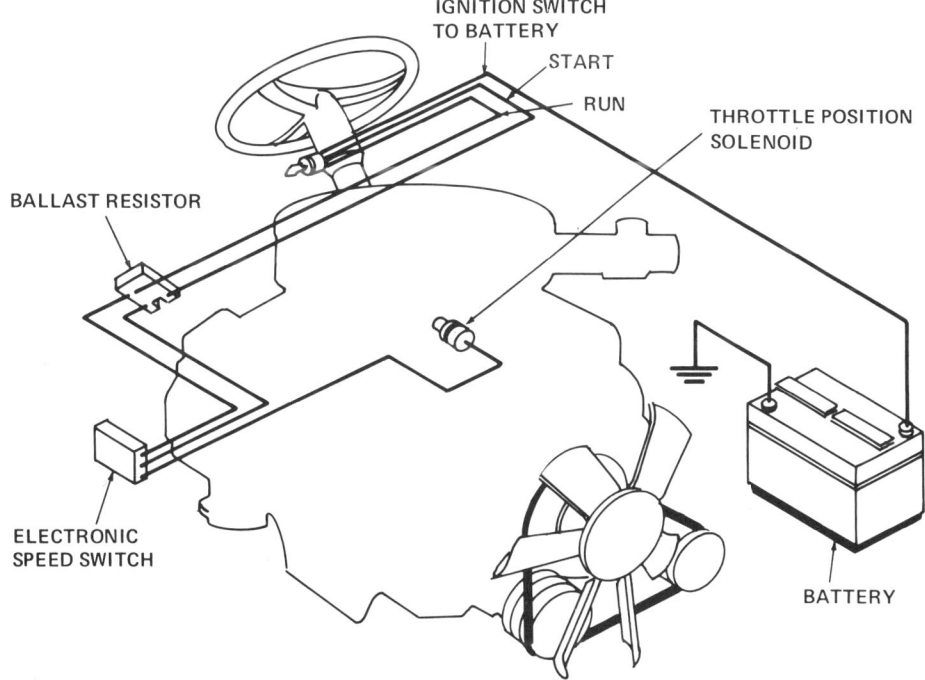

Figure 17-12. Chrysler's catalytic protection subsystem.

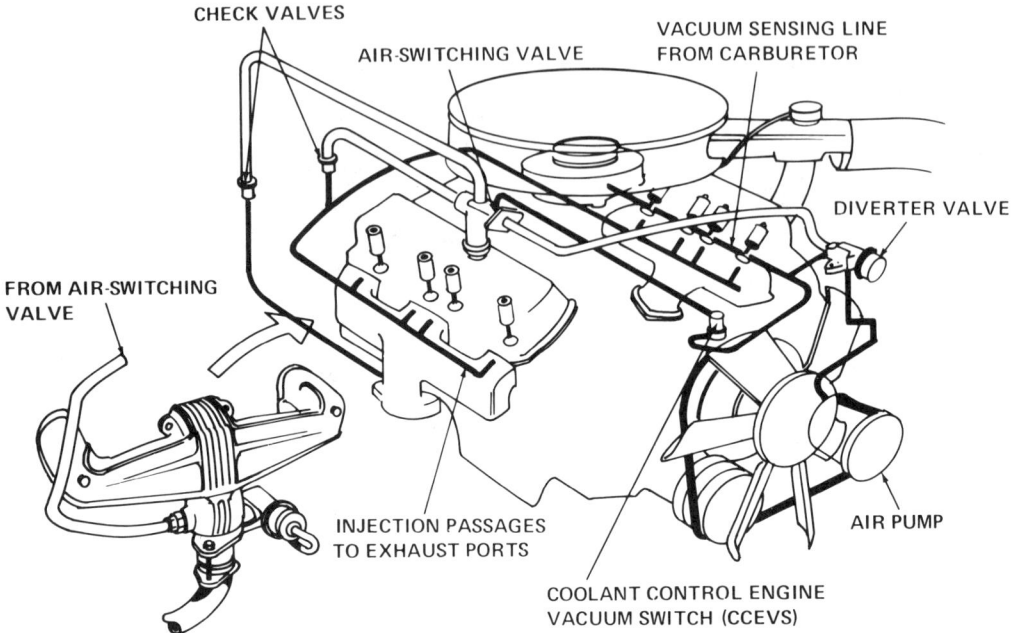

Figure 17-13. Typical Chrysler air injection system used in conjunction with the catalytic converter.

harmless nitrogen while the rear portion oxidizes the HC and CO compounds, thus changing them into harmless carbon dioxide and water vapor.

The Chrysler three-way converter receives additional air in a similar manner as the unit illustrated in Figure 17-8. In other words, the Chrysler air injection system pumps air into the rear half of the converter to assist the oxidation process.

Ford Catalytic Converter Systems

Ford Motor Company also uses a monolithic converter on its vehicles (Fig. 17-14). Before 1977, all Ford converters had a single monolithic substrate. However, some 1977 and later automobiles have converters with two monolithic substrates (Fig. 17-15).

Due to this monolithic design, a Ford catalytic converter is also not serviceable. Consequently, the entire unit requires replacement after about 50,000 miles or in case of damage.

Since the monolithic converter transmits a high level of heat through its housing, Ford vehicles have various forms of solid and perforated heat shields. These shields are welded or clamped in place, according to the particular model, in order to protect given vehicle components.

Ford Secondary Air Supply

A Ford catalytic converter does require a secondary air supply. This additional air contains the oxygen that will further ox-

FORD CATALYTIC CONVERTER SYSTEMS

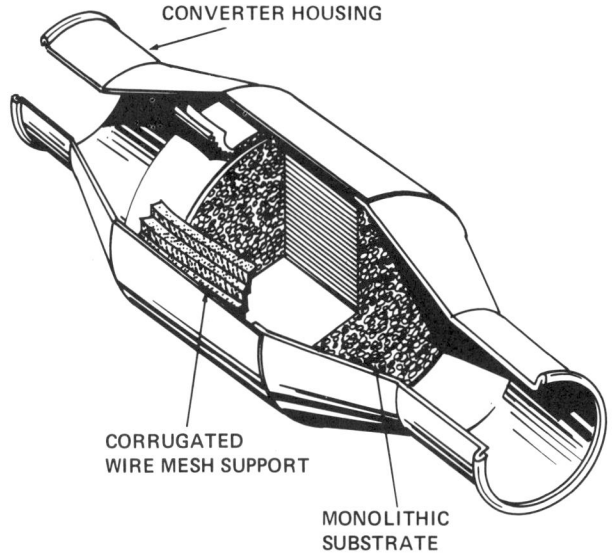

Figure 17-14. Ford single-substrate converter.

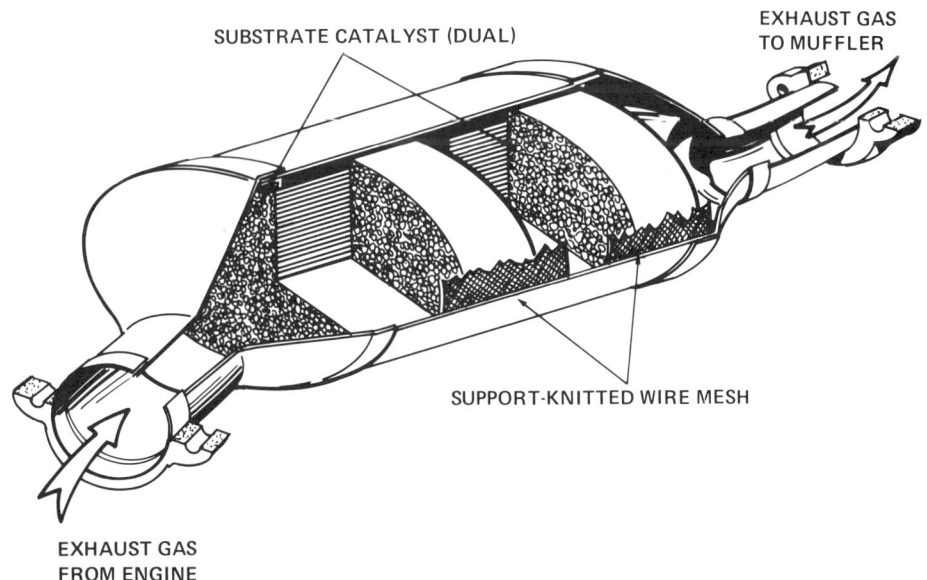

Figure 17-15. Ford dual-substrate converter.

idize the HC and CO compounds within the catalytic converter.

The secondary air supply for the converter comes from Ford's Thermactor air injection system (Fig. 17-16). (Chapter 13 of this text describes the design and operation of this system in detail.)

Ford Catalytic Protection System

The basic Thermactor air injection system does require some additional components to protect the catalytic converter from overheating. These include the solenoid vacuum valve and PVS electric switch. The solenoid vacuum valve controls the operation of the air bypass valve. The solenoid valve accomplishes this task by opening or closing the vacuum circuit to the vacuum differential valve according to an electrical signal.

When power is cut off to the solenoid, it deenergizes. As a result, the vacuum signal cannot pass through the valve to the vacuum differential valve (Fig. 17-17a).

When the electrical circuit is complete to the solenoid, its valve opens (Fig. 17-17b). Since the valve portion of the solenoid is now open, the vacuum signal can now pass to the vacuum differential valve (VDV). In later systems, Ford eliminated the differential vacuum valve. Consequently, when the solenoid valve is open, a ported vacuum signal can pass through the valve to directly operate the air bypass valve.

Depending on the vehicle, Ford utilizes several methods to energize the solenoid valve. For example, some vehicles use a temperature switch within the air cleaner. However, those vehicles using a catalytic protection have either a floor-mounted switch or an electric ported vacuum switch.

Figure 17-18 illustrates a typical Ford electric ported vacuum switch. This switch can be either normally opened or normally closed, depending on its design. In either case, the switch signals the vacuum solenoid valve if coolant temperature reaches or exceeds 235 °F. The solenoid valve then interrupts the vacuum signal to the VDV or air bypass valve. Consequently, air under pressure from the injection pump dumps into the atmosphere. This action prevents the catalyst within the converter from overheating, especially when the vehicle idles for long periods of time, such as in heavy traffic conditions.

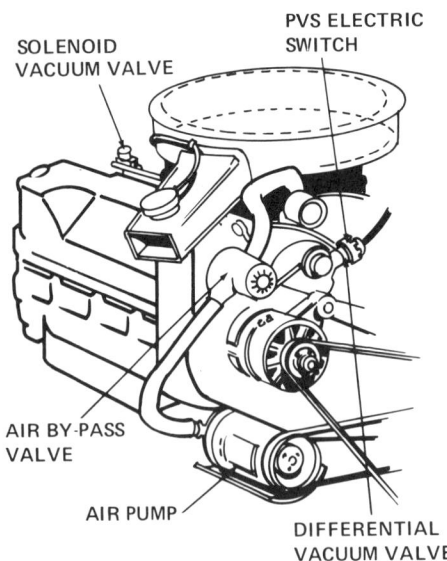

Figure 17-16. Typical Ford air injection system.

Ford Three-Way Converter System

Beginning in 1978, Ford installed a three-way converter and a closed-loop feedback emission control system on many

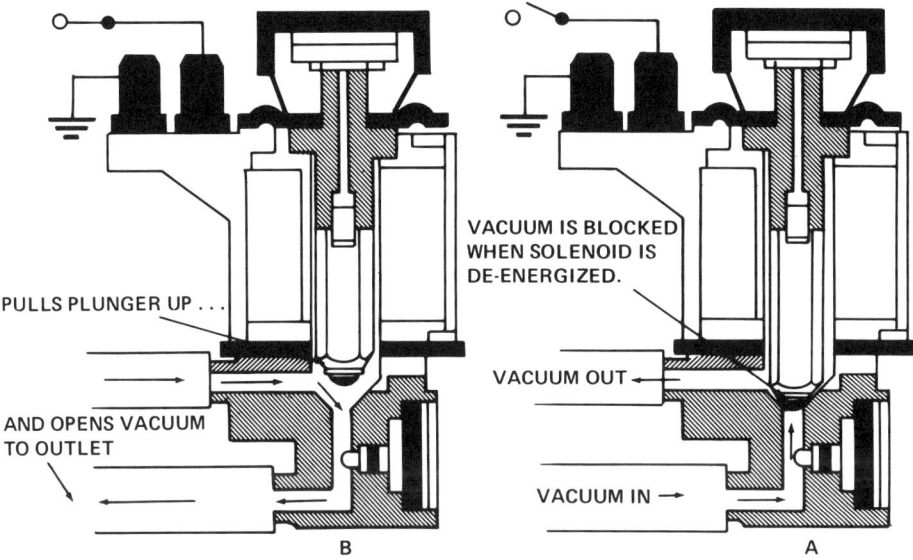

Figure 17–17. Design and operation of a Ford solenoid vacuum valve.

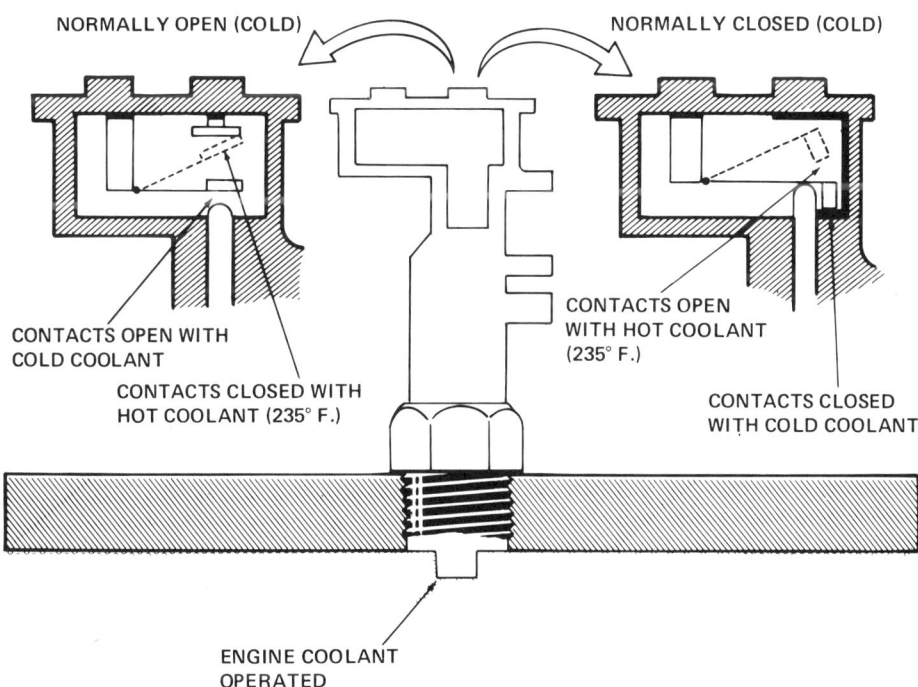

Figure 17–18. Electric ported vacuum switch used on some Ford catalytic air injection systems.

of its compact models (Fig. 17–19). The two-stage three-way converter receives additional air from the Thermactor air injection system, as explained in Chapter 13.

The closed-loop feedback system consists of several sensors and an electronic control module. For instance, an oxygen sensor continuously monitors the unburned air/fuel ratio in the exhaust and directs a signal to the control module. The module also receives signals from a throttle-angle switch, two vacuum switches, and the ignition control side of the coil. The module then directs a signal to the vacuum solenoid regulator that controls the vacuum supply to the carburetor's power valve. This action provides precise control of the air/fuel ratio required by the two-stage converter to reduce the exhaust emission levels.

General Motors Catalytic Converter Systems

General Motors installs a pellet-type converter on it vehicles. Since this converter is serviceable, with special equipment a technician can remove and replace the pellets after 50,000 miles of service or if the material becomes damaged.

Since the pellet-type converter has internal insulation, its skin remains at about the same temperature as the outside of an ordinary muffler. Thus, only a few vehicles such as the Corvette require any form of heat shield.

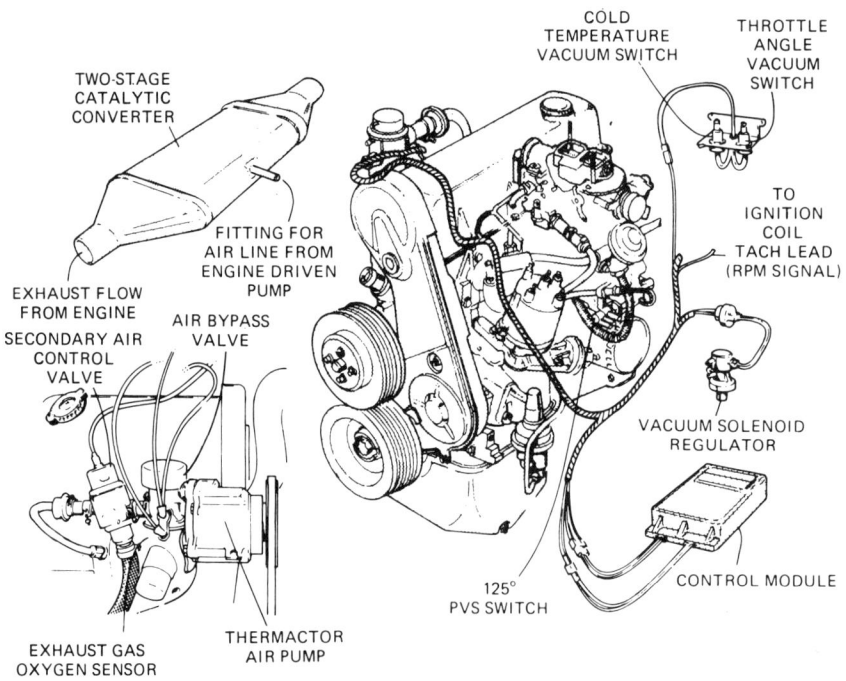

Figure 17–19. Typical Ford three-way converter used in conjunction with an electronic emission control system.

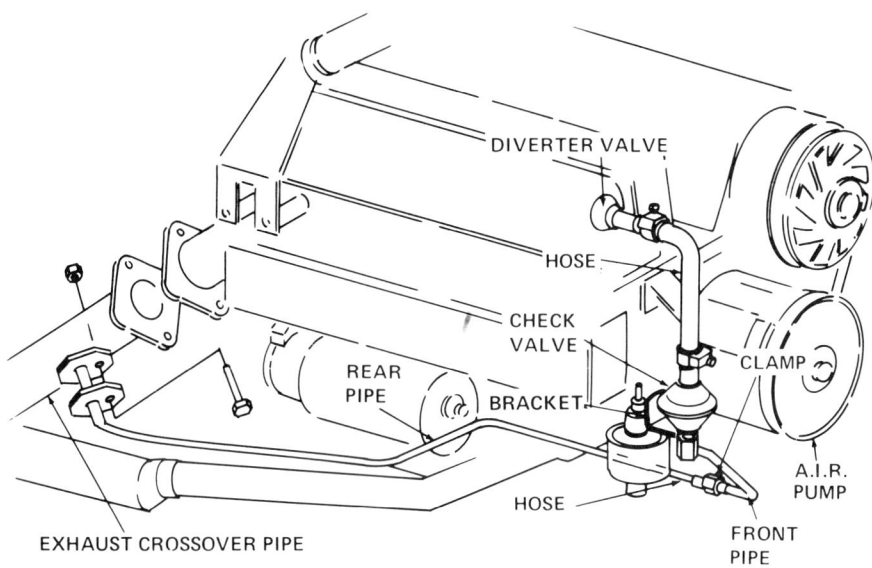

Figure 17-20. Air injection system used by General Motors on a vehicle with a catalytic converter.

General Motors Secondary Air Supply System

General Motors also uses an air injection system to supply the necessary oxygen to the catalytic converter in order for it to perform its function (Fig. 17-20). As explained in Chapter 13, about the only difference in the catalytic converter injection system over earlier designs is where the injected air enters the exhaust system. In the catalytic system, pumped air enters to the exhaust pipe downstream from the exhaust manifold.

General Motors Three-Way Converter Installation

In 1978 General Motors also began to use a closed-loop feedback emission control system in conjunction with a three-way converter on some of its compact models. This C-4 system, as it is commonly called, is illustrated in Figure 17-21.

Along with the three-way convertor, the C-4 system has an oxygen sensor, temperature switch, electronic control unit, and a control air/fuel ratio carburetor. The oxygen sensor fits into the exhaust manifold or pipe. This device develops a voltage, which varies correspondingly with the volume of oxygen present in the exhaust gas. A rise in the voltage indicates a decrease in the oxygen content, and a drop in voltage indicates an increase in the oxygen content. The temperature switch directs a signal to the electron control unit (ECU) when cooling system temperature is higher than 90 °F. This signal is necessary before the ECU can begin regulating the carburetor's air/fuel ratio.

The ECU receives the signal from the various sensors. Then, the ECU computes the signal and informs the vacuum modulator as to how much vacuum to feed to

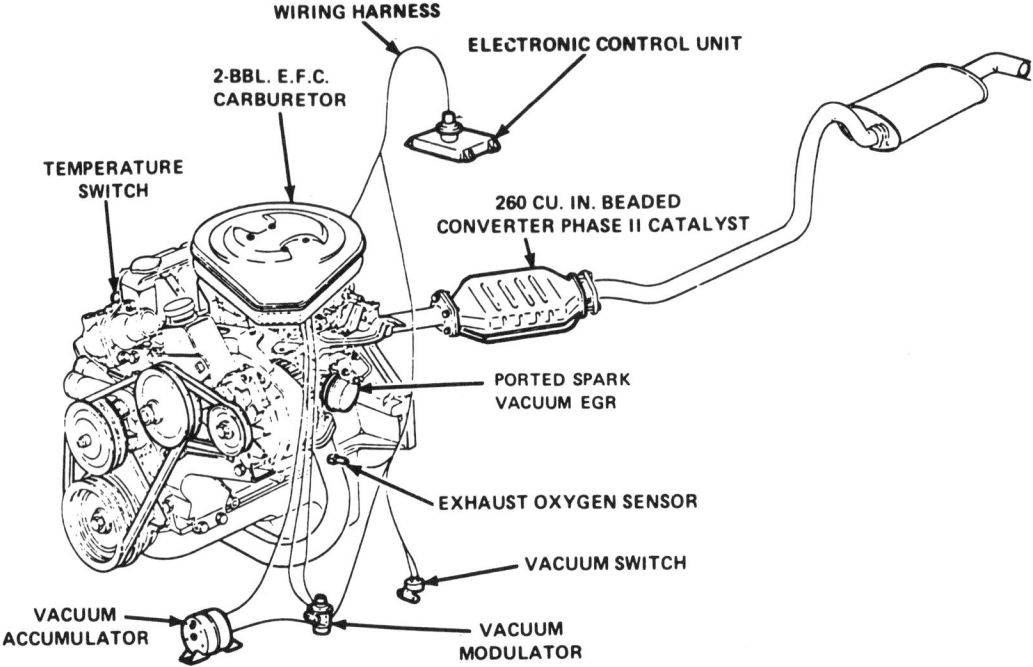

Figure 17-21. Closed-loop electronic system used on some GM vehicles with a three-way converter (Courtesy of General Motors Corp.).

the vacuum diaphragm within the carburetor's idle and main metering system. This action provides the precise control of the air/fuel ratio from the carburetor for adequate emission control by the catalytic converter.

Summary

1. One of the more successful methods of reducing harmful emissions treats the exhaust gas after it leaves the combustion chambers.
2. Manufacturers began to use catalytic converters in 1975.
3. The catalytic converter has two main functions.
4. The catalytic converter fits into the exhaust system between the exhaust manifold and muffler.
5. Although the catalytic converter looks like another muffler, there are some obvious differences in design.
6. Many catalytic converter installations require a number of heat shields.
7. A catalyst is a substance that causes a chemical reaction without being changed by the reaction.
8. There are three kinds of catalysts used in converters: platinum, palladium, and rhodium.
9. Manufacturers use two basic types of converters: oxidation and reduction.
10. During the oxidation process within the catalytic converter, the hydrocarbons are broken apart into hydrogen and carbon atoms.
11. Since the converter oxidizes the HC and CO pollutants, the process generates considerable heat.

12. The actual operating temperature within the converter varies during different phases of engine performance.
13. In order for the catalyst to oxidize the harmful compounds, additional air is necessary.
14. An oxidation converter has little effect on NOx levels because the control of this pollutant requires a separate reaction called reduction.
15. The reduction converter provides a reaction that changes NOx to harmless nitrogen and carbon dioxide.
16. The monolithic converter consists of a shell, honeycomb monolith, flow diffuser, and stainless steel mesh.
17. The pellet-style converter has a substrate, upper and lower baffles, and insulation.
18. Both types of oxidation converters perform efficiently but both have advantages and disadvantages.
19. Beginning in 1978, Ford and General Motors started using a two-stage three-way converter.
20. Chrysler uses the monolithic-type converter on its vehicles.
21. Some 1977 and later Chrysler vehicles use a small oxidation-type converter welded into the engine's exhaust pipe.
22. Some 1975 Chrysler models have a catalytic protection system.
23. Ford also uses a monolithic converter on its vehicles.
24. Ford vehicles have various forms of perforated and solid heat shields.
25. A typical Ford catalytic protection system includes a solenoid vacuum valve an a PVS electric switch.
26. Beginning in 1978, Ford started to install a three-way converter and a closed-loop feedback emission control system on many of its compact vehicles.
27. General Motors installs a pellet-type converter on its vehicles.
28. Chrysler, Ford, and General Motors all use an air injection system to supply the necessary oxygen to the catalytic converter.
29. In 1978 General Motors also started to use a three-way converter on some of its compact models.

Review Questions

The questions listed below will assist you in determining how well you remember the material contained in this chapter. Read each question carefully before choosing the answer. If you cannot answer the question, review the section in the chapter that covers the material.

1. Manufacturers began to use catalytic converters in _____.
   a. 1973
   b. 1975
   c. 1977
   d. 1970

2. The shell of the catalytic converter is made of _____.
   a. stainless steel
   b. iron
   c. aluminum
   d. brass

3. An oxidation converter changes _____ _____ to harmless gases.
   a. hydrocarbons and carbon dioxide
   b. hydrocarbons and oxides of nitrogen
   c. hydrocarbons and carbon monoxide
   d. oxides of nitrogen and carbon monoxide

4. A reduction converter changes _____ _____ to harmless gases.
   a. oxides of nitrogen and carbon dioxide
   b. oxides of nitrogen and carbon monoxide
   c. oxides of nitrogen and hydrocarbons
   d. hydrocarbons and carbon monoxide

5. The pellets in a catalytic converter are formed of _____ coated with platinum or palladium.
   a. stainless steel
   b. copper oxide
   c. iron oxide
   d. aluminum oxide

6. Chrysler uses a _____ type of catalytic converter.
   a. two-stage
   b. three-way
   c. pellet
   d. monolithic

7. After 1977 Chrysler used _____ oxidation converter(s) on some vehicles.
   a. one
   b. two
   c. three
   d. no

8. After 1977 Ford converters have _____ monolithic substrate(s).
   a. zero
   b. two
   c. one
   d. three

9. Ford began to use a three-way converter in _____.
   a. 1976
   b. 1977
   c. 1978
   d. 1979

10. The _____ in the Ford and General Motors closed-loop feedback emission control system informs the electronic control unit as to the amount of O_2 remaining within the exhaust gases.
    a. oxygen sensor
    b. temperature switch
    c. vacuum switch
    d. control carburetor

For the answers to these review questions, turn to the Appendix.

Chapter 18

Catalytic Converter Service

Factors Reducing Service Life of Catalytic Converters

Federal law now requires that a catalytic converter provide at least 50,000 miles of normal use before service is necessary. However, certain factors reduce the normal life of the converter and require its premature replacement or servicing. Other than physical damage, these factors include the use of the wrong fuel, poor engine condition, or improper service procedures.

Unleaded Fuel

Vehicles with catalytic converters must utilize an unleaded fuel. The lead additive in certain fuel greatly reduces the efficiency of the catalyst. The lead itself plates the catalyst to form a coating that prevents the exhaust gas pollutants from reaching and, therefore, reacting with the catalyst. Consequently, the continuous use of a leaded fuel eventually causes the converter catalyst to lose its ability to promote the oxidation of HC and CO compounds.

Even in a situation of the emergency use of leaded fuel in a vehicle, the converter sustains some damage. The return to the use of unleaded fuel permits the converter to regain some, but not all, of its effieiency. In reality, how much efficiency the converter does retain depends on how long the leaded fuel was used. The problem in dealing with this particular situation is that there is no known procedure for testing the remaining converter efficiency.

Manufacturers prevent the use of leaded fuel in a vehicle by the design of the filler tube leading down to the fuel tank. Vehicles requiring leaded fuel have a special filler tube, which has a restriction placed at its opening (Fig. 18-1). This restriction prevents the entry of the larger

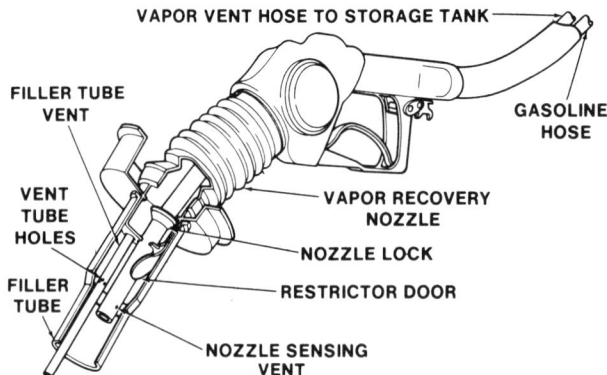

Figure 18–1. A restriction placed in the filler tube that permits only the smaller unleaded fuel nozzles to enter (Courtesy of Chrysler Corp.).

leaded-fuel delivery nozzles at the gasoline pumps. In addition, to designate the vehicle fuel requirements further, the manufacturer usually installs a decal reading "unleaded fuel only" beside the filler cap and sometimes on the instrument panel.

Engine Condition

Proper engine condition and maintenance are necessary to prevent overheating the catalyst. Excessively high catalyst temperature reduces the converter's service life. On the average, converters operate well at internal temperatures up to 1,500 °F. However, at temperatures above this, the catalyst begins to melt or break down.

When any engine is in poor mechanical condition or requires a tune-up, the exhaust gas passing into the converter will contain excessive amounts of hydrocarbons. These excessive amounts change the converter into a catalytic furnace, where the temperatures are high enough to destroy the catalyst.

Even when an engine is in good condition and tuned properly, its improper use can create excessive converter temperatures. For instance, operating the engine at idle for long periods is one of the worst operating conditions for converter life. At prolonged idle, the engine develops more heat than at normal highway speeds. Therefore, never operate an engine with a converter more than 10 minutes at idle.

Service Precautions

To prevent fuel vapors from reaching the converter and causing high temperatures during vehicle maintenance, follow these general rules:

1. Use an oscilloscope, if available, for locating misfiring cylinders, instead of shorting or removing spark plug wires from an operating engine.

2. If an oscilloscope is not available, never operate an engine more than 30 seconds with one spark plug wire removed or shorted.

3. Do not crank an engine over for more than 60 seconds when flooded or fir-

ing intermittently. In the case of a flooded engine, repair the cause of the flooding condition first. Then, dry out the engine as necessary before starting by removing the spark plugs.

4. Never operate the engine more than 10 minutes at idle.

5. Do not attempt to start a vehicle with a standard transmission by pushing it. Instead, use a spare battery and jumper cables.

6. With a vehicle in motion, never turn off the ignition switch.

7. Always repair substandard operating conditions such as dieseling, heavy surging, backfiring, or repeated stalling. These conditions can lead to premature converter failure.

8. It is not recommended that you use liquid engine or carburetor cleaners, which inject directly into the carburetor.

9. Only use unleaded gasoline in the vehicle. Never use low-lead or leaded gasoline.

10. Avoid running out of fuel while the engine is operating or while driving on the highway, especially at high speeds. This may damage the converter.

11. Do not use engine or ignition replacement parts that are not certified, recommended, or approved as being equivalent to original equipment.

12. Do not pump the accelerator pedal to start a hot engine that has stalled.

13. When raising or lowering a vehicle on a hoist, be sure all the hoist arms and other equipment are properly positioned to avoid damaging the converter and other components. If the hoist makes contact with any portion of the vehicle other than the proper lift points, check all the underbody components for physical damage and operating clearance before lowering the vehicle.

Converter System Inspections

To physically check the condition of the catalytic converter system (Fig. 18-2):

1. Inspect the converter for physical damage such as a ripped skin, holes, and a crushed or bulged shell.

2. Check the exhaust pipe leading to the converter for holes or damage.

3. Inspect the muffler and tail pipe for holes or damage.

4. On American Motors and General Motors vehicles, examine the opening in the tail pipe. Pellets coming out of the tail pipe are an indication that the internal stainless steel basket or baffle assembly containing the pellets is breaking up. This condition is usually due to excessively high temperatures, which cause a catalyst support to distort. This opens holes through which the exhaust gas may blow the beads. Although the pellet-type of converter is rechargable with new beads, there is no way to open up the converter and repair a damaged support assembly. In addition, before installing the new converter find the cause of the problem, which resulted in the high temperatures within the converter. Otherwise, the new converter will fail prematurely.

5. If the outside of the converter, exhaust pipe, muffler, or tail pipe show any sign of damage, replace the affected component.

6. Inspect all heat shields to make certain that they are secure around the com-

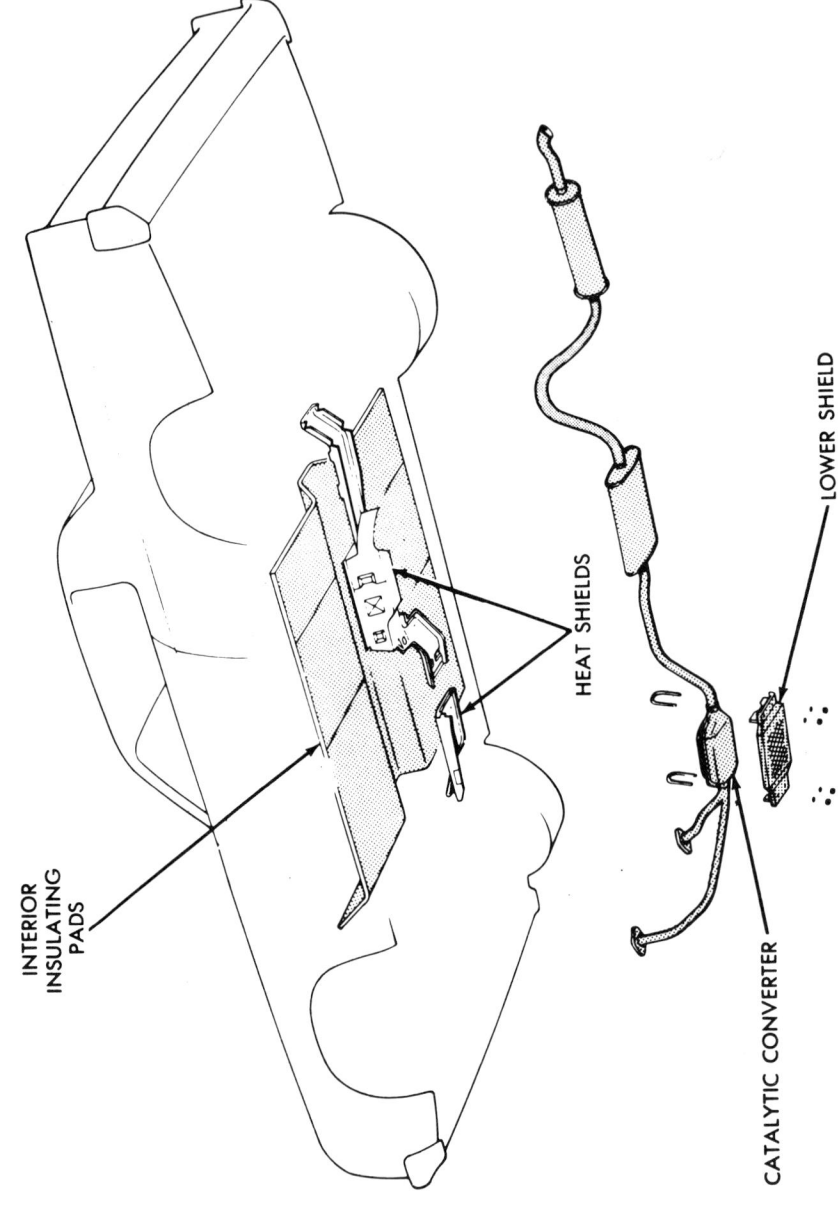

Figure 18-2. Inspecting a catalytic converter system (Courtesy of Chrysler Corp.).

ponents they protect and also for physical damage.

Testing Catalytic Converters

There are two common methods of testing catalytic converters. These include the checking of a converter for restrictions with a vacuum gauge and testing the converter's efficiency in reducing HC and CO emissions with an infrared analyzer.

Several abnormal factors can cause restrictions within the catalytic converter. For example, excessive engine oil consumption can partially plug the converter with carbon. Also, a leaking automatic transmission vacuum modulator may permit so much automatic transmission fluid to enter the combustion chambers that the spark plugs foul. Should this occur, excessive hydrocarbons pass through the engine and into the catalytic converter.

Operation of an engine with foul spark plugs, ignition malfunction, or improper air/fuel mixture raises the temperature within the catalytic converter. At these high temperatures, the converter cover may bulge or distort. Furthermore, inside the converter the high temperatures melt the substrate. If this melts, even partially, normal exhaust gas flow is blocked and catalytic action is lost. In addition, any restriction to exhaust gas flow causes poor engine performance.

To test the converter and exhaust system for restrictions with a vacuum gauge:

1. Connect the vacuum gauge, using a sufficient length of hose to a nonrestricted port on the intake manifold (Fig. 18-3), where the gauge measures the total

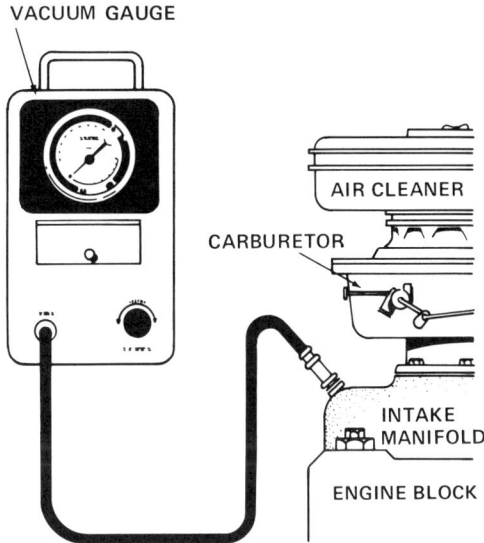

Figure 18-3. Proper connection of a vacuum gauge into the intake manifold.

vacuum within the manifold. The length of hose, about three feet, should be long enough to dampen excessive vibrations from the gauge needle. However, under certain conditions, it may be necessary to dampen the pointer action by placing a small clamp around the hose in order to slightly restrict its passageway.

2. Connect a tachometer to the engine.

3. Warm up the engine until it reaches normal operating temperature.

4. Slowly accelerate the engine until it reaches 2,000 rpm.

5. Note the reading on the vacuum gauge. The needle should drop slightly and then rise quickly to a reading that is 3 to 5 inches higher than the normal idle vacuum.

6. Quickly close the throttle; the gauge needle should now return to the idle reading just as rapidly as it rose.

Results and Indications

If the gauge needle at first reaches a normal reading at idle and at 2,000 rpm, but it begins to drop towards zero (Fig. 18-4) and then rise slowly to a below-normal reading at 2,000 rpm, some form of restriction exists within the converter or exhaust system.

In the case of some large displacement engines, it may be necessary to perform this test while actually driving the vehicle until it reaches the road speed where the engine loses power. In order to perform this test under actual driving conditions:

1. Connect the vacuum gauge to the intake manifold with a long enough hose so that you can position the gauge itself inside the vehicle.

2. Connect a tachometer to the engine and route its wires so this gauge is also inside the vehicle.

3. Drive the vehicle through the road speed range where engine performance begins to drop off.

4. Note the reading on both gauges. The tachometer reading will most likely be different than the figure mentioned earlier due to the size of the engine, size of the restriction, and the actual load on the engine. But regardless of the actual reading, it should be about the same each time a loss of power occurs, if a restriction exists within the converter or exhaust system.

The vacuum gauge needle will react normally at first, but as the engine begins to lose power, it will begin to drop off towards zero. The amount of needle deflection toward zero will depend on the amount of restriction. In some cases, the needle never reaches zero but remains at the very low figure until the driver reduces engine load.

Testing the Converter Efficiency With an Infrared Analyzer

To test a catalytic converter's efficiency in reducing HC and CO emissions, follow these steps:

1. Calibrate an infrared analyzer and insert its probe into the vehicle's tail pipe.

2. Thoroughly tune the engine, making all operational adjustments necessary to meet manufacturer's as well as the appropriate emission control specifications.

3. After making all required adjustments, check the HC and CO levels on the infrared analyzer. If the engine along with the ignition, fuel, and emission control systems are functioning properly but the analyzer still indicates HC and CO levels that are too high (Fig. 18-5), the catalytic con-

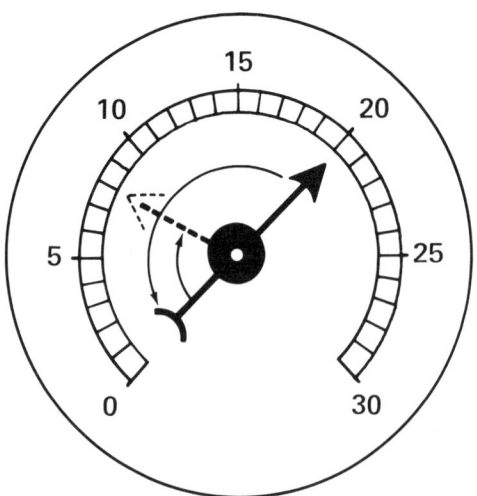

Figure 18-4. The vacuum gauge needle will drop toward zero and rebound to a low reading if the converter has a restriction.

CATALYTIC CONVERTER SERVICE

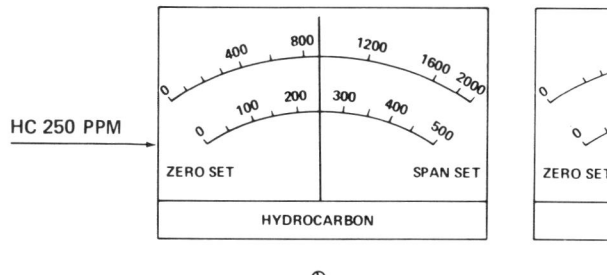
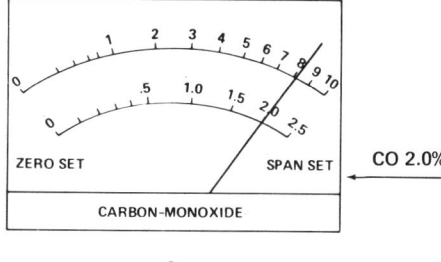

Figure 18-5. Typical infrared analyzer readings from a vehicle with a defective catalytic converter.

verter is most likely defective and requires replacement or servicing.

Catalytic Converter Service

If a pellet-type converter, such as the one made by AC-Delco and used by GM and AMC, is no longer reducing HC and CO emissions, you can refill it with a new charge of catalyst. To perform this procedure:

1. Raise the vehicle on a hoist.

2. Attach an aspirator (Fig. 18-6) to the vehicle's tail pipe. If the vehicle has a dual-exhaust system, plug the other tail pipe.

3. Connect a compressed air-line coupling to the fitting on the aspirator. This action creates a vacuum within the converter that holds the pellets in position.

4. Remove the catalytic converter's drain-and-fill plug:

   a. For a pressed-end plug, drive a small chisel between the plug and the converter shell to deform it. Next, remove the plug with a pair of pliers.

   b. For a threaded plug, use a three-fourths-inch hex wrench to remove it.

5. Clamp the vibrator and catalyst container into position over the converter fill opening (Fig. 18-7). Note: A special adaptor is necessary if the converter has a pressed-in plug.

6. Disconnect the compressed air-line coupling from the aspirator and snap it in place on the fitting of the vibrator. This action drains the catalyst pellets from the converter housing.

7. After the process removes all the pellets from the converter, disconnect the air-line coupling. Then, remove the container from the vibrator unit.

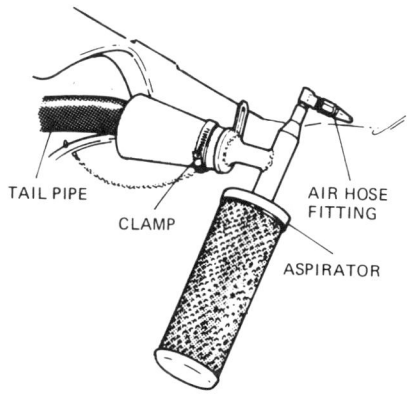

Figure 18-6. Attaching an aspirator to the tail pipe.

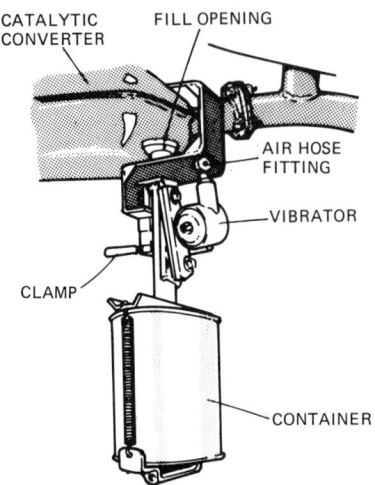

Figure 18-7. Attaching the vibrator and catalyst container in place over the converter fill opening.

8. Dispose of the used catalyst pellets.

9. Fill the converter with the new replacement catalyst pellets.

10. Install the fill-tube extension to the vibrator unit. Note: An adaptor is necessary on those converter types using a pressed-in plug.

11. Connect a compressed air-line coupling to both the vibrator unit and the aspirator.

12. Connect the catalyst container to the vibrator unit.

13. Turn on the air supply. The pellets will move and pack themselves into the converter housing.

14. When the vibrator container is empty, disconnect the air line at the vibrator.

15. Remove the vibrator unit. Then, make sure that all the catalyst from the container has entered the converter, and it is full.

16. Wipe the threaded fill plug with an antiseize compound. Next, thread it into the converter housing and torque it to specifications. Note: A special service plug with a bridge is necessary with the converter originally equipped with a pressed-in plug. Torque this plug also to specifications.

17. Disconnect the air-line coupling from the aspirator and remove it from the tail pipe.

18. Lower the vehicle sufficiently to start the engine. Then, check the converter drain plug for leaks.

19. Check the emission levels with an infrared analyzer to make sure that they are now within specifications.

Catalytic Converter Replacement

A Chrysler or Ford monolithic-type catalytic converter requires replacement when defective or damaged. Because catalytic converter installations do differ somewhat from one vehicle to another, it is impossible in the space provided to cover all the procedures. However, listed below are some general removal and replacement procedures adaptable for use on most vehicle designs.

Converter Removal

To remove a typical converter (Fig. 18-8):

1. Raise the vehicle on a hoist to a comfortable working height.

2. If so equipped, remove the lower heat shield. It is usually held in place by U-bolts or straps.

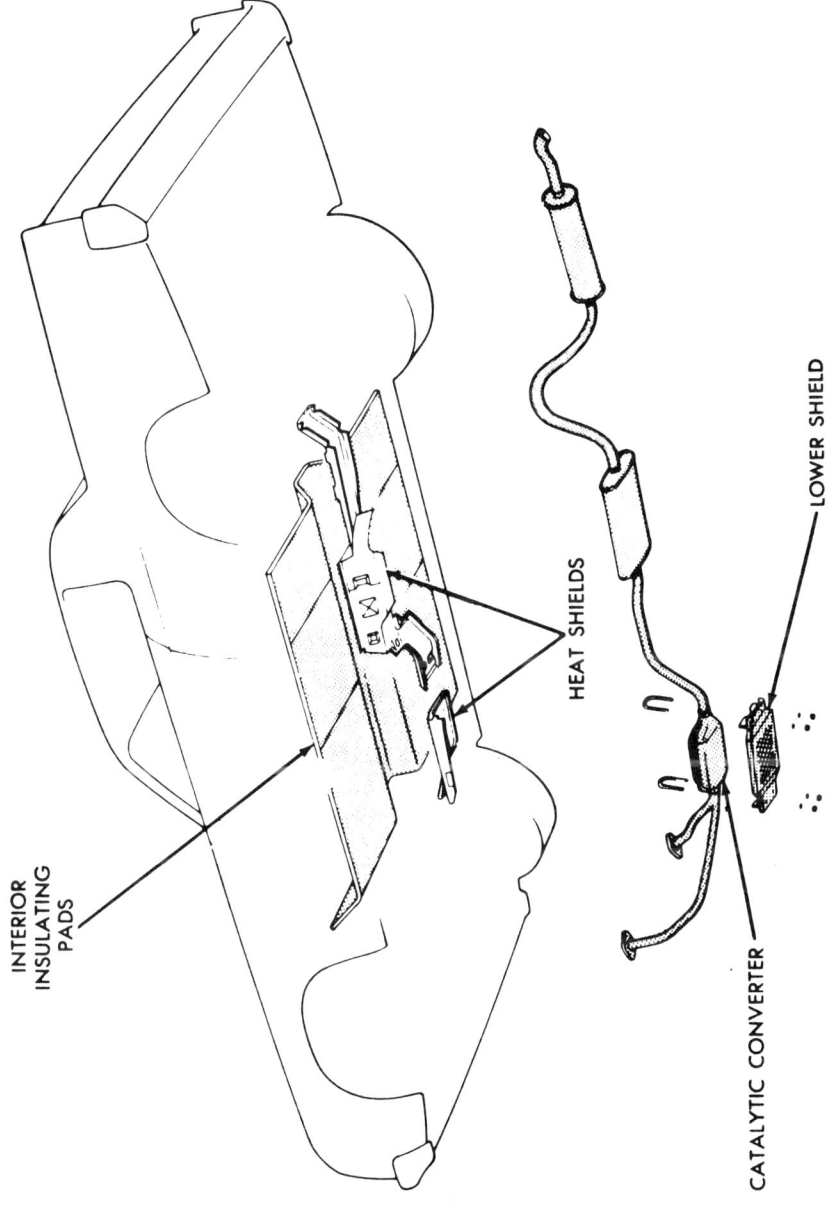

Figure 18-8. Typical catalytic converter installation (Courtesy of Chrysler Corp.).

3. Remove the converter's flange fastener at the front and rear of the unit. Note: Some converters incorporate a short length of pipe on each end of the unit. In such cases, it will also be necessary to loosen the connection on these pipes.

4. Separate the old converter from the exhaust system pipes and remove it from the vehicle.

Converter Installation

To install a new catalytic converter:

1. Position the new converter to the inlet exhaust pipe. Note: If gaskets are necessary at the converter flange, always replace them with new ones during the installation process.

2. As necessary, install the front flange attaching bolts loosely.

3. Connect the outlet flange to the pipe leading to the muffler. Next, install the rear attaching bolts or nuts loosely.

4. Install the lower heat shield into position, and align the converter, heat shield, and any other attached components according to manufacturer's instructions.

5. Tighten all nuts and bolts to factory specifications.

6. Lower the vehicle. Then, start the engine and check the exhaust system for leakage.

Summary

1. Federal law now requires that a catalytic converter provide at least 50,000 miles of normal use before service is necessary.

2. Vehicles with catalytic converters must utilize an unleaded fuel.

3. The use of leaded fuel damages the catalyst within the converter.

4. Manufacturers prevent the use of leaded fuel in a vehicle by the design of the fuel tank's filler tube.

5. Proper engine condition and maintenance are necessary to prevent overheating the catalyst.

6. When an engine is operating poorly, the exhaust gas passing into the converter contains excessive amounts of hydrocarbons.

7. During vehicle maintenance, there are certain general rules you must follow to prevent fuel vapors from reaching the converter in excessive amounts.

8. Physically check the condition of the catalytic converter system following the steps outlined in this chapter.

9. There are two common methods of testing a catalytic converter.

10. Several abnormal factors can cause restrictions within the catalytic converter.

11. Follow the procedure outlined in this text to check a converter for restrictions.

12. To test a catalytic converter's efficiency in reducing HC and CO emissions, use the procedure given in this chapter.

13. Service a pellet-type converter following manufacturer's instructions or those provided within this text.

14. Replace a Chrysler or Ford monolithic-type catalytic converter when defective or damaged following manufacturer's instructions or those given in this chapter.

Review Questions

The questions listed below will assist you in determining how well you remember the material contained in this chapter. Read each question carefully before choosing the answer. If you cannot answer the question, review the section in the chapter that covers the material.

1. Vehicles with a catalytic converter must _____.
   a. use leaded fuel
   b. use unleaded fuel
   c. use a combination of leaded and unleaded fuel
   d. none of these

2. If an engine is in poor mechanical condition or in need of a tune-up, excessive amounts of _____ will pass into the converter.
   a. hydrocarbons
   b. carbon monoxide
   c. oxides of nitrogen
   d. carbon dioxide

3. Never operate an engine with a catalytic converter more than _____ seconds with a plug shorted out.
   a. 40
   b. 10
   c. 20
   d. 30

4. Never operate an engine with a catalytic converter more than _____ minutes at idle.
   a. 1
   b. 5
   c. 10
   d. 15

5. If pellets are coming out of the tail pipe, replace the _____.
   a. tail pipe
   b. converter
   c. muffler
   d. catalyst

6. An engine that burns excessive amounts of oil can plug the _____.
   a. converter
   b. tail pipe
   c. exhaust pipe
   d. muffler

7. To test a catalytic converter for restrictions, use a _____.
   a. vacuum pump
   b. engine analyzer
   c. vacuum gauge
   d. infrared analyzer

8. Use a _____ to check a converter's efficiency in reducing HC and CO emissions.
   a. vacuum pump
   b. engine analyzer
   c. vacuum gauge
   d. infrared analyzer

9. A _____ catalytic converter is serviceable by replacing its catalyst.
   a. Ford
   b. GM
   c. Chrysler
   d. none of these

10. A _____ converter requires replacement if its catalyst is defective.
    a. monolithic
    b. AMC
    c. GM
    d. both b and c

For the answers to these review questions, turn to the Appendix.

Appendix
Answers to Review Questions

Chapter 1

1. b
2. a
3. c
4. a
5. d
6. c
7. b
8. c
9. a
10. d

Chapter 2

1. b
2. a
3. c
4. d
5. c
6. a
7. d
8. b
9. a
10. b

Chapter 3

1. b
2. a
3. c
4. d
5. b
6. a
7. c
8. b

Chapter 4

1. a
2. d
3. b
4. c
5. d
6. a
7. a
8. c
9. b
10. d
11. c
12. c
13. a
14. b
15. c
16. d
17. a
18. b
19. a
20. d
21. a
22. c

Chapter 5

1. b	4. b	7. c	9. d
2. a	5. d	8. b	10. a
3. c	6. a		

Chapter 6

| 1. b | 3. c | 5. c | 7. c |
| 2. a | 4. a | 6. d | 8. b |

Chapter 7

1. a	4. c	7. a	10. c
2. d	5. b	8. d	11. a
3. b	6. c	9. b	12. c

Chapter 8

| 1. a | 3. c | 5. a | 7. c |
| 2. d | 4. b | 6. b | |

Chapter 9

1. a	5. b	9. d	13. d
2. d	6. a	10. c	14. a
3. c	7. c	11. a	15. b
4. b	8. b	12. b	16. a

Chapter 10

1. c	4. d	7. c	9. d
2. b	5. b	8. a	10. c
3. a	6. a		

Chapter 11

1. b
2. a
3. c
4. b
5. d
6. a
7. c
8. b
9. d
10. a

Chapter 12

1. b
2. a
3. d
4. c
5. a
6. c
7. b
8. d

Chapter 13

1. b
2. a
3. d
4. c
5. b
6. a
7. c
8. c
9. d
10. a

Chapter 14

1. a
2. d
3. b
4. c
5. b
6. a
7. d
8. c

Chapter 15

1. b
2. a
3. d
4. c
5. b
6. a
7. c
8. d
9. b
10. c
11. a
12. d

Chapter 16

1. b
2. a
3. b
4. d
5. c
6. b
7. a
8. c
9. d
10. b
11. a
12. c
13. b
14. b

Chapter 17

1. b
2. a
3. a
4. b
5. d
6. d
7. b
8. b
9. c
10. a

Chapter 18

1. b
2. a
3. d
4. c
5. b
6. a
7. c
8. d
9. b
10. a

Index

A

Acceleration
  exhaust gas back pressure during, 292–294, 304
  PCV valve efficiency during, 89, 92, 95–96
Additives in fuel, 19
Advanced basic ignition timing, 28–30, 67, 209–211
  see also Spark-timing control systems; Timing, ignition
AIR, see Air Injection Reactor
Air
  atmospheric pressure, 20–21, 35
  composition, 17–18
Air cleaners, 174–177, 201–202, 204–205
Air Injection Reactor (AIR), 249, 264
Air injection systems
  aspirator type, 266–268, 283–284
  and catalytic converters, 255–266
  Chrysler catalytic air injection system, 256–257, 277–279
  description, 249–255
  with dual catalytic converters, 263–264
  Ford
  —catalytic air injection system, 258–264, 279–283
  —catalytic protection system, 352
  General Motors catalytic air injection systems, 264–266, 283
  infrared analyzer use, 65, 274
  pulse air systems, 268–270
Air pollution
  contributing factors, 3–4
  defined, 1
  man-made, 2–3
  natural, 2
  research and legislation, 9–12
  versus smog, 3
Air pump
  description, 250–253
  infrared analyzer use, 64–65
  testing, 274–275
  thermactor, 258, 352
Air-conditioning idle solenoid, 171–172, 201
Air/fuel ratio
  adjusting, 54–55
  emission levels, effect of, 25–26
  in ideal conditions, 17
  importance, 5–6
  lean mixture, 26, 68–69
  measurement of, 38–39
  rich mixture, 23, 70
Altitude compensation, 161
American Motors PCV system, 93
Aneroid bellows, 161

INDEX

Antiknock additive, see Tetraethyllead
Aromatic hydrocarbons, see Ring hydrocarbons
Aspirator-air injection systems, 266–268, 283–284
Atmospheric pressure, 20–21, 35, 161

B

Back-pressure transducer, 292–294, 304–306, 309, 328–330
BDC, see Bottom dead center
Benzene, 18
Benzopyrene, 6, 23, 24
Blow-by gases, 85–86
Bottom dead center (BDC), 21
Branched-chain hydrocarbons, 18, 19

C

C-4 system, see Closed-loop feedback emission control system
CAI, see Converter Air Injection
CAIR, see Converter Air Injector Reactor
Calibration, infrared analyzer, 55–58
California Air Resources Board, 11
California Institute of Technology studies, 2, 9
Carbon dioxide (CO_2), 7, 24, 26
Carbon elements, 127, 140
Carbon monoxide (CO)
  and air/fuel ratio, 26–29
  causes and diagnosis of excessive emissions, 76–78
  CO meter, 38
  description and effects of, 7–8, 24
Carburetor
  fuel bowl venting, 124–126
  incorrect adjustments, 74, 77–78
  modifications
    —air-conditioning idle solenoid, 171–172, 201
    —air/fuel ratio adjustment, 186, 188–191
    —antidieseling solenoid, 171, 199–201

Carburetor (contd.)
  —choke circuit, 161–165
  —choke vacuum brake, 167–168
  —choke-delay valve, 168–169, 198–199
  —dashpot, 170–171, 199
  —deceleration valve, 169–170, 217
  —electronically controlled, 165–166
  —high-speed light-load circuit, 160–161
  —idle circuit, 156–160
  —jet air control valve system, 173–174, 186–187
  —propane enrichment equipment use, 191–198
  PCV passages, cleaning, 107
Carter Thermo Quad carburetor, 161
Catalyst materials, 339
Catalytic converters
  and air injection systems, 255–266
  Chrysler vehicles, 345–350
  CO, testing equipment for, 41
  description, 337–338
  dual, air injection systems for, 263–264
  Ford vehicles, 350–354
  General Motors vehicles, 354–356
  and idling, 360
  infrared analyzer use, 64–65, 66–67, 364–365
  inspections and testing, 361–365
  oxidation converters, 339, 341
  —monolith design, 341–343, 344–345
  —pellet design, 342, 344–345
  recharging catalyst, 365–366
  reduction converters, 341
  replacement, 366–368
  sulfuric acid mist problem, 9
  three-way, 345, 349–350, 352, 354, 355
  unleaded fuel, use with, 359–360
CCEV switch, see Coolant-controlled engine vacuum switch
CHAM-CV, see Hot air modifier check valve
Charcoal canisters, 126–127, 140–141
Charge temperature switch (CTS), 297
Check valve, 255, 276

Choke
  circuit modifications, 161–165
  delay valve, 168–169, 198–199
  vacuum brake, 167–168
  valve malfunction, 76–77
Chrysler Corp.
  catalytic air injection system, 256–257, 277–279
  catalytic converter systems, 345–350
  catalytic protection system, 347
  EEC systems, 113
  EGR systems
  —floor jet system, 288, 295, 315–316, 318
  —inspection, testing and servicing, 315–321
  —maintenance reminder system, 298–299, 321
  —ported system, 316, 318–319
  —time-delay systems, 297, 320
  —venturi vacuum system, 316, 319–320
  electronic spark control system, 224–228, 242–246
  power heat control valve, use, 178
  propane enrichment adjustment procedures, 197–198
  Pulse Air Feeder system, 268–270
Clear Air Act of 1963, 9–10
Closed-loop feedback emission control (C–4) system, 355–356
CO, see Carbon monoxide
CO_2, see Carbon dioxide
Combustion chamber
  deposits, 5
  engine modifications, 146–147, 150–156
  hemispheric, 153
  jet air system, 153–156
  stratified charge, 151–153
  and temperature, 23, 30
Compression
  in emission modified engine, 147
  malfunctions, 75–76
  role in combustion, 24–25
  stroke, 21–22
  testing for loss, 50–51

Converter Air Injection (CAI), 265
Converter Air Injector Reactor (CAIR), 265
Coolant-assisted choke mechanisms, 163–164
Coolant-controlled engine vacuum (CCEV) switch, 256, 279
Coolant-controlled vacuum valves, 294
Crankcase emission control, see Positive crankcase ventilation systems
Crankcase vapors, see Blow-by gases
CTS, see Charge temperature switch
CVCC engine design, 151–152

D

Dashpot, 170–171, 199
Deceleration valve, 169–170, 217
Dieseling, 171
Distributor tester, 72–73
Diverter valve, 253–254, 258–263, 275–276, 279–280
Dual-diaphragm EGR valves, 294–295, 310
Dual-diaphragm vacuum advance, 212, 213, 233–235
Dual-exhaust systems, 64
Dump diaphragm, 291–292

E

Early fuel evaporation (EFE) valve, 178
EEC systems, see Evaporation emission control systems
EFE valve, see Early fuel evaporation valve
EGR systems, see Exhaust gas recirculation systems
Electrical assist choke system, 162–163
Electrically operated vacuum pumps, 43
Electronically controlled timing, 223–228, 242–246
Emissions
  air/fuel ratio, effect of, 25–26
  basic ignition timing, effect on, 28–30, 209–211
  CO
  —and air/fuel ratio, 26–29

Emissions (contd.)
—causes and diagnosis, 76–78
—description and effects of, 7–8, 24
controls, effect on fuel consumption, 12
HC
—causes and diagnosis, 71–76
—description and effects of, 4–6
idle speed, effect on, 27
legislative standards for, 9–10, 11–12
NOx, 6–7, 18, 29–30
particulate, 8
SOx, 8–9
spark plug misfires, effect on, 30
temperature, effect on, 30
testing with infrared analyzer, 62–70
Engine
air injection, see Air injection systems
carburetor, see Carburetor
compression
—in emission modified engine, 147
—loss, testing for, 50–51
—malfunctions, 75–76
—role in combustion, 24–25
condition for catalytic converter efficiency, 360
EGR systems, see Exhaust gas recirculation systems
exhaust system leaks, effects of, 62–64
ignition system malfunctions, 5, 71
intake system malfunctions, 73–74
modifications, see Engine modification systems
operating cycle
—compression stroke, 21–22
—exhaust stroke, 23–24
—intake stroke, 20–21
—power stroke, 22–23
PCV systems
—filter cleaning and replacement, 108–110
—hose and valve replacement, 107–108
—inspection and testing, 102–107
—malfunction symptoms, 74, 101–102
—type 1 - open system, 88–89
—type 2 - restricted system, 89–92

Engine (contd.)
—type 3 - tube-to-air cleaner system, 92–93
—type 4 - closed system, 93–96
spark-timing control, see Spark-timing control systems
testing for problems
—distributor tester, 72–73
—ignition oscilloscope, 71
—infrared analyzer, see Infrared analyzer
—tachometer, 53–54
—timing light, 51–53
—vacuum gauge, 35–36, 47–51, 363–364
timing system malfunctions, 72–73
vacuum leaks, 67–68, 73–74
Engine modification systems
camshafts, 148
carburetor
—modifications, 156–174
—see also Carburetor
choke circuit, 161–165
combustion chambers, 146–147
compression ratio reduction, 147
function, 145–146
intake manifold, 148–150
operating temperatures, increase, 177–178
pistons, 146–147
spark-advance control devices, 177
thermostatically controlled air cleaners, 174–177, 201–202
vacuum-operated heat-riser valves, 178–179
valve ports and seats, 147–148
Environmental Protection Agency (EPA), 10–11
Equipment, testing
distributor tester, 72–73
ignition oscilloscope, 71
infrared analyzer, see Infrared analyzer
PCV tester, 105–106
propane enrichment equipment, 41, 79, 191–198

Sun MAG-81 magnetic timing unit, 242–243, 245
tachometer, 37, 53–55, 106
timing light, 36–37, 51–53, 72
vacuum gauge, 35–36, 47–51, 363–364
vacuum pumps, 41–43, 79–80
Evaporation emission control (EEC) systems
  carbon elements, 127, 140
  charcoal canisters, 126–127, 140–141
  component replacement, 140–142
  filler caps, 114–116, 137–138
  filter separators, 123–124
  inspection of, 137–139
  liquid-vapor separators, 119–122
  maintenance intervals, 135
  operation, 127–132
  overfill-limiting devices, 117–119, 138–139
  purging methods, 128–132
  rollover leakage protection devices, 122–123, 141–142
  types, 113–114
  vapor vent hose replacement, 140–141
Exhaust gas recirculation (EGR) systems
  Chrysler vehicles
  —floor jet system, 288, 295, 315–316, 318
  —maintenance reminder system, 298–299, 321
  —ported system, 316, 318–319
  —time-delay systems, 297, 320
  —venturi vacuum system, 316, 319–320
  description, 287–288
  EGR valve, 289–290, 314–315, 318, 323–324, 332–333
  Ford vehicles
  —back-pressure transducer subsystem 304–306, 328–330
  —electronically controlled system, 306–309, 330–332
  —high-speed modulating subsystem, 302–303, 325–328
  —ported system, 322–323
  —testing, 321–332
  —valves, 299–302, 323–324

EGR systems (*contd.*)
  —venturi vacuum subsystem, 303–304, 324–325
  General Motors vehicles, 308–310, 332–333
  maintenance, importance, 313
  malfunctions, 69, 74–75
  modulating devices, 292–295
  system design, 288–291
Exhaust stroke, 23–24
Exhaust system leaks, 62–64
Expansion tanks, 118
EXXON gasoline, properties of, 19

F

Federal regulation, 10–11
Filler caps, 114–116, 137–138
Filters
  EEC system, 140–141
  infrared analyzer, 59–62
  PCV inlet, 108–110
Flexible sampling hose, 61
Flooding, engine, 123, 360–361
Floor jet EGR system, 288, 295, 315–316, 318
Ford Motor Co.
  catalytic air injection system, 258–264, 279–283
  catalytic converter systems, 350–354
  EGR systems
  —back-pressure transducer subsystem, 304–306, 328–330
  —electronically controlled system, 306–309, 330–332
  —high-speed modulating subsystem, 302–303, 325–328
  —ported system, 322–323
  —testing, 321–332
  —valves, 299–302, 323–324
  —venturi vacuum subsystem, 303–304, 324–325
  HCV valve, use, 178
  PCV system, 93–94
  propane enrichment adjustment procedures, 195–197

Ford Motor Co. (*contd.*)
  spark-delay valve, 215
Fuel, *see* Gasoline

G

Gas calibration check of infrared analyzer, 58
Gasoline
  composition of, 18
  EXXON, properties of, 19
  octane rating, 18, 19
  unleaded, use with catalytic converters, 359–360
  vapor control, *see* Evaporation emission control systems
General Motors Corp.
  catalytic air injection systems, 264–266, 283
  catalytic converter systems, 354–356
  EFE valve, use, 178
  EGR systems, 308–310, 332–333
  propane enrichment adjustment procedures, 192–195

H

Hand-operated vacuum pumps, 41–43
HC, *see* Hydrocarbons
HCV valve, *see* Vacuum-operated heat control valve
Health, Education and Welfare, Department of, 9
Heat-riser valves, 178–179
Hemispheric combustion chamber, 153
High-speed light-load circuit, 160–161
Honda Compound Vortex Controlled Combustion (CVCC) design, 151–152
Hot air modifier check valve (CHAM-CV), 164
Hydrocarbon meter, 38
Hydrocarbons (HC)
  as auto emissions, 4–6
  defined, 2
  excessive, causes and diagnosis, 71
  in gasoline, 18–20

I

Idle
  circuit modifications, 156–158
  effect on catalytic converter, 360
  enrichment systems, 158–160
Idle speed
  adjusting, 53–54
  effect on HC emissions, 27
  infrared test results, 67–70
Ignition oscilloscope, 71
Ignition system malfunctions, 5, 71
Ignition timing, *see* Timing, ignition
Inert gas, 287
Infrared analyzer
  air/fuel ratio adjustment, 186, 188–191
  calibration, 55–58
  description, 37–44, 55
  EEC system testing, 139–140
  maintenance, 58–62
  PCV systems, testing, 106–107
  test procedures, 65–67, 76, 78–79, 274, 364–365
  use of, 62–70
Intake stroke, 20–21
Intake system malfunctions, 73–74
Inversion, 4
Isooctane, *see* 2-2-4-trimethylpentane

J

Jet air control valve system, 173–174, 186–187
Jet air system, 153–156

K

Knocking, engine, 21

L

Lead bromide, 23
Lead bromochloride, 23
Lead chloride, 23
Lead (Pb) particulates, 8, 19
Lean misfire, 26, 68–69
Liquid-vapor filter separators, 123–124

Liquid-vapor fuel separator, 119–122
Los Angeles, California, 1, 3–4

M

MAI, see Manifold Air Injection
Maintenance reminder system, Chrysler, 298–299, 321
Manifold Air Injection (MAI), 249, 264–265
Manifold, injection, 255, 276
Manifold, intake
  EGR system design, 288–289
  modifications, 148–150
Man-made air pollution, 2–3
Methane hydrocarbons, 2
Methylcyclopentadienyl manganese tricarbonyl (MMT), 19
Misfire, lean, 26, 68–69
MMT, see Methylcyclopentadienyl manganese tricarbonyl
Modulator ball, 90–92
Monolith design catalytic converters, 341–343, 344–345
Mount St. Helens volcano, 2
Mufflers, 337–338

N

National Emissions Standard Act, 9–10
Natural air pollution, 2
N-heptane, 19
N-hexane gas, 56–57
Nitric oxide (NO), 7
Nitrogen, 6, 18
Nitrogen dioxide (NO_2), 7
NO, see Nitric oxide
NO_2, see Nitrogen dioxide
NOx, see Oxides of nitrogen
N-pentane, 18

O

O_2, see Oxygen
O_3, see Ozone
Octane rating of fuel, 18–19

Oregon Department of Environmental Quality, 11
Orifice spark advance control (OSAC) valve, 216–217, 239
Oxidation converters, 339, 341
  monolith design, 341–343, 344–345
  pellet design, 342, 344–345
Oxides of nitrogen (NOx)
  composition, 7
  emissions, 6, 18, 29–30
  and high temperatures, 25
Oxides of sulfur (SOx), 8–9
Oxygen (O_2), 7
  for engine combustion, 6, 17
Ozone (O_3), 7

P

PAF, see Pulse Air Feeder
Palladium as catalyst, 339
Particulate emissions, 8
PCV systems, see Positive crankcase ventilation systems
Pellet-type catalytic converters, 342, 344–345
Photochemical smog, 3, 6, 7
Piston
  engine modifications, 146–147, 150–156
  in operating cycle, 20–24
Plants, air pollution from, 2
Platinum as catalyst, 339
Pollution, see Air pollution
Ported vacuum, 290–291, 316, 318–319, 322–323, 332–333
Ported vacuum switch (PVS), 235
Positive crankcase ventilation (PCV) systems
  filter cleaning and replacement, 108–110
  history, 85–87, 101
  hose and valve replacement, 107–108
  inspection and testing, 102–107
  malfunction symptoms, 74, 101–102
  retrofit devices, 96–97
  type 1 - open system, 88–89
  type 2 - restricted system, 89–92

PCV systems (*contd.*)
  type 3 - tube-to-air cleaner system, 92–93
  type 4 - closed system, 93–96
Post emission control devices, 64
Power circuit test, 78–79
Power heat control valve, 178
Power stroke, 22–23
Powered timing light, 36–37, 52–53, 72
Propane enrichment equipment
  carburetor adjustments, 79, 191–198
  description and use, 41
Pulse Air Feeder (PAF), 268–270
Pulse-air injection systems, *see* Aspirator-air injection systems
Purge control valve, 138
Purging methods, 128–132
PVS, *see* Ported vacuum switch

Q

Quenching
  effect of, 4–5, 23
  engine modifications for, 146–147, 150–151

R

Relief valve, 254–255
Retrofit crankcase emissions systems, 96–97
Rhodium as catalyst, 339
Rich air/fuel ratio, 23, 70
Ring hydrocarbons, 18, 19
Road-draft ventilation system, 86–87
Rollover leakage protection devices, 122–123, 141–142

S

Saginaw air pumps, 250
Sealed system, 93
Smog, 3, 6, 7
Solenoid, 124–126, 220–221
  air-conditioning idle, 171–172, 201
  antidiesel, 171, 172, 199–201

SOx, *see* Oxides of sulfur
Spark plug, 22
  malfunctions, 71
  misfires, effect on HC emissions, 30
Spark-advance control devices, 177
Spark-timing control systems
  dual-diaphragm vacuum advance, 212, 213, 233–235
  effect on emissions, 209–211
  electronically controlled timing, 223–228, 242–246
  OSAC valves, 216–217, 239
  spark-delay valves, 214–216, 238
  temperature-sensitive vacuum control valves, 212, 214, 235–238
  transmission controlled spark, 219–223, 241–242
  vacuum control valves, 217–219, 239–241
Stalling, 170–171
State regulation, 11
Straight-chain hydrocarbons, 18, 19
Stratified-charge combustion chamber, 151–153
Suction-air injection systems, *see* Aspirator-air injection systems
Sulfuric acid mist, 9
Sun MAG-81 magnetic timing unit, 242–243, 245

T

Tachometer, 37, 53–55, 107
Tamperproof automatic chokes, 165
TCS system, *see* Transmission-controlled spark system
TDC, *see* Top dead center
TEL, *see* Tetraethyllead
Temperature
  air cleaner sensor, 204–205
  in air/fuel mixture, 5
  combustion chamber, 23, 30
  and compression pressure, 25
  during compression stroke, 21
  effect on catalytic converter, 360, 363

Temperature (contd.)
　minimum operating, 177–178
　sensitive vacuum control valves, 212, 214, 235–238
Terpene hydrocarbons, 2
Testing equipment, see Equipment, testing
Tetraethyllead (TEL), 8, 19, 21–22
Thermactor system, 258, 352
Thermo ignition control (TIC), 235
Thermo vacuum switch (TVS), 235
Thermo vacuum valve (TVV), 235
Thermo valve, 173–174
Thermostat malfunction indication, 67
Thermostatic vacuum switch, Ford, 324–325
Thermostatically controlled air cleaners, 174–177, 201–202
Three-way catalytic converters, 345, 349–350, 352, 354, 355
TIC, see Thermo ignition control
Timed air bypass valves, 258
Time-delay EGR systems, 297, 320
Timing, ignition, 28–30
　effects on HC emissions, 209–211
　malfunctions, 72–73
　overadvanced, symptoms, 67, 72
　testing using timing light, 51–53
　see also Spark-timing control systems
Timing light, 36–37, 51–53, 72
Top dead center (TDC), 20
Transmission-controlled spark system, 219–223, 241–242
Trees, air pollution from, 2
TVS, see Thermo vacuum switch
TVV, see Thermo vacuum valve
Two
　2-methylbutane, 18
　2-2-4-trimethylpentane, 19

V

Vacuum control valves, 217–219, 239–241
Vacuum differential valve (VDV), 259
Vacuum gauge, 35–36, 47–51, 363–364
Vacuum leaks, 67–68, 73–74
Vacuum motor diaphragm, 178–179, 202–204
Vacuum pumps, 41–43, 79–80
Vacuum sources for EGR valve, 290–291, 316, 318–319, 322–323, 332–333
Vacuum test for PCV system, 103–104
Vacuum-operated heat control (HCV) valve, 178
Vacuum-operated heat-riser valves, 178–179
Valve ports and seats, modifications, 147–148
Vapor lock, 123
VDV, see Vacuum differential valve
Ventilation, engine, see Positive crankcase ventilation systems
Venturi vacuum control system, 291, 316, 319–320, 324–325
Volcanos, 2

W

World War II, 1